Predicative Forms in Natural Language and in Lexical Knowledge Bases

Text, Speech and Language Technology

VOLUME 6

The titles published in this series are listed at the end of this volume.

Predicative Forms in Natural Language and in Lexical Knowledge Bases

Edited by

Patrick Saint-Dizier

IRIT-CNRS, Toulouse, France

SPRINGER-SCIENCE+BUSINESS MEDIA, B.V.

A C.I.P. Catalogue record for this book is available from the Library of Congress.

DOI 10.1007/978-94-017-2746-4

Printed on acid-free paper

Table of Contents

Foreword

This volume is a selection of papers presented at a workshop entitled *Predicative Forms in Natural Language and in Lexical Knowledge Bases* organized in Toulouse in August 1996. A predicate is a named relation that exists among one or more arguments. In natural language, predicates are realized as verbs, prepositions, nouns and adjectives, to cite the most frequent ones.

Research on the identification, organization, and semantic representation of predicates in artificial intelligence and in language processing is a very active research field. The emergence of new paradigms in theoretical language processing, the definition of new problems and the important evolution of applications have, in fact, stimulated much interest and debate on the role and nature of predicates in natural langage. From a broad theoretical perspective, the notion of predicate is central to research on the syntax-semantics interface, the generative lexicon, the definition of ontology-based semantic representations, and the formation of verb semantic classes.

From a computational perspective, the notion of predicate plays a central role in a number of applications including the design of lexical knowledge bases, the development of automatic indexing systems for the extraction of structured semantic representations, and the creation of interlingual forms in machine translation.

In this volume, the notion of predicate is studied from several perspectives. Since it is not possible to be exhaustive in this broad domain, the aim is to address a number of current most important issues that can be classified as follows:

- linguistic aspects,
- the Generative lexicon,
- corpus-based analysis,
- verb semantic classes formation,
- WordNet and relations among verbs,
- knowledge representation and reasoning issues.

Since most articles often address several of these issues at once, the structure of the volume reflects two major 'orthogonal' perspectives: the first seven papers deal with core issues in predication, and the last five present different theoretical prespectives.

The Introduction provides an outline of the major issues and models of lexical semantics which are refered to in the subsequent articles. The models are compared more explicitely in the second article, around the specific example of the verb pair *buy/sell*, which was the topic of a panel session, summarized here. The article by Christiane Fellbaum describes the foundations and the organization of the WordNet lexical base for verbs. The article by Achim Stein focuses on verb semantic classes in Italian and their use in disambiguation, while the article by Patrick Saint-Dizier focuses rather on techniques for the formation of verb semantic classes. The article by Evelyne Viegas et al. is a presentation of a large-size project: Mikrokosmos with an emphasis on verbs describing actions. Finally, the article by Dimitrios Kokkinakis introduces a method for corpus-based acquisition of verb syntax and semantics.

The second part of the volume, starts with an article by Martha Palmer et al. which presents a treatment of the relations between verb classes and syntax within the Tree Adjoining Grammar theory. The article by Charles Jones focusses on argument structure and introduces new perspectives on this much debated topic. The article by Jayez et al. introduces a particular form of predicates: connectors; their pragmatic aspects, in terms of inferences, are studied. The next article, by Maria Angeles Zarco, presents a detailed study of the interligual representation of complex predicates. Finally, the article by Federica Busa deals with the treatment of agentive nominals, a topic rarely addressed, within the framework of generative lexicon theory.

I would like to thank the members of the programme committee for their help in making the workshop an interesting and productive exchange, and the anonymous reviewers for their contribution in preparing this volume. The programme committee of the workshop was composed of: *Sabine Bergler, Ted Briscoe, Nicoletta Calzolari, Bonnie Dorr, Danièle Dubois, Christiane Fellbaum, Danièle Godard, Marie-Luce Herviou, Jacques Jayez, Judith Klavans, Beth Levin, Martha Palmer, James Pustejovsky and Frederique Segond.* The workshop was in part sponsored by Rank Xerox, Meylan, France.

Patrick Saint-Dizier

AN INTRODUCTION TO THE LEXICAL SEMANTICS OF PREDICATIVE FORMS

A Computational Perspective

PATRICK SAINT-DIZIER
IRIT - CNRS
Université Paul Sabatier
118 route de Narbonne
31062 TOULOUSE Cedex FRANCE

Abstract. This introduction presents the main paradigms considered in the papers of this volume: notions of argument structure and thematic roles, the notion of a network of word-senses, as embodied in WordNet, approaches for constructing verb semantic classes, the Lexical Conceptual Structure and the Generative Lexicon.

In this introductory chapter, we present some of the main lexical semantics paradigms used in the chapters that follow. Most of the elements presented here come from different linguistic traditions. To make the text homogeneous, we have somewhat reformulated or simplified some of them. In the next chapters, as shall be seen, these linguistic elements have also been adapted or rephrased to be better adapted to a computational perspective.

We successively present the notions of argument structure and thematic roles, the notion of a network of word-senses, as embodied in WordNet, approaches for constructing verb semantic class, the Lexical Conceptual Structure and the Generative Lexicon.

1. Argument Structure and Thematic Relations

A large number of studies across languages in lexico-semantics, syntax and morpho-syntax have shown very robust regularities and strong generalizations around the notion of argument structure. Semantic labels, often called thematic roles, have been proposed to name the semantic relation which holds between an argument and its predicative term. Complex relations,

P. Saint-Dizier (ed.), Predicative Forms in Natural Language and in Lexical Knowledge Bases, 1–52.

called linking rules, have been established between grammatical functions (subject, object, indirect object) in sentences and arguments. Subsequent relations have been established between arguments, labeled by thematic roles, and represented by means of variables, and deeper semantic representations, such as the Lexical Conceptual Structure. Most of the studies on argument structure and thematic roles tend to converge on concrete elements (e.g. arity of a predicate), while they may diverge on more theoretical considerations. As we will see below, these divergences may be more or less deep and may be motivated by different perspectives, in particular on the use and the utility of thematic roles. Besides a relatively neutral presentation of the notions of argument structure and thematic roles, some crucial debates are touched on here in order to better outline both the foundations and the practical utility of these notions.

1.1. ARGUMENT STRUCTURE

The development of research on argument structure has in large part been motivated by the need to go beyond mere enumerations of possible combinations (e.g. subcategorization frames) of predicates with arguments. Different investigations have revealed the necessity of a more organized representation of predicate-argument relationships, according to very abstract principles. Lexical knowledge has been shown by a number of authors to be organized around families and classes of predicates, as shown in e.g. (Levin 93). This organization enables us to syntactically and semantically characterize very precise and subtle combinations of arguments with a predicate. In particular, a number of syntactic behaviors are determined by the semantics of the compound predicate-arguments.

Several aspects of lexical knowledge being most probably hierarchically organized, one of the main tasks is to identify general knowledge and constraints which are valid for a whole family of predicates. More specific knowledge and, ultimately, idiosyncratic knowledge, will be described at lower levels in the hierarchy.

A hierarchical organization is also a way to avoid redundancies and to characterize phenomena at an appropriate level of abstraction. The use of the predicate-argument structure viewed as expressing the compositional relations which hold between a verb and its arguments, combined with the Projection Principle (Chomsky 65, 81) (and a few other principles) allows the production or the recognition of well-formed propositions. These ideas are further elaborated upon in (Stowell 81), (Pesetsky 82), (Marantz 84), (Rappaport and Levin 88), (Pinker 89), (Williams 94), (Goldberg 94) and (Levin and Rappaport 95).

Every predicate is assumed (or is postulated) to have an argument structure, although assigning one to a given verb-sense may turn out to be a very difficult task. The argument structure specifies the number of arguments the predicate has, the arguments being the participants which are minimally required for the activity or state described by the predicate to be understandable. This minimality is subject to debate. Indeed, from a cognitive and AI perspective, the minimal number of arguments may be slightly larger, by including arguments which play a very important role in the proposition, e.g. some instruments or localizations, which would not be included in a stricter linguistic perspective.

An argument structure is often associated with a subcategorization frame and a thematic grid, but an argument structure does not *a priori* say anything about syntactic categories. It specifies in a very abstract way the arity (i.e. the number of arguments) of a predicate. The syntactic aspects are dealt with by additional, independent principles. For example, the Projection Principle (and its variations) describes how arguments are realized at D-structure. Alternations (Levin 93) show which arguments, for semantic classes of verbs, can e.g. be moved to other grammatical positions, be deleted, or be pronominalized.

For example, a verb such as *give* is of arity 3, exemplified in 'X gives Y to Z'. Additional examples are given here:

run, verb, arity 1.
cut, verb, arity 2 (or 3 if one wants to include the instrument, however this argument is unlikely to be part of the argument structure).
disturb, verb, arity 2 or 3 if the cause of the disturbance is included into the argument structure (which is likely).
mix, verb, arity 2, notice that the mixed elements are all included into a single argument, which may have the form of a set of conjoined NPs. This argument is often said to be *plural.*
between, preposition, arity 3 (X is between Y and Z).
on, preposition, arity 2.
envious, adjective, arity 2.
algorithm, noun, arity 1 (algorithm of X).

Even if this is not totally necessary, some approaches combine argument structure with elements from the subcategorization frame. In that case, syntactic categories are mentioned usually as XPs, as in, e.g.:

run, NP_1.
give, NP_1, NP_2, PP_3. (This latter PP can also be an NP in the case of a ditransitive construction.)
envious, NP_1, PP_2.

The number in index indicates the argument position at D-structure, or at any other structure termed a 'basic form'. In Lexical Functional Grammars

(Kaplan and Bresnan 82), grammatical functions are given instead of syntactic categories, with SUBJ and OBJ being usually NPs and OBL(iques) being usually PPs:

run, (SUBJ).
say, (SUBJ, OBJ, OBL).

Finally, another notation, defined in (Rappaport and Levin 88) and based on the notion of D-structure, introduces variables representing arguments and a typing of the preposition:

go: X $<$ Prep_{loc} Y $>$
push: X $<$ Y $\text{Prep}_{direction}$ Z $>$.

In this notation, the variable X represents the subject (called the external argument, which motivates it being outside the bracketed list). The other arguments are called the internal arguments; they are represented within the bracketed structure. Prepositions introducing PPs, are noted as Prep; they are respectively typed, in these examples, loc(alization) and direction. Verbs which do not have any external argument, such as unaccusatives, e.g., *fall*, do not have any variable outside the bracketed list:
fall: $<$ X $>$.
In languages (e.g. Romance except French) where there is no compulsory realized subject, the argument is realized as an object, as in the following Italian sentence:
è caduto Giovanni. (John fell) In French, this argument will have to be realized as a subject for the sentence to be well-formed:
Jean est tombé.
Argument structure is then not isomorphic to D-structure and to c-structure (constituent structure, in terms of arguments) in Lexical Functional Grammar.

From a methodological point of view, assigning an argument structure to a predicate may be a difficult task. It may indeed, for example, be quite difficult to identify the 'basic' form and structure of a verb-sense with respect to to the slight sense variations and the metonymies this verb-sense accepts. For example, the French verb *afficher* has a standard use:
afficher quelque chose sur une surface, (to post something on a wall), and
afficher son orgueil, (to exhibit one's pride.)
If these two uses may be felt to be related to the same sense of *afficher* (even if it has different translations in English), then the problem is to identify the arity. It is clearly 3 since, in the second example, the pride is 'exhibited' on one's face or in one's attitude, but that third argument remains usually implicit. The notion of a verb-sense is itself subject to intense debates: one can have a very large or restricted view of what a sense is (an illustration

of this problem is given in the sections devoted to WordNet and to the generative lexicon).

From this point of view, incorporation (Baker 88) raises difficult problems for argument structure. For example, for a number of movement verbs, the preposition may or may not be present:
to cross [$_{NP}$ *the street*] vs *to cross* [$_{PP}$ *over the river*]
to climb [$_{NP}$ *the mountain*] vs *to* [$_{V}$ *climb down*] [$_{NP}$ *the mountain*].
Consequently, the arity of each of these verbs is 2, but in the subcategorization list we have either an NP or a PP. In the second example, the preposition is an integral part of the verb:
to (climb down) the mountain.
In that case, the object remains an NP. A quite similar problem arises with the incorporation of the object as in:
John is fishing.
?*John is fishing fish.*
John is fishing cod.
In that case the arity may be either 1 or 2, since the object cannot be considered a modifier. From a 'cognitive' perspective, it would be 2 since the object is cognitively present in the statement, even if it is not explicit.

1.2. THEMATIC ROLES

Thematic roles are abstract labels that characterize the semantic relations between predicates and their arguments. Each argument of a predicate is marked with a thematic role, which indicates, in a very general way, the 'semantic' role played by the argument with respect to the predicate. Thematic roles have been subject to many controversies, and there is still little agreement on their nature, definition, and role in linguistic theories (Gruber 67), (Jackendoff 87), (Rappaport and Levin 88), (Roca 92), (Ravin 90).

1.2.1. *Definition of thematic roles*

Before studying in more depth the status and role of thematic roles, we provide a partial list of roles, which is generally agreed upon:

- **agent**: the entity who intentionally initiates, makes or originates the action described by the predicate,
- **patient**: the entity that undergoes the action described by the predicate, it is often an animate entity,
- **theme**: the entity moved (in a very general sense) as a consequence of the action expressed by the predicate, it is often a non-animate entity,
- **experiencer**: the entity that experiences some psychological state resulting from the predicate,

- **goal, source, location**: are roles related to spatial, temporal or abstract fields, expressing respectively the goal, the source or the position of a temporal, spatial or abstract entity.

Thematic roles are postulated by a number of authors to be universal, non-ambiguous, and to cover the whole spectrum of the predicate-argument relationships. This is certainly somewhat optimistic. Thematic roles are essentially assigned to NPs, by verbs, prepositions and VPs via predication. Their uses and meaning may be either direct or metaphorical.

Predicates, and in particular verbs, are usually associated with a thematic grid, which contains in general one role per argument position. However, since a verb usually may accept a large variety of arguments, grids remain often quite general. For example, the following predicates have the following grids:

give: [Agent, Theme, Goal].
buy: [Agent, Theme, Source].
love: [Experiencer, Patient].
toward: [Theme∨Agent, Goal].

These simple examples call for several remarks. First, roles being very general, the relations they describe remain superficial. However, even with this high degree of generality, there are many verbs for which it is not possible to define a thematic grid using, e.g. the above roles. For example a number of communication verbs such as *negotiate* or verbs expressing a quantity such as *cost, weight* as in *costs 30 Francs*.

Some theories, such as Government and Binding, via the θ-criterion, require that each argument position be assigned one and only one thematic role, and that each argument position be assigned a different role. This constraint clearly depends on the form and nature of the roles (using very refined roles will entail a lower probability to mark two argument positions with the same role), but it has been shown to be false in a number of cases (see, e.g. Jackendoff 87)

Next, there may be several candidate roles for a given argument, for two reasons: the distinction between roles is not as straightforward as it may seem at first glance, and there is, in fact, a kind of continuum between roles. For example, there are verbs where the first argument (which is often an agent, and in general the subject in the basic form or at D-structure) has a very weak degree of volition, as in: *John thinks about Mary*
where the first argument can either be an agent or an experiencer. In fact, depending on what John is thinking about (e.g. Mary or a math problem), its degree of volition may be different. The second reason is that arguments may play different 'roles'. For example, for the verb *give*, the first argument is clearly an agent, but it is also the source of the transfer of possession. One may consider that, for that verb, the agent role is more prominent than

the source, but this is not necessarily always the case. This problem can be resolved by associating with each argument a cluster of roles, possibly ordered from the most to the least prominent.

The next remark concerns the thematic grid of the preposition. In general, a preposition assigns a thematic role to its 'object' arguments, i.e. the arguments in the scope of the PP it heads. Therefore, *toward* assigns goal to its object NP. A preposition being a relation, it is also necessary to take into account the other argument, a kind of 'external' argument, that the preposition shares with the verb or the VP in which it appears. However, that external argument may bear different roles:
The arrow moves toward the target: the external argument is a theme,
John runs toward the restaurant: John is an agent.
This problem can be solved in two ways: by leaving the first role underspecified or by listing the different possibilities (as in the above example, the symbol / being interpreted as a disjunction).

The next remark concerns the introduction of disjunction of roles. Some predicates may indeed assign different roles to a given argument position depending on the argument. For example, for the French verb *alerter*, we may have:
Jean alerte le directeur du danger. (John warns the director of the danger) where John is an agent and,
Le bruit alerte le surveillant. (The noise warns/alerts the guard), where the noise is a theme (probably a causal theme).

The last remark that we want to formulate is that other categories than NPs (and by extension, PPs) may be assigned thematic roles: adjectives, propositions, PROs can be assigned roles. For example, in:
To drink alcohol is prohibited before driving a car,
the proposition *to drink alcohol* is an entire proposition, which receives a role, probably theme. Similarly, the PRO in:
$[_S$ PRO driving a car $]$
is marked (unexpressed) Agent.

1.2.2. *About the status of thematic roles*

Thematic roles are a very old concept. The first traces can be found in ancient traditions, as old as Panini's *Karakas*. More recently, thematic roles have been introduced by several authors for different purposes: Gruber's thematic relations (Gruber 65), Fillmore's deep cases in his Case Grammar (Fillmore 68), Davidson's event logic (Davidson 67). Slightly more recently, they have been developed by Jackendoff (Jackendoff 72, 83), Talmy (Talmy 85), Cruse (Cruse 73) and within the GB tradition, as θ-roles. In spite of all these investigations, no one has proposed, or even attempted to propose, a comprehensive and non-ambiguous list of thematic roles.

Thematic roles can be considered either from a syntactic or from a semantic and cognitive perspective. GB syntacticians and others have included θ-roles in grammatical rules and principles. The θ-criterion ensures that each argument is assigned a thematic role and that two different arguments are marked with two dictinct roles. Thematic roles have also been used to resolve some anaphors or adjectival passives (see (Roca 92)).

A radically different point of view is presented in (Jackendoff 83), where the author claims that thematic relations are part of a level of semantic/conceptual structure, not part of syntax. Variables in the Lexical Conceptual Structure (LCS, see section 4) can be marked with thematic roles. The same variable can be marked with a different role in different parts of the LCS formula (e.g. X in X give Y to Z) is marked Agent and Source in the LCS). Roles are then viewed as mnemonics, not as argument positions. They are however not considered as primitives, but as structural notions, comparable to subject and object (see also (Ravin 90)).

Thematic grids do not form, alone, a semantic representation because most roles cover too many notions (see 1.2.3), but combined with semantic classes (or families) of verbs (e.g. those of WordNet such as possession, communication, or cognition), they can be viewed as defining a minimal, coherent, and relatively comprehensive, form of semantic representation.

A compromise between these two positions consists in viewing thematic roles as a bridge between syntax and semantics. On the syntactic side, thematic roles are linked to the argument structure and to the subcategorization frame. They also play a prominent role for defining grammatical functions of each argument of a predicate. A number of more or less overlapping hierarchies of thematic roles have been defined, according to a prominence criterion (Grimshaw 91), (Rappaport and Levin 88). One of the most commonly admitted is the following (found, e.g. in (Ravin 90)):
Agent < *Experiencer* < *Goal, Source, Location* < *Theme, Patient.*
Where Theme and Patient are the most prominent roles. They correspond to the grammatical function Object (in 'basic' forms), which is the closest argument to the predicate, then comes the oblique and the subject. Consequently, if there is an argument marked as Agent, then it will naturally be the subject of the sentence. If there is none, then an argument marked as Experiencer is a potential candidate to subjecthood. On the semantic side, thematic roles bind arguments to conceptual variables in the LCS or to more more abstract and possibly underspecified positions.

Finally, a fourth point of view on thematic roles is that they are unnecessary. This position is adopted in (Rappaport and Levin 88), with several clear and insightful arguments. A first class of arguments is based on the fact that thematic roles can't be defined for any situation and that there are many irregularities in syntax and in semantic representations, which

strongly affect their usefulness. The relation between the LCS and the syntactic structure (e.g. a D-structure) is simply mediated by the predicate-argument structure, which does encode the compositional relations between the verb and its arguments. Another class of arguments is related to meaning: a thematic grid is not a viable representation because it hides, to a certain extent, the possible cross-classifications of verbs with respect to the meaning components that they may share, and possibly reflected in their syntactic properties (Levin 93), among which the alternations shared by those verbs. The unstable character of thematic grids is also advocated in (Levin 93), and exemplified by means of the locative alternation (see section 3). For example, for the verb *load*, we have the following grid:
[Agent, Theme, Goal]
and the d-structure form:
Load: X, < Y, Prep_{loc} Z >, corresponding to:
John loads hay onto the truck.
The alternation called 'Spray-Load', produces the form:
Load: X, < Y, Prep_{with} Z >, corresponding to:
John loads the truck with hay.
where the roles given in the thematic grid are no longer appropriate, they are not stable under alternations (changes in syntactic form). To overcome this difficulty, the authors propose either to allow several roles to be assigned to a given argument position or to introduce a feature analysis of roles, slightly similar to the proto-role philosophy (see section 1.2.3):
Load: [+Agent, [+Goal, +Affected, + Theme], +Theme].

1.2.3. *Thematic proto-roles, roles as prototypes*

In this section, we summarize Dowty's approach to thematic roles, which is an attempt to give more precise definitions to thematic roles and to integrate them into a deeper semantic perspective than previous works. The basic idea (Dowty 89), which is true for most types of semantic representations, is that a semantic theory of predicates, and of verbs in particular, must enable us to draw a set of consequences on the evolution of the 'current' world from their lexical semantics representation. For example, from the form 'X gives Y to Z', the theory must be able to produce the inferences that state that (1) Y no longer belongs to X, but to Z, and that (2) X can perform concrete actions. Another important point worth mentioning is that a semantic distinction is "visible" at the thematic role level if and only if it is possible to show that this distinction is relevant for argument selection and lexicalization.

Roles as types and prototypes.
Dowty's hypothesis is then that a thematic role could be viewed as a 'cluster' of presuppositions and implications, e.g. an Agent is a rational being

with a certain degree of volition. In order to define these presuppositions and implications, Dowty first defines the notion of *individual thematic role*, attached to a given predicate δ:
$\Box[\delta(x_1, x_2, ..., x_i, ..., x_n) \Rightarrow \alpha(x_i)]$.
This expression says that the individual thematic role of x_i is the set of the properties α such that the above implication is true.

Then, a type of thematic role can be defined. For a given argument position, playing a specific role in any predicate $\delta \in T$, a set of predicates, a *thematic role type* can be defined as the intersection of the individual thematic roles of each δ. Therefore, a certain thematic role type τ is the intersection of all the properties $\alpha_\delta(x_i)$ of the predicates δ. From that perspective, τ summarizes the characteristics of a given thematic role type, it also allows the definition of thematic roles as a set of implications.

Such a set of thematic role types can be defined for any random set of individual thematic roles. Therefore an infinite number of types can be defined. Among them, Dowty distinguishes what he calls *L-thematic roles types* which are relevant types defining roles such as Agent or Patient. Unfortunately, these types remain very vague, and it does not seem possible to precisely specify sets of entailments for each relevant role such as Agent or Patient. For example, the role Theme occurs in too many verbs to allow for the definition of a precise set of entailments, using the above method. Similarly, it is quite difficult to establish the properties and entailments associated with the notion of agenthood. Dowty then concludes that this approach does not lead to sufficiently convincing results.

Thematic proto-roles

Thematic proto-role originated from the fact that thematic roles cannot satisfactorily be defined in terms of discrete roles. Rather, roles should be defined from semantic properties. This idea is first found without much detail in (Cruse 86) where the following decomposition of the notion of agent is proposed:
Agent: *+volitional, +agentive, +effective, +initiative.*
Proto-roles are abstractions defined over several concepts or properties. Dowty suggests in (Dowty 91) that there are only two thematic role-like concepts which have a real role in argument selection: the Agent proto-role and the Patient proto-role. Here are the specific properties associated with each of these proto-roles, from (Dowty 91):
The Agent proto-role:

1. volitional involvment in the event or state described by the verb (Mary eats cakes),
2. sentience and/or perception (Mary knows about it),
3. causes an event or change of state in another participant (Phone calls cause interruptions),

4. causes a movement, relative to a position or to an entity (The wind blew down the barn).

The Patient proto-role:

1. undergoes a change of state (John opens the window),
2. incremental theme: a theme affected in its integrity (Ann writes a book),
3. causally affected by another participant (smoking provokes cancer),
4. stationary relative to movement or to another participant (the truck entered the garage).

Also, the object denoted by the Agent proto-role exists independently of the event described by the verb, whereas the object denoted by the Patient proto-role may not exist.

Dowty suggests that the properties of the proto-roles are essential in the acquisition of lexical meaning. Given a predicate with n arguments, then the role Agent is assigned to the argument which has the greatest number of properties of the Agent proto-role, and similarly for the assignment of the Patient role. If two arguments have approximately the same number of properties of a proto-role, then they can both be either Agent or Patient, and therefore either subject or object. When three arguments are at stake, the argument that recieves the smaller number of proto-Agent or proto-Patient properties will be lexicalized as the oblique.

Compared to Patients and Agents, the other thematic roles given above are felt to be more basic, but still not primitives. A last word about Themes, which are decomposed in (Dowty 91) into:

- holistic themes: denote entities not affected in their integrity, they may move or change of possession,
- incremental themes: denote entities affected in their integrity.

1.2.4. *Representation of thematic roles*

Traditionally, i.e. in most linguistic approaches (LFG, HPSG), thematic grids are associated with argument structure and subcategorization frames:
Give: [SUBCAT: $< NP_1, NP_2, PP_3[to] >$,
ROLES: $< agent, theme, destination >$].
The linear order reflects the degree of obliqueness of arguments.

A neo-Davidsonian representation allows us to focus on the event. Arguments as well as modifiers are predicates operating over events denoted by verbs, a verb is associated with a variable denoting an event, and the different arguments and modifiers are expressed separately:
$\forall e\ \Box[giving(e) \Rightarrow\ \exists x\ agent(x,e) \wedge \exists y\ source(y,e)\ \exists z\ goal(z,e)]$.
This formalism allows the specification of various types of constraints, such as in the following example:
$\forall e,\ \forall x\ \Box[[flying(e) \wedge agent(x,e)]\ \Rightarrow has-wings(x)]$.

Similarly, it is also possible to state more general constraints on the form of the formula. For example, we may want to say that a given thematic role, e.g. agent, is assigned to only one argument. The following constraint can then be defined:
$\forall e,\ \forall x\ \Box[agent(x,e) \Rightarrow\ \forall y\ [agent(y,e) \Rightarrow\ x = y]]$.

1.2.5. *Thematic roles in NLP*

Thematic roles are used in natural language processing with different goals in mind. They can first be labels of semantic nets, frames or conceptual graphs, in the tradition initiated by Fillmore. They can also be used to mark the roles in predicate-argument structures (Pugeault et al. 92). They are also used in large lexical knowledge bases to make more explicit the meaning of the predicates (Dorr 95). This is, for example, the case with the frames being introduced in WordNet (see section 2) which resemble thematic-roles.

In this volume, several papers deal with thematic roles, or use them at different levels: papers by C. Jones, A. Stein, D. Kokkinakis, P. Saint-Dizier and E. Viegas et al. The reader will discover different definitions for thematic roles, which largely overlap, and different useful uses.

2. Lexical Semantics Relations and WordNet

Lexical semantics relations introduce a different perspective on semantics (see (Cruse 86) for a detailed presentation) in the sense that they introduce semantic relations between lexical entries. The basic assumption is that the semantic properties of a lexical item are fully reflected in a number of relations associated with different types of contexts (the environments in which sentences utter). Since relations may be very diverse (e.g. in thesauri, or in semantic nets devote to applications), let us limit the notion of context to linguistic contexts, which should cover the whole spectrum of senses of an item, in particular if we assume that extra-linguistic contexts can in principle be mirrored linguistically. Linguistic contexts are also easier to control, analyse and manipulate. It is then of much interest to take into account and to study the convergence of grammatical properties and semantic properties of lexical items, and how they are realized in utterances.

Let us now assume that contexts can be defined by either isolated words or by expressions. Then it is possible to analyse the affinities of a word with respect to a context. Contrasts will then enable us to define semantic distinctions. There are two ways of analyzing affinities: a *syntagmatic* way, using an analytical approach based on the description of properties associated with lexical items and contexts, and a *paradigmatic* way, by stating relations between words or between a word and a context. These relations

The number of levels in a taxonomy is subject to debate. Whereas taxonomies of natural elements (e.g. in biology) do not tend to go beyond 5 levels (Cruse 86), it seems that more technical ones (i.e. related to human artefacts) can go up to 7 levels. A taxonomy may also have holes, if there are nodes for which there are no words in a given language. Taxonomies may be related to a technical domain (and possibly part of a thesaurus) or of general purpose. For this latter type, there are now efforts to normalize the top-level nodes so that similar typing (or selectional restrictions) can be used by different users and in different languages. Such an effort, is carried out e.g. in WordNet and in the LDOCE dictionary (Longman Dictionary of Contemporary English). Finally, let us notice that decomposing an entity into more specific sub-entities is a non-trivial task. It is more a matter of intuition than of systematic methodology to decompose an entity along a certain dimension rather than along another one. Indeed, there exist few hierarchies of properties which would guide this decomposition.

Meronomy describes the part-whole relation. This relation is fairly complex; it attempts to describe various forms and degrees of differentiation of parts with respect to a whole. Forms of differentiations may indeed be quite varied, e.g. spatial, functional. Meronomies are mainly used for nouns. Different types of meronomies for nouns are described in (Winston and Chaffin 87), a summary is given in (Saint-Dizier and Viegas 95). However, combined with a partial temporal precedence relations, meronomies can be used to describe the subevents associated with a complex event. For example, *give a talk* can more or less sequentially (and recursively) be decomposed into: *introduce, expose, develop, comment on, summarize, conclude.*

Proportional series are non-branching hierarchies. They allow for the (strict) ordering of lexical items according to a certain property or dimension. This mode of classification does not require the specification of precise values for the property considered in the classification. It simply introduces a precedence relation. This approach is particularly useful if one does not want or is unable to assign precise values to properties, or if this assignment requires the introduction of complex systems (e.g. a Euclidean system for movement verbs to characterize the speed of movement). Proportional series are also very useful to compare lexical items from different languages according to a property or to a dimension. For example, a classification of movement verbs according to speed can be carried out separately for French and English and then correspondences can be established.

We distinguish among three kinds of proportional series:

1. series encoding an *a priori* continuous hierarchy, where limits between elements are somewhat fuzzy, as in: small-average-large, stroll-walk-run-rush,

2. non-continuous and gradable (i.e. related to a certain measurable property) hierarchy: tiptoe-jump, implorer- demander- requérir- ordonner, (implore- ask- order- request), vaporiser- liquéfier- solidifier, etc.
3. non-continuous and non-gradable hierarchies coding e.g. institutional or technical hierarchies. These are more appropriate for nouns.

2.1.3. *Non-hierarchical Relations*

Non-hierarchical relations are binary or ternary. They basically structure lexical items in terms of synonyms and various forms of oppositions. Other types of relations, such as 'level of language' are found in thesauri to distinguish between e.g. technical or scientific terms, usual terms and slang terms; they are not usually described as lexical semantics relations.

Synonymy characterizes two lexical items which have a very high degree of overlap in terms of usage, denotations or similar property-value pairs, but the degree of synonymy is not directly reflected by the importance of the overlap. Very few terms are absolute synonyms, but often lexical items are synonyms in some precise contexts (Cruse 86). For example, in the context of *Teach*, *remember* and *memorize* are synonyms. This introduces a kind of ternary relation, which includes the context.

Antonyms and opposites cover a large variety of phenomena. A simple definition would be that L1 and L2 are opposites if they have a lot of chraracteristics which are identical, but they also differ quite notably in a few other characteristics. Like synonyms, antonyms and opposites are often largely contextual. There are different forms and degrees of opposition. Antonyms refer to gradable properties while opposites refer to non-gradable (i.e. boolean) ones.

For example, in the context of *to arrive*, *to leave* and *to stay* are opposites, similarly for *try - fail / succeed*, whereas *to grant* and *to refuse* may be considered as antonyms (considering that there are alternatives). There are several forms of antonyms which are analyzed in-depth in (Cruse 86). An interesting study of antonymy in corpora is reported in (Fellbaum 95).

2.2. AN OVERVIEW OF WORDNET

WordNet is a large project in English, developed at Princeton University, which started in the early eightees at the initiative of G. Miller and C. Fellbaum (Miller and Fellbaum 91). WordNet is deliberately 'limited' to paradigmatic relations (although it introduces in its last version frame-based descriptions). Relations structure the different senses of a lexical item. The WordNet WWW interface is available at: *www.cogsci.princeton.edu.*

2.2.1. *Main Organization*

WordNet makes detailed and accurate sense distinctions for a given lexeme, based on usage, resulting sometimes in more than 20 senses for a verb (e.g. *give* has 27 senses, but there are now trends to reduce this number, in particular in the EuroNet project, The European WordNet). This reflects in a quite extensional way the complexity and the creativity of language. In WordNet, the different syntactic categories are studied independently, for technical and methodological reasons. The authors are of course aware of that sooner or later the relations between the different categories will have to be studied and developed.

Conventional dictionaries as well as most thesauri are built around words. In WordNet, the smallest unit is called a *synset*, which is a set that contains all the senses of different lexical items that express the same 'concept' or 'notion', e.g. [give, offer]. The notion of concept is of course not an absolute one and may receive different levels of granularity. However, in a synset, all the words, with the sense considered (and often paraphrased by a natural language expression in the synset definition to make it explicit), behave identically in natural language utterances. If a word is polysemous, then it appears in several synsets. Consequently, distinctions and synsets are based essentially on usage, not (fully) on the lexicographer's intuition. Lexical semantics relations are established between synsets. About 17,000 verbs have been studied in WordNet, which is very large. About 11,500 synsets have been defined. WordNet is now accessible in different versions and environments (the original database is in standard Prolog) with a very well-designed graphical interface.

Verb semantic classifications have received little attention in the past, compared to the efforts devoted to nouns. However, a first partition of verbs into large 'families' (or semantic domains, also called "files" in WordNet, but this term is vague) is necessary to allow for a better and more 'modular' analysis of senses and relations between senses. A verb may appear in several families, if it is polysemous or if has different 'facets' that naturally fall into different classes.

The analysis of verbs by family is based on the observation that most, if not all, relations operate within quite precise semantic domains. In (Fellbaum 93, 97), verbs are first divided into states (this set is relatively small) and actions. This latter set is divided into 14 families, which may be quite large, based on classes discussed in (Miller and Johnson-Laird 76). These classes are the following: *motion, possession, perception, contact, communication, competition, change, cognition, consumption, creation, emotion (or psychological), bodily care and function, social behavior and interactions, and expression of weather.* These main classes are then further divided into subclasses, according to different dimensions, properties and domains (e.g.

family, law, education etc.). It is clear that, for each level of classification, there are verbs which are more prototypical than others, around which others may aggregate. However, once again, in this domain much work has been carried out for nouns (e.g. (Rosch 75)), but very little for verbs.

Families are headed by 'unique beginners', whose definition and justification is not an easy task, WordNet includes e.g.: *act, move, get, become, be, make.* To some extent, one may consider that these unique beginners may correspond to primitives, as those developed in the Lexical Conceptual Structure approach (Jackendoff 90). In that case, primitives could be combined with semantic domains (e.g. possession, space), corresponding to the fields of the Lexical Conceptual Structure (see section 4 below).

A number of experiments have been carried out to explore the psychological validity of the relations used in WordNet (Chaffin et al. 94). Results are good and they confirm on a larger scale the intuitions at the origin of WordNet. Interestingly enough, troponymy and opposition turned out to be the most frequently considered, indicating therefore a certain psychological salience of these relations. The results also suggest that there is a kind of 'homomorphism' between the verb organizations in human's mind and in WordNet.

2.2.2. *The Paradigmatic Relations in WordNet*

The number of relations developed in WordNet for each type of lexical category has been deliberately kept small, for various reasons, explained in (Fellbaum 97). Only the most appropriate, unambiguous, and productive relations have been considered for each category. We briefly review them here for verbs, examples are often those given in (Fellbaum 97). It should be noted that a relation between two synsets holds for any item in these synsets (this may explain why senses in WordNet are very accurate, and therefore quite numerous for productive items).

Entailment relates two verbs V1 and V2 where V1 necessarily entails V2 in our current world. For example, *snoring* entails *sleeping.* As can be seen from this example, this relation is not symmetric: *sleeping* (fortunately) does not entail *snoring.* Notice also that *snoring* and *sleeping* are not related by an isa relation, as is *moving* which subsumes *running.* In that case, it is clear that the lower verb in the hierarchy (running) entails the higher one (moving). However, how far this 'deduction' can be carried out remains an open problem, and depends on how verbs are structured in the isa hierarchy.

The relation of entailment is also often related to a temporal dimension. Snoring is temporally included into sleeping. Similarly, there is a kind of part-of relation holding between the cause and the entailed verb: if *eating* entails *chewing* then one can say that chewing is a part of the more complex action of eating, similar to *ingesting.*

Troponymy relates two verbs in a generic-specific way. For example, since *running is a way of moving*, it follows that run and move also stand in a hyponymy relation. Similarly, disassembling or disconnecting are ways of detaching, and sawing is a way of cutting. It is not very easy to organize verbs into taxonomies. The hyponymy relation can however be specialized for verbs: precise criteria can be considered such as the reference to a means, a cause, a certain property and its associated potential values, a conveyance, etc. For example, disconnecting is a means of detaching, cycling is a conveyance for moving or riding, shouting is a way of speaking very loudly. The troponymy relation gathers all these distinctions in stating that V1 is in a troponymy relation with V2 if 'V1 is to V2 in a certain manner'. As noted in the observations above, troponymy can be viewed as a certain form of entailment.

Opposition covers, as indicated above, a large spectrum of phenomena. Oppositions enable us to make actions which are complementary such as screw-unscrew, pack-unpack (the 'negative' morphological mark is not always there), or antagonistic actions: leave-stay. Oppositions in WordNet are grouped under a single relation. Its denotation is much smaller than synonyms. For example, *give* has 27 synonyms/senses and only one opposite (*take, takeaway*).

The **cause** relation relates causative and resultative verbs such as *heat-melt, push-fall.* WordNet has also a Cause pointer from causative, transitive verbs to anticausative, intransitive sense of the same verb (*ring, whiten, desintegrate*).

2.2.3. *Syntactic Constructions and Lexical Semantics Relations*

Several studies, among which the most exhaustive for English is probably (Levin 93), have shown that there is a relation between a verb meaning and its syntactic forms (see section 3.3 below). The semantic classes formed from syntactic criteria are very different in form and size form those of WordNet. However, there are more relations between the two approaches that it seems at first glance.

First, it seems that the verb-senses of a given synset share the same syntactic behaviors. This has been studied on a large scale in (Dorr and Jones 96) where most verbs in Levin's classes have been tagged with WordNet synset numbers. Next, some classifications motivated by syntactic distinctions correspond to 'natural' WordNet classifications, where verbs with different syntactic behaviors appear under different taxonomies. For example, the distinction between *create from raw material* and *create mentally* is motivated from different factors in WordNet and in Levin's verb semantic classes. It should be noted, however, that this distinction is language dependent, for example, it can't be made in French, where the alternation system is less rich (see Saint-Dizier, this volume).

2.2.4. *The example of the verb Give*
Let us illustrate the above relations for the verb *to give*, from WordNet version 1.5.

Give has 27 senses, here is a sample:
1. give, cause to have (give her trouble),
2. yield, give, afford,
3. give: transfer of possession of a concrete or abstract object,
4. give, give an idea, an opinion,
5. hold, make, have (give a reception). etc.

It therefore appears in 27 different synsets.

Give has only one antonym, for sense 3, which is take, take away. The other senses don't have any.

The hyperonym trees are, for example:
sense 2: supply, then give (sense 3 above) and then the abstract notion of transfer,
sense 4: state, say, tell, then express, utter,
sense 5: direct (= be in charge of), then manage, then control, command.

There is no entailment relation. The verb *say* has, for sense 6 (*pronounce, articulate*) an entailment link with *talk, speak, utter*, etc., meaning express in speech.

Two senses are subject to a cause relation. For example, sense 3 is associated with *have, have got, hold*, expressing the change of possession.

Finally, the sentence frames just differentiate between 'somebody' and 'something', e.g. *Give something to someone*, and *give someone something.* Frames are under developement and may become more useful in the near future.

WordNet is used or evoked in a majority of the articles presented in this book. Consider in particular the papers by C. Fellbaum, P. Saint-Dizier and the panel paper by F. Busa *et al.*.

3. Syntactic Alternations and Verb Semantic Classes

In her book, Beth Levin (Levin 93) shows, for a large set of English verbs (about 3200), the correlations between the semantics of verbs and their syntactic behavior. More precisely, she shows that some facets of the semantics of verbs have strong correlations with the syntactic behavior of these verbs and with the interpretation of their arguments.

Her book has several aims. She first precisely delimits the different forms of verb syntactic behavior. Each of these forms is described by one or more *alternation* (e.g. alternations describe passive forms, there-insertions and reflexive forms). Then, she proposes an analysis of English verbs according to these alternations: each verb is associated with the set of alternations it undergoes. A preliminary investigation showed that there are sufficient correlations between some facets of the semantics of verbs and their syntactic behavior to allow for the formation of classes. From these observations, Beth Levin has then defined about 200 verb semantic classes, where, in each class, verbs share a certain number of alternations. In other words, classes are based on one or more alternations, and the verbs in a class accept most of these alternations: there are quite numerous exceptions, but there is sufficient evidence to motivate the construction of these classes. The non-acceptance of an alternation may also be a classificatory criterion.

This very important work emerged from the synthesis of specific investigations on particular sets of verbs (e.g. movement verbs), on specific syntactic behaviors and on various types of information extracted form corpora. Other authors have studied in detail the semantics conveyed by alternations, e.g. (Pinker 89) and the links between them (Goldberg 94).

3.1. THE ALTERNATION SYSTEM

An alternation, roughly speaking, describes a change in the realization of the argument structure of a verb. The scope of an alternation is the proposition. Modifiers are considered in some cases, but the main structures remain the arguments and the verb. Arguments may be deleted or 'moved', NPs may become PPs or vice-versa, and some PPs may be introduced by a new preposition. Alternations may also be restricted by means of constraints on their arguments.

Beth Levin has defined 79 alternations for English. They basically describe 'transformations' from a 'basic' form. However, these alternations have *a priori* little to do with the assumptions of Government and Binding theory and Movement theory, in spite of some similitudes. The form assumed to be basic usually corresponds to the direct realization of the argument structure, although this point of view may clearly be subject to debate. Here are now a few types of alternations, among the most common ones, references about works establishing these relations can be found in (Levin 93).

The *Transitivity alternations* introduce a change in the verb's transitivity. In a number of these alternations the subject NP is deleted and one of the objects becomes the subject, which must be realized in English. The *Middle alternation* is typical of this change:
John cuts the cake → *The cake cuts easily.*
As can be noticed, it is often necessary to add an adverb to make the sentence acceptable. The *Causative/inchoative alternation* concerns a different

set of verbs:
Edith broke the window $\rightarrow$ *The window broke.*
Verbs undergoing this alternation can roughly be characterized as verbs of change of state or position.

Under the transitivity alternations fall also alternations where an object is unexpressed. This is the case of the *Unexpressed object alternation* where the object1 is not realized. A number of verbs undergo this alternation. In most cases, the 'typical' object is somewhat 'implicit' or 'incorporated' into the verb, or deducible from the subject and the verb. This is the case, e.g., for the *Characteristic property of agent alternation*:
This dog bites people $\rightarrow$ *This dog bites.*
We also find alternations that change the object NP into a PP, as in the *conative alternation*:
Edith cuts the bread $\rightarrow$ *Edith cuts at the bread.*

A second set of alternations deals with changes within the arguments of the VP. One of the most popular alternations is certainly the *Dative alternation* which concerns verbs of giving, of future having, of transfer, etc., as in:
Edith hands the baby a toy $\leftrightarrow$ *Edith hands a toy to the baby*
(we use the symbol $\leftrightarrow$ for some examples when we feel that both forms have equal status, i.e. one or the other could be considered as basic). The same phenomenon occurs for the *Benefactive alternation*:
I carve a toy for the baby $\leftrightarrow$ *I carve the baby a toy.*
The *Spray/Load alternation* involves the permutation of the arguments in the VP and the preposition alternation on $\leftrightarrow$ with:
to spray paint on the wall $\leftrightarrow$ *to spray the wall with paint.*
English is particularly rich in this type of phenomenon. Let us note also the *Material / product alternation*:
Martha carves a toy out of a piece of wood $\leftrightarrow$ *Martha carves a piece of wood into a toy,*
and the *With / Against alternation*:
to hit a stick against the fence $\leftrightarrow$ *to hit the fence with a stick.*

Other sets of alternations include the introduction of oblique complements, reflexives, passives, there-insertion, different forms of inversions and the introduction of specific words such as the way-construction.

It is clear that these alternations are specific to English. They are not universal, even though some are shared by several languages (e.g. the passive alternation). Every language has its own alternation system, and has a more or less important number of alternations. The characteristics of the language, such as case marking, are also important factors of variation of the form, status and number of alternations. English seems to have a quite large number of alternations; this is also the case e.g. for ancient languages such as ancient Greek. French and Romance languages in general have much fewer alternations, their syntax is, in a certain way, more rigid. The number of alternations also depends on the way they are defined, in particular the degree of generality via constraints imposed on context elements is a major factor of variation (see remarks on this point in (Saint-Dizier, this volume)).

3.2. CONSTRUCTION OF VERB SEMANTIC CLASSES

As explained above, verb semantic classes are then constructed from verbs, modulo exceptions, which undergo a certain number of alternations. From this classification, a set of verb semantic classes is organized. We have, for example, the classes of verbs of putting, which include Put verbs, Funnel Verbs, Verbs of putting in a specified direction, Pour verbs, Coil verbs, etc. Other sets of classes include Verbs of removing, Verbs of Carrying and Sending, Verbs of Throwing, Hold and Keep verbs, Verbs of contact by impact, Image creation verbs, Verbs of creation and transformation, Verbs with predicative complements, Verbs of perception, Verbs of desire, Verbs of communication, Verbs of social interaction, etc. As can be noticed, these classes only partially overlap with the classification adopted in WordNet (section 2.2.1 above). This is not surprising since the classification criteria are very different.

Let us now look in more depth at a few classes and somewhat evaluate the use of such classes for natural language applications (note that several research projects make an intensive use of B. Levin's classes, among which several papers in this volume). A first thing to note is that when we speak of verbs, we in fact think in terms of verb-senses and consider a given sense, whatever this may mean. Next, experience shows that acceptability decisions aren't very easy to make for a quite large number of alternations (this is not unique to English), which in fact explains the introduction of exceptions in classes. We may consider that the lack of flexibility in the acceptability decisions may be compensated by the introduction of exceptions in classes. Next, with respect to WordNet, the classes obtained via alternations are much less hierarchically structured, which shows that the two approaches are really orthogonal.

There are other aspects which may weaken the practical use of this approach, in spite of its obvious high linguistic interest, from both theoretical and practical viewpoints. The first point is that the semantic definition of some classes is somewhat fuzzy and does not really summarize the semantics of the verbs it contains. An alternative would be to characterize a class by a set of features, shared to various extents by the verbs they are composed of (this is done e.g. for defining thematic roles via the notion of proto-role, see section 1.2.3). Next, with respect to the semantic characterization of the class, there are some verbs which seem to be really outside the class. Also, as illustrated below, a set of classes (such as movement verbs) does not include all the 'natural' types of classes one may expect (but 'completeness' or exhaustiveness has never been claimed to be one of the objectives of this research). This may explain the unexpected presence of some verbs in a class. Finally, distinctions between classes are sometimes hard to make, and this is reinforced by the fact that classes may unexpectedly have several verbs in common. Let us illustrate these observations on two very representative sets of classes: verbs of motion and verbs of transfer of possession (notice that a few other classes of transfer of possession, e.g. deprivation, are in the set of classes of Remove verbs).

Verbs of *Motion* include 9 classes:

- Inherently directed motion (arrive, go,...),
- Leave verbs,
- Manner of motion:
    - Roll verbs (bounce, float, move, ...),
    - Run verbs (bounce, float, jump, ...),
- Manner of motion using a vehicle:
    - Vehicle name verbs (bike, ...),
    - Verbs not associated with vehicle names (fly,..),
- Waltz verbs (boogie, polka, ...),
- Chase verbs (follow, pursue, ...),
- Accompany verbs.

Note that the labels 'Roll' and 'Run' do not totally cover the semantics of the verbs in the corresponding class. They are however useful mnemonics for these classes. Also, the difference between the two classes is not very clear. Waltz and chase verbs are interesting examples of very specific classes which can be constructed from alternations. However, few domains are represented, and major ones are missing or under-represented (e.g. type of movement, medium of movement, manner of motion, etc.). However, it should be stressed that full coverage was not Beth Levin's objective; her aim was rather to identify those meaning components which have an influence on syntax and which can originate verb semantic classes. Fromthat point of view, results are of much interest.

Verbs of *transfer of possession* include 9 classes:

- Give verbs (feed, give, lease, ...),
- Contribute verbs (distribute, donate, submit, ...),
- Providing:
    - Fulfilling verbs (credit, provide, ...),

    - Equip verbs (arm, invest, ...),
- Obtaining:
    - Get (book, buy, call, cash, order, phone, ...),
    - Obtain (accept, accumulate, seize, ...),
- Future having verbs (advance, assign, ...),
- Exchange verbs,
- Berry verbs (nest, clam, ...).

In this example, the difficulty of defining the semantics of a class is evident, e.g.: fulfilling, future having: these terms do not exactly characterize the class. Note also the Get class is very large and contains very diverse verbs. Domain descriptions (family, education, law, etc.) as well as moral judgements on the transfer (legal, illegal, robbery) are not accounted for in this classification.

3.3. ABOUT THE UNDERLYING SEMANTICS OF ALTERNATIONS

It is of much interest to analyze in depth the set of verbs which undergo an alternation. It is also interesting to analyze exceptions, i.e. verbs not associated with an alternation but which are closely related to verbs which are associated with it, in order to narrow down the semantic characterization of this alternation.

There is, first, a principle of non-synonymy of grammatical forms (Goldberg 94): 'a difference in syntactic form always spells a difference in meaning'. We have, for example, the following syntactic forms with their associated Lexical Semantic Template (Goldberg 94):

- Ditransitive: X CAUSES Y to RECEIVE Z,
- Caused Motion: X CAUSES Y to MOVE Z,
- Resultative: X CAUSES Y to BECOME Z,
- Intransitive Motion: X MOVES Y,
- Conative: X DIRECTS action AT Y.

From these general observations, we see that form and meaning cannot be considered apart. From the point of view of the principle of compositionality, the meaning of a sentence should not only be derived from the meaning of its components, but it should also include the implicit, partial semantics associated with the syntactic construction (e.g. (Goldberg 94)). Let us now consider several examples.

3.3.1. *About the identification of relevant meaning components*

(Pinker 89, pp. 104 and following), explains that in the conative construction, where the transitive verb takes an oblique object introduced by the preposition *at* instead of a direct NP, there is the idea that the subject is

attempting to affect the oblique object, but may not succeed. But the conative alternation applies to much narrower sets of verbs than those whose actions could be just attempted and not realized. For example, verbs of cutting and verbs of hitting all undergo the alternation, but verbs of touching and verbs of breaking do not:
John touches at Mary's hand.
John breaks at Mary's car.

It turns out, in fact, that verbs accepting the conative construction describe a certain type of motion and a certain type of contact.

The same situation occurs for the *Part-possessor ascension alternation* (*Ann cuts John's arm ↔ Ann cuts John on the arm*) which is also accepted by verbs of motion followed by contact. Here verbs of breaking do not participate in that alternation whereas verbs of hitting and touching do.

Finally, the *Middle alternation*, which specifies the ease with which an action can be performed on a theme, is accepted only by verbs that entail a real effect, regardless of whether they involve motion or contact. Therefore, verbs of beaking and of cutting undergo this alternation whereas verbs of touching do not.

As can be seen from these examples, a common set of elements of meaning can be defined for a set of alternations, such as motion, contact and effect, which contributes to differentiating the semantics conveyed by alternations, and therefore to characterizing quite precisely verbs which potentially undergo and alternation or not. Therefore, membership of a verb to a class depends on some aspects of meaning that the semantic representation of the verb constrains. These aspects may moreover be surprisingly subtle and refined, and difficult to identify and to describe in a formal system. These observations reinforces the arguments in favor of a certain autonomy of lexical semantics.

Let us now survey the semantic components conveyed by some major alternations.

3.3.2. *The dative alternation*

The dative alternation applies to a number of verbs of transfer of possession, but the semantic components which account for the difference between verbs which do accept it and those which do not are really subtle. This alternation conveys the idea of X CAUSE Y to HAVE Z. However, as noted by (Pinker 89), while the class of verb of *instantaneous imparting of force causing a ballistic motion* (throw, flip, slap) allow the dative alternation, the *verbs of continuous imparting of force in some manner causing accompanied motion* do not (pull, push, lift).

Similarly, verbs where "X commits himself that Y will get Z in the future" allow the dative alternation (offer, promise, allocate, allot, assign).

There are also verb classes which accept either one or the other form of the dative alternation (with or without the preposition *to*). Verbs of 'long-distance' communication (fax, telephone) also accept this alternation.

From these examples, it is possible to deduce that the dative alternation is accepted by verbs where the actor acts on a recipient (or a destination) in such a way that causes him to possess something. This is opposed to the actor acting on an object so that it causes it to go to someone. For example, in verbs like *push*, the actor has not *a priori* in mind the destination, but just the object being pushed. On the contrary, *ask* accepts the dative alternation because when someone is asking something he has (first) in mind the way the listener will react, the 'physical' transfer of the information is in general less important.

3.3.3. *The location alternations*

The location alternations (a family of alternation which involve a permutation of object1 and object2 and a preposition change) are also of much interest. The participation to certain of these alternations allows one to predict the type of motion and the nature of the end state. Verbs which focus only either on the motion (e.g. pour) or on the resulting state (e.g. fill) do not alternate. Verbs that alternate constrain in some manner both motion and end state. Let us now specify in more depth these constraints, since in fact quite few verbs do alternate.

For example, let us consider the *into/with* alternation. (Pinker 89) differentiates among verbs which more naturally accept the *into* form as their basic form and which alternate with a *with* form. Their general form is:
Verb NP(+theme) onto NP(+destination), and they alternate in:
Verb NP(+destination) with NP(+theme).
These verbs naturally take the theme as object (e.g. pile). Other verbs more naturally take the location/container as object (e.g. stuff), their basic form is more naturally:
Verb NP(location) with NP(+theme), and alternate in:
Verb NP(+theme) onto NP(+destination).
For these two types of constructions, only a very few verbs require the obligatory presence of the two objects.

If we now consider the first set of verbs, those whose basic form is more naturally the 'into/onto' form, then verbs which have one of the following properties alternate: simultaneous forceful contact and motion of a mass against a surface (brush, spread, ...), vertical arrangement on a horizontal surface (heap, pile, stack), force is imparted to a mass, causing ballistic motion along a certain trajectory (inject, spray, spatter), etc. Those which do not alternate have for example one of the following properties: a mass is enabled to move via gravity (spill, drip, spill), flexible object extended in one direction put around another object (coil, spin, twist, wind), mass is expelled from inside an entity (emit, expectorate, vomit). As can be seen here, the properties at stake are very precise and their identification is not totally trivial, especially for verbs which can be used in a variety of utterances, with some slight meaning variations.

These properties are derived from the observation of syntactic behaviors. While some properties seem to have a clear ontological status, others seem to be much more difficult to characterize. They seem to be a conglomeration of some form of more basic properties.

3.3.4. *The causative and the passive alternations*

Among the many other alternations which have a clear correlation with semantic properties, let us briefly mention causatives/anti-causatives and passives.

First, not all intransitive verbs can be transformed into causative transitives and vice-versa. (Pinker 89) shows that there are three main classes of intransitives which causativize. First there are the verbs which describe an instrinsic change of physical state such as *melt, shrank*, but also *close* and *open*. We don't think that these verbs form a real class, we rather prefer to say that these verbs share, to some extent, a certain property or a set of properties. A second class concerns verbs of motion taking place in a certain manner (slid, float, roll). The last class is more complex, it includes verbs describing a manner of locomotion (walk, race, gallop) or directly a means of locomotion (cycle, ferry, boat, bus). Noncausativizable classes include e.g. verbs of volition or internally caused actions (eat, drink, sing, run), verbs of coming into existence and of sound emission (buzz, chatter), etc.

Many verbs can passivize. In fact, most verbs which clearly have an agent and a theme (or patient) do passivize. There are no clear classes with agent and patient arguments which do not passivize, but there are other classes with no agents or themes which undergo passivization. We won't go through all these classes since they are too numerous. Note that verbs like *open* and *receive*, which describe concrete actions passivize with e.g. a subject of type instrument:

The door is opened by the key,

or other types of verbs such as:

Major financial restrictions were justified by the economical situation.

The first example above may be subject to debates. We indeed believe that the basic form of open is something like:

Edith opens the door with a key.

Now since the instrument *key* turns out to play a major role in the action of opening a door, compared to the subject, the *key* can 'metaphorically' be assigned an agent role, which then explains the passive form. Probably the same is true for the second example.

3.4. SEMANTICS OF THE VERB AND SEMANTICS OF THE CONSTRUCTION

Let us now consider the combination of a verb, with its own semantics, with a syntactic construction. The *Construction Grammar* approach (Goldberg 94) sheds a particularly clear and insightful light on this interaction; let us present here some of its aspects, relevant to the verb semantic class system. The first point concerns the nature of the verb semantics, the nature of the semantics of a construction and the characterization of the interactions between these two elements. The second point concerns the meaning relations between constructions.

Verbs usually have a central use, characterized by a specific syntactic form, but they may also be used with a large variety of other syntactic forms. In this case, the meaning of the proposition may be quite remote from the initial meaning of the verb. Let us consider a few illustrative cases. In:
Edith baked Patrick a cake.
the initial sense of *bake* becomes somewhat marginal, in favor of a more global meaning:
'Edith INTENDS to CAUSE Patrick TO HAVE cake'.
There is not here a special sense of bake which is used, but *bake* describes a kind of 'manner' of giving Patrick a cake.

Consider now the case of *slide*, suggested by B. Levin. From the following sentences:
Edith slid Susan
Edith slid the door the present.
Edith slid the present to Susan
Edith slid the present to the door.
one might conclude that there are two senses for slide (probably very close). The first sense would constrain the goal to be animate while the second would have no constraint. Now, if we impose, in the ditransitive construction, that the goal must be animate, then we can postulate just one sense for *slide*, which is intuitively more conçeptually appropriate. We then need to posit constraints in the alternations on the nature of the arguments which would then allow only those verbs which meet the constraints to undergo that alternation. As noticed very early by Lakoff, a verb alone (and its associated lexical semantics) cannot be used to determine whether a construction is acceptable, it is necessary to take into account the semantics of the arguments.

Depending on the construction and on the verb, the verb may either play an important part in the elaboration of the semantics of the proposition or may simply express the means, the manner, the circumstances or the result of the action, while the construction describes the 'central' meaning. In fact, the meanings of verbs and of constructions often interact in very subtle ways. One might conclude then that there is no longer a clear separation between lexical rules and syntactic rules.

The difficulty is then to identify and describe the syntactically relevant aspects of verb meaning, i.e. those aspects which are relevant for determining the syntactic expression of arguments, via linking rules. Pinker (Pinker 89) notes that these aspects should exist in small number, since they resemble characteristics of closed-classes. This is not very surprising, since syntactic alternations form a set of closed elements. These syntactic aspects, from a generative point of view, allow us to better account for novel uses of verbs in particular constructions by allow possibly unexpeted combinations of verbs with constructions. However, the crucial task is then to be able to effectively interpret these combinations.

From the examples given above, we observe that alternations are not, in general associated with a single abstract meaning, but that they are rather associated with a family of closely related meanings. (Goldberg 94) shows that, for the ditransitive construction, many of the verb classes which can be associated with that construction give rise to slightly different interpretations. In:
Edith baked Patrick a cake.
Edith baked a cake with the intention of giving it to Patrick. In the case of verbs of future having, such as:
John assigns Mary a task.
John acts to cause Mary to get a task at some point in the future, but Mary has no obligation to carry out the task. Similarly for verbs of permission, as in:
Ann allows John an ice-cream.
Ann enables John to have an ice-cream, but nothing indicates that John will effectively have an ice-cream. From that point of view, since the difference in meaning are quite small and can be represented by a set of related meanings, we can view the ditransitive construction in English as a form of *constructional polysemy*: a particular form can generate closely related, but different, senses. This perspective gives a more central status to alternations. If we pursue this direction, we can then say that probably the central sense of the alternation (which is also the most frequent) is the successful transfer and that the other senses are derived or are extensions from that initial sense.

3.5. RELATIONS AMONG ALTERNATIONS

The last point to study about alternations is the semantic characterization of relations between alternations. Via a number of principles and within an object-oriented approach with inheritance, (Goldberg 94) has elaborated several links which hold between alternations: polysemy, subpart, instance and metaphorical extension.

We present here a few examples that she has developed. The *polysemy link* relates an alternation, with a precise meaning with one of its extensions, such as e.g. the ditransitive and the benefactive ditransitive:
Edith gave Ann an apple → Edith permitted Ann an apple.
which can be characterized by the introduction of an intended causation:
CAUSE-RECEIVE(agent, recpt, patient) →
INTEND CAUSE-RECEIVE(agent, recpt, patient).
The polysemy lies in the fact that the meaning associated with the ditransitive is viewed as a kernel meaning from which more elaborate meanings can be derived via other alternations.

The *subpart link* relates an alternation to another one that describes a proper part of the former. This is for example the case of the caused motion construction which is linked to the intransitive motion (the theme is then the subject of the clause since the object is removed), or the resultative linked to the intransitive resultative. The *instance link* allows us to relate an alternation with one of its instances and the *metaphorical extension* links more remote constructions having to do with metaphorical extensions, e.g. change of location and change of state with the same verb:
kick the ball out of the room → kick the ball damaged.

In this volume, the alternation system and verb semantic classes are further explored in the articles by A. Stein, M.A. Zarco, P. Saint-Dizier, M. Palmer et al. and in the panel paper.

4. The Lexical Conceptual Structure

Let us now introduce the Lexical Conceptual Structure (LCS), which is an elaborated form of semantic representation, with a strong cognitive dimension. The LCS came in part from the Lexical Semantics Templates (see above) and from a large number of observations such as those of (Gruber 65). The present form of the LCS, under which it gained its popularity, is due to Jackendoff (Jackendoff 83, 90). The LCS was designed within a linguistic and cognitive perspective, it has some similarities, but also major differences, with approaches closer to Artificial Intelligence such as semantic nets or conceptual graphs (which are very similar). The LCS is basically designed to represent the meaning of predicative elements and the semantics of propositions, it is therefore substantially different from frames and scripts, which describe situations in the world like going to a restaurant or been cured from a desease. It is not specifically oriented toward communication acts or toward the representation of abstract objects of the world (by means of e.g. state of affairs, complex indeterminates), represented as objects, as in Situation Semantics (e.g. a complex indeterminate to model a person who utters an sentence).

4.1. MAIN PRINCIPLES AND CHARACTERISTICS

The LCS is mainly organized around the notion of *motion*, other semantic/cognitive fields being derived from motion by analogy (e.g. change of possession, change of property). This analogy is fine in a number of cases, as shall be seen below, but turns out to be unnatural in a number of others. From that point of view, the LCS should be considered both as a semantic model providing a representational framework and a language of primitives on the one hand, and as a methodology on the other hand, allowing for the introduction of new primitives to the language, whenever justified. Another important characteristics of the LCS is the close relations it has with syntax, allowing the implementation of a comprehensive system of semantic composition rules. From that point of view one often compares the LCS with a kind of X-bar semantics.

The LCS attemps at satisfying a number of principles, among which:

- *expressivity*: the theory must be capable of expressing all the relevant semantic distinctions of a natural language,
- *universality*: semantic representations must be sufficiently abstract to be language independent,
- *compositionality*: the theory must provide means to compute the meaning of a sentence from the individual meaning of words and from the syntactic structure,

In the LCS, a single level of semantic representation is postulated: the conceptual structure, where all types of relevant information are represented. Representations are based on the following principles:

- each major syntactic constituent maps into a conceptual constituent of the appropriate type,
- each conceptual category encodes linguistic and possibly sensory information,
- most categories support the type-token (class-object or type-instance) distinction,
- many categories support quantification, although it is not explicit in the LCS (but may be specified on top of it),

– each conceptual category has a representation based on a function-argument structure, and each argument is itself a conceptual constituent of some category.
– the conceptual structure of a lexical item has zero or more open places for arguments, allowing compositionality to operate.

4.2. THE CONSTITUENTS OF THE LCS

Let us now introduce the different elements of the LCS language. They are mainly: conceptual categories, semantic fields and primitives. Other elements are conceptual variables, semantic features (similar to selectional restrictions, e.g. such as eatable entity, liquid), constants (representing non-decomposable concepts like e.g. money, butter) and lexical functions (which play minor roles). (Pinker 89) also introduces relations between events (e.g. means, effect). These latter elements are not presented formally, but by means of examples.

4.2.1. *Conceptual Categories*

(Jackendoff 83) introduces the notion of conceptual constituent defined from a small set of ontological categories (also called conceptual parts of speech), among which the most important are: *thing, event, state, place, path, property, purpose, manner, amount, time.* These categories may subsume more specific ones, e.g. the category *thing* subsumes: *human, animal, object.* These categories may be viewed as the roots of a selectional restriction system.

The assignment of a conceptual category to a lexical item often depends on its context of utterance, for example the noun *meeting* is assigned the category *time* in:
after the meeting
while it is assigned the category *event* in:
the meeting will be held at noon in room B34.
There are constraints on the types of conceptual categories which can be assigned to a lexical item. For example, a color will never be assigned categories such as event or distance.

Conceptual categories are represented as an indice to a bracketed structure:
$[_{<semantic\ category>}\ ]$
where the contents of that structure has the type denoted by the semantic category. Here are a few syntactic realizations of conceptual categories:
$[_{thing}$ *Mozart* $]$ *is* $[_{property}$ *famous* $]$.
He composed $[_{amount}$ *many* $[_{thing}$ *symphonies* $]]$.
$[_{event}$ *The meeting* $]$ *starts at* $[_{time}$ *2 PM* $]$.

Ann switched off the electricity $[_{purpose}$ *to prevent a fire* $]$.
The light is $[_{state}$ *red* $]$.
$[_{event}$ *Edith begins a new program* $]$.

4.2.2. *Conceptual primitives*
The LCS is based on a small number of conceptual primitives. The main ones are BE, which represents a state, and GO, which represents any event. Other primitives include: STAY (a BE with an idea of duration), CAUSE (for expressing causality), INCH (for inchoative interpretations of events), EXT (spatial extension along something), REACT, EXCH (exchange), ORIENT (orientation of an object), etc. Their number remains small, while covering a quite large number of concepts. A second set of primitives, slightly larger (about 50) describes prepositions: AT, IN, ON, TOWARD, FROM, TO, BEHIND, UNDER, VIA, etc. These primitives are 'lower' in the primitive hierarchy and their number is *a priori* fixed once for all.

4.2.3. *Semantic Fields*
The LCS uses some principles put forward in (Gruber 65), namely that the primitives used to represent concepts of localization and movement can be transposed to other fields by analogy, and generalized. This statement has some obvious limits, but it has been used as far as possible in the LCS. It is clear that there are many similarities across semantic fields, for example, similar prepositions can be used for the expression of movement, localization, time and possession.

The main fields considered in the LCS are the following: localization (+loc), time (+temp), possession (+poss) and expression of characteristics of an entity, its properties (+char,+ident) or its material composition (+char,+comp). (Pinker 89) introduces additional fields such as: epistemic (+epist) and psychological (+psy).

Primitives can then be specialized to a field, e.g. GO_{+loc} describes a change of location, GO_{+temp} a change of time, GO_{+poss} a change of possession, and $GO_{+char,+ident}$ a change in the value of a property (e.g. weight, color). Similarly, BE_{+loc} describes a fixed localization, and BE_{+poss} the possession of something by someone (ommitting the arguments). All combinations of a primitive with a semantic field are not relevant.

4.3. CONSTRUCTION OF LCS REPRESENTATIONS

The LCS is then constructed from conceptual domains (D) composed of conceptual categories (T) and semantic fields (S). A primitive p is then a partial function:
$p \in P,\ p : D_1 \times D_2 \times \ldots \times D_n \rightarrow D_0$

where $D_i \in D$. D_0 is the type of the primitive p. We have, for example, the following definitions:
$GO: thing \times path \rightarrow event$
$TO: thing \vee place \rightarrow path$
$GO_{+loc}: thing \times (path \wedge loc) \rightarrow event.$
Notice that conceptual primitives are stable under semantic fields restrictions.

LCS representations are constructed formally as follows:

- (1) $p \in P$, $v \in Var$, $semantic-field \in S$, $p_{+semantic-field}$ are well-formed LCS representations,
- (2) $p_{+semantic-field}(C_1, C_2, ..., C_n)$ is a wff LCS if the C_i are wff LCS representations (semantic fields may be underspecified or absent),
- (3) $[_e\, C_1, C_2, ..., C_n\,]$ is a wff LCS if e is a conceptual category ($e \in T$) and if the C_i are defined as in (2).

Condition (2) describes the instantiation of arguments by variable substitution while condition (3) describes the creation of the most external structure of an LCS and the treatment of modifiers by concatenation (or adjunction).

The most important structures are the following (where semantic fields have not been included for the sake of readability since there are several possible combinations):

1. $PLACE \rightarrow [_{place}\, PLACE-FUNCTION([_{thing}\,])]$
2. $PATH \rightarrow [_{path}\, TO/\, FROM/\, TOWARD/\, AWAY-FROM/$
 $VIA([_{thing/place}\,])]$
3. $EVENT \rightarrow [_{event}\, GO([_{thing}\,], [_{path}\,])]\ /$
 $[_{event}\, STAY([_{thing}\,], [_{place}\,])]\ /$
 $[_{cause}\, CAUSE([_{thing/event}\,], [_{event}\,])]$
4. $STATE \rightarrow [_{state}\, BE([_{thing}\,], [_{place}\,])]\ /$
 $[_{state}\, ORIENT([_{thing}\,], [_{path}\,])]\ /$
 $[_{state}\, EXT([_{thing}\,], [_{path}\,])]$

PLACE-FUNCTIONS are symbols such as ON, UNDER, ABOVE, related to the expression of localization. Note that STAY is treated as a form of event.

4.4. A FEW EXAMPLES

Writing the LCS representation of a lexical item (and of a proposition) requires to analyze the type of event/state described, the existence of a cause, and then to analyze the role of possible arguments and modifiers such as trajectory and means. This latter point may be quite difficult to represent, and depends on the proposition been treated. A verb like *run* alone does not say anything about the trajectory of its subject: it is left underspecified. The trajectory becomes instantiated in a sentence such as *John runs into the store* to the value: TO_{+loc} path. More specific verbs such as *leave* naturally include (or incorporate) the idea of an origin, they will therefore contain the mention of a source in their semantics:

run: $[_{event}\ GO_{+loc}([_{thing}\ ],\ [_{path}\ ])]$
into: $[_{path}\ TO_{+loc}([_{place}\ IN_{+loc}([_{thing}\ ])])]$
run into: $[_{event}\ GO_{+loc}([_{thing}\ john\ ],\ [_{path}\ TO_{+loc}([_{place}\ IN_{+loc}([_{thing}\ ])])])]$
leave: $[_{event}\ GO_{+loc}([_{thing}\ john\ ],\ [_{path}\ FROM_{+loc}([_{place/thing}\ ])])]$

To make composition of lexical items more clear and to establish an implicit link with syntax, we can use λ-abstractions (possibly typed by LCS forms), as follows:
$\lambda I, P\ [_{event}\ GO_{+loc}([_{thing}\ I\ ],\ [_{path}\ P\ ])]$
where I and P stand for arguments. These notations for arguments are often used in the LCS to indicate the argument positions: conventionally I is the subject, J the first object and K the second object when it exists. These arguments may appear several times, at different places, as e.g. in transfer of possession verbs such as *give* where the subject is both the cause of the transfer (agent) and the source:

$$\lambda I, J, K,\ [_{event}\ CAUSE([_{thing}\ I\ ],\ [_{event}\ GO_{+poss}([_{thing}\ J\ ],\ [_{path}\ FROM_{+poss}([_{thing}\ I\ ]),\ TO_{+poss}([_{thing}\ K\ ])])])].$$

Note that this form of representation motivates the assignment of several thematic roles to arguments: I is the agent (in the cause) and the source (in the path). Jackendoff and others have demonstrated that thematic roles are not primitive notions, but that they can be deduced from LCS representations.

Some arguments may be *incorporated*, i.e. not explicit in the syntax, as for example in the verb *butter* the argument *butter* is incorporated. In *sell*, the money used to pay for the object being sold is also incorporated. The LCS makes it explicit, in a argument not realized in the syntax (it has no index associated):

$$\lambda I, J, K,\ [_{event}\ CAUSE([_{thing}\ I\ ],\ [_{event}\ GO_{+poss}([_{thing}\ J\ ],\ [_{path}\ FROM_{+poss}([_{thing}\ I\ ]),\ TO_{+poss}([_{thing}\ K\ ])]),\ GO_{+poss}([_{thing}\ MONEY\ ],\ [_{path}\ FROM_{+poss}([_{thing}\ K\ ]),\ TO_{+poss}([_{thing}\ I\ ])])])])]$$

MONEY refers informally to the value of the object. The representation of *buy* would be the same except for the variables I and K whose role smust be reversed in the paths, I remaining the cause.

The contents of an argument can also be restricted by some forms of selectional restrictions. For example, for the verb *drink*, the entity being drunk must be a liquid. The representation of *drink* is the following, where the verb is seem essentially as denoting a movement:

$$[_{event}\ CAUSE([_{thing}\ I\ ],$$
$$[_{event}\ GO_{+loc}([_{thing}\ J = LIQUID],$$
$$[_{path} TO_{+loc}([_{place}\ IN_{+loc}([_{thing}\ MOUTH - OF([_{thing}\ I\ ])])])])])]$$

MOUTH-OF is a specific function that refers to or determines the position of the mouth of I. LCS in fact abounds of these more or less ad'hoc functions which are necessary to describe the semantics of verbs.

The LCS is clearly well-adapted to describe the semantics of relational elements, in particular verbs and prepositions. More examples can be found in the literature, in particular in (Jackendoff 90), and (Dorr 93) where LCS is used as interlingua forms in machine translation. It is not designed for nouns and adjectives, and probably also not for adverbs which require more complex forms of predication. The LCS cannot account for a number of 'scalar' properties associated with verbs. For example, it cannot describe the difference between *run* and *walk*. The only simple means which is used is to add an manner, as a modifier, indicating that the movement is realized 'runningly' or 'walkingly' as in:

run: $[_{event}\ GO_{+loc}([_{thing}\ ], [_{path}\ ], [_{manner}\ RUNNINGLY])]$

Let us consider now a few more complex examples which illustrate the genericity and the limits of LCS forms, other examples can be found e.g. in (Pinker 93), where, in addition, relations between events are introduced. To make LCS descriptions more homogeneous, one can proceed by sets of semantically related verbs (verbs of construction, of sound emmission, of movement toward a target, etc.), for which LCS forms are very close, if not identical.

The construction verb class includes verbs like *build, construct, compose*, etc. Let us concentrate on the verb *build*, which includes usages such as:

build a house, a circle, a project, a relation.

The general representation of this verb is:

$$\lambda J, I, [_{event}\ CAUSE([_{thing}\ I\ ],$$
$$[_{event}\ GO_{+char,+ident}([_{path}\ FROM_{+char,+ident}($$
$$[_{+prop}\ EPIST - STATUS(J) = non - exist\ ]),$$
$$TO_{+char,+ident}([_{+prop}\ EPIST - STATUS(J) = exist\ ]),$$
$$FROM_{+char,+ident}(part - of(J)),$$
$$VIA_{+char,+ident}(definition - constitutive(J))])])].$$

which describes the coming into being of J. Two functions, related to lexical data, are used: part-of(J) which gets the parts of J, and definition-constitutive(J) which gets the definition of J (e.g. a circle is a set of points equidistant from a particular point: the center). The function EPIST- STATUS(J) says whether its argument J is known or not. *Build* is probably the generic verb of the construction verb class. If we consider the following sense of the verb *compose*: *compose a sonata*, which is basically restricted to musical pieces (imposed by constraints proper to the verb), we get exactly the same representation, but with a restriction on the object J=MUSICAL-PIECE.

Let us now consider the French verb *supporter* which is highly polysemic. The basic sense, which is the most concrete, conveys the idea of holding up / maintaining something in a certain position, as in:
La poutre supporte le toit.
The object of the verb is a physical-object, often an element of a structure. A second sense of *supporter* is accept, admit, endure, tolerate, as in the following usages:
supporter la violence, la paresse, les insultes, une personne = accept, tolerate
supporter les épices, le froid, le gel, le feu = resist.
In this case, the type of the object argument is either an NP of type 'psychological attitude' or a PP of type 'aggressive physical phenomenon'. This verb also has a sense related to its English counterpart to support:
supporter quelqu'un = soutenir, or *supporter un logiciel, une architecture* = accepter (to accept/support a certain software).
The type of the object is here quite specific, either 'computer element', 'human organization', or 'human'. Finally, there is a more restricted sense, meaning to take on, to take over or to assume, as in:
assumer des dépenses, une tâche, un rôle, un commandement.

Ley us introduce the function PSY-ATTITUDE-WRT(J), whose associated values depend on the semantic nature of the object J. If we have the following two main types of arguments:
- $mental - attitude \vee human$ then the value is accept as in *supporter la paresse* (accept lazyness),
- $agressive - physical - object$ then the value is resist.
Notice that accept and resist may be represented themselves as LCS formula, or they may be elements of an ontology of psychological attitudes, not decomposed any further at this level.

The minimal LCS form for these verbs is the following (where the conceptual category +psy is borrowed from (Pinker 93):

$$\lambda I, J, \ [_{state}\ BE_{+psy}([_{thing}\ I\ ], \\ [_{path}\ TO_{+psy}([_{prop}\ PSY - ATTITUDE - WRT(J)\ ])])]$$

The LCS is considered in a number of papers in this volume, including papers by F. Busa, M.A. Zarco, and the panel paper.

5. Sense Variation and Generativity in Lexical Semantics

5.1. INTRODUCTION

Investigations within the generative perspective aim at modelling, by means of a small number of rules, principles and constraints, linguistic phenomena (either morphological, syntactic or semantic) at a high level of abstraction, level that seems to be appropriate for research on multi-linguism and language learning. These works, among other things, attempt at modelling a certain form of 'creativity' in language: from a limited number of linguistic resources, a potentially infinite set of surface forms can be generated.

Among works within the generative perspective, let us concentrate on the Generative Lexicon (Pustejovsky 91, 95), which has settled in the past years one of the most innovative perspectives in lexical semantics. This approach introduces an abstract model radically opposed to the 'flat' sense enumeration lexicons. This approach, which is now well-known, is based:

1. on the close cooperation of three lexical semantic structures: the argument structure (including selectional restrictions), the aspectual structure and the Qualia structure,
2. on a detailed type theory and a type coercion procedure and
3. on a refined theory of compositionality.

These are briefly presented below. The Generative Lexicon (GL) investigates the problem of the multiplicity of usages of a sense of a lexeme and shows how they can be analyzed in terms of possible type shiftings with respect to the type expected by a central or core sense. Type shifting is modelled by a specific inference mechanism: type coercion. It shows in particular the inter-dependence of arguments and predicates to define the meaning of an expression.

5.2. USAGE RELATIONS IN THE PREDICATE-ARGUMENT PAIR

In this section, we survey the different types of usage relations between an argument and a predicate on a general level. By usage relation, we mean the analysis of the relation between a predicative term and one of its arguments in a language utterance. These usage relations are not all treated by the GL, but we think itis important to survey them to give a more comprehensive view of sense variation.

The relational perspective taken here is mathematically richer than a functional one, allowing us to introduce ambiguity and non-determinism. Finally, it gives an equal role to the predicate and to the argument. We have the following forms of usage relations:

- **Selection**: the type of the argument is subsumed by the type (possibly polymorphic or 'dotted') expected by the predicate for that argument.

In general, the sense conveyed by the verb, which may be largely underspecified, is made more precise by the type of the argument by means of more or less complex compositional rules. For example, for the verb *construire* (construct), the expected object is, for example, of type 'physical-object $\vee$ abstract-construction'. Selection is called selective binding in the GL. Selection is clearly a more frequent phenomenon for general purpose (generic) verbs than for verbs with a more restricted meaning.

- **Metonymy**: in an argument, an entity is used to refer to another one, via lexical relations, among which mainly the part-whole relation (and its different forms (Winston et al. 87)), but also forms such as producer-product or institution-person relations. A number of metonymies have a high degree of systematicity in language (Lakoff 80), possibly over different languages. There are well-known examples of metonymies such as *buses are on strike to-day* (object for user or controller) or *I bought a Ford* (make for object). Most of these forms are treated by the GL via the constitutive or formal roles, but in a rather unconstrained way (except for the constraints implicit in the Qualia).
We can somewhat structure metonymies by considering 3 major forms:
    - metonymies based on the part-of relation,
    - metonymies based on various forms of argument type shiftings such as the make for object illustrated above (*I bought a Ford/ a Sun*). There is always a semantic link such as possession between the expected and the realized arguments. Another link is, e.g. physical object $\rightarrow$ monetary value.
    - metonymies introducing an emphasis on properties of the argument related to actions and to its uses. From the argument's properties, argument type shiftings (also called semantic alternations) can be defined such as the alternation physical object $\rightarrow$ event. The ambiguous example often used in the GL: 'begin a book' is treated by this type of metonymy, using the general rule that any physical object was once created and can be used in a number of ways.
- **Metaphor**: this phenomenon is more complex than metonymy. It is however possible to isolate several types of metaphors which are relatively concrete, and which also have a good degree of systematicity. These metaphors refer usually to partial homomorphisms between ontologies of different conceptual domains, with a focus on some aspects (or semes) carried prominently by the argument. General schemas have been defined, e.g. in (Lakoff 80), such as: *Ideas are objects, ideas are food, creation is birth, activity is a substance*, etc. which can, roughly

speaking and within the GL perspective, allow a type to be changed into another. Metaphors are not, in general, treated by the GL, at least in a simple way.
Rules for metaphors are based on type shiftings; their scope must undoubtly be restricted to some groups of nouns. The semantic interpretation of a metaphorical expression is in general not a trivial matter and much remains to be done in this area.

- **Co-composition** deals with the emergence of unexpected new senses, often with an important sense variation, from the combination of an argument with a predicate. This phenomenon usually goes somewhat beyond compositionality and may be at the origin of semi-fixed or fixed forms. Compared to the definition given in (Pustejovsky 91), we can give here a more restricted view of co-composition. Co-composition occurs when there is an important change in meaning, not simply when an argument has influence over the verb's semantics. We do not think, in fact, that an argument may modify the meaning of a verb, it can just specialize it, but this specialization remains compositional. Situations of co-composition occur when the meaning of a proposition is not just a certain composition of the meaning of its parts, but something different.
Examples abound of this phenomena, such as: *produire des témoins, se payer un zéro, gagner sa vie* (literally: to produce witnesses, to pay a zero, to earn one's living). Co-composition is particularly difficult to analyze. Let us note two points. First, in a number of co-composition relations, a few, usually secondary, features are strongly highlighted, producing the feeling of 'unexpected' use. Secondly, these forms are not formed at random and do not acquire a meaning also at random. Some of them may be the remnants of productive forms, perfectly compositional, several centuries ago. Fixed forms may then be considered as an impoverishment of previously compositional forms.

These forms of variations are claimed (e.g. in (Lakoff 80)) to have a quite good degree of systematicity among verbs. This systematicity may clearly be altered by inherent constraints of verbs. For example, one may say: *construire une maison, un cercle* (to construct a house, a circle) but: *bâtir une maison, *un cercle* (to build a house, a circle), and both verbs are in the building verb semantic class, but *bâtir* accepts only certain types of objects.

5.3. THE PROBLEM OF SENSE DELIMITATION

5.3.1. *Preliminaries*

Let us say very informally that we assume that a sense (more or less large and constrained) has a basic form and one (or more) basic usage(s) (a surface form reflecting the basic sense). The basic sense expects some types of arguments but it also may originate derived usages (via the above generative operations), which are more or less constrained and limited. Sense delimitation is then defining boundaries to derived usages and therefore to sense variation. Note that we can only observe linguistic realizations of these derived usages. Meaning is then assigned to such linguistic expressions from the meaning of their constituents and by taking into account the semantic aspects introduced by the generative operation at stake. Taking this into account, sense variations, more or less important, can be associated with the basic sense of a verb, via a decompositional approach of meaning.

Sense delimitation is largely an open problem. It is indeed almost impossible to state precise and general principles that characterize the boundaries of different senses of a lexeme and what a sense exactly is. To make this approach workable, let us assume that senses exist independently of linguistic realizations, whatever they are called (e.g. concepts), and that it is possible to formally represent them. Sense delimitation largely depends on the theoretical and practical perspectives and, to a large extent, on intuition. The difficulty is then to elaborate a coherent system of sense delimitation and sense/usage variation. Extreme solutions have been proposed, which are not totally satisfactory. For example WordNet (see above) tends to introduce one sense per usage, where a usage is characterized by the semantic features of the arguments. For example, WordNet has 27 different senses for the verb give. Distinctions between senses are often very subtle and somewhat hard to represent in a formal semantic representation. This approach is very useful in the sense that it provides a very detailed description of the usages of a large number of words in English, but we think it lacks generalizations about language which are often useful for NLP systems to work efficiently. On the other extreme, we find AI systems which tend to postulate a unique sense for a lexeme and very complex derivation procedures, involving complex logical systems, to produce different sub-senses. This may be explained by the fact that most AI approaches are not concerned with usages in language but just by meaning representation.

The perspective of the GL is in-between these two extremes, where the different usages of a word are organized around a small, relatively generic, number of senses. From these senses, generative procedures produce or recognize derived usages, with possible slight changes in meaning.

A second problem is the definition of the nature and the form of constraints related to sense delimitation. For example, there are meaning components within a given sense which can be more or less constrained or opened. Within a given sense, modalities of the action often realized by means of adjuncts and which are often more peripheral, such as: manner, means, goals, existence of an effect or telicity, may be left unconstrained whereas elements such as the type of agent, the existence of a cause, or the type of the first object (e.g. physical object or not, as in *vendre un objet* = to sell versus *vendre quelqu'un* = to denounce) may be much more constrained.

Then, given a set of usages assumed to be related to a unique sense, on what basis should the basic usages/sense be defined ? We will see that it is often the most usual usage which is the most primitive, probably the most concrete one, often one of the most widely used, and also possibly historically and ontogenetically the oldest. The basic usage must then, from its semantic properties and representation, be able to allow and possibly to explain the derived usages. We therefore do not postulate an abstract sense from which usages are derived, but rather a concrete usage which is primitive, to which corresponds a basic sense, and derived usages with possible sense variations.

Finally, notice that it is not always easy to identify a basic usage with respect to a derived one, there is indeed in some cases a kind of continuum between basic and derived usages. In the following example, it is quite hard to precisely say where the metaphor starts:
créer une maison, un décor, une mise en scene, une pièce de théatre, une théorie, une école de pensée.
(to create a house, a scenery, the scenery of a play , a theater play, a theory, a school of thought).

5.3.2. *Some principles of sense delimitation within a generative perspective*

As we said above, sense delimitation remains largely intuitive, and depends on the theoretical or the practical perspective. This has been discussed at length in the literature, within a general linguistic and cognitive perspective. The task becomes more complicated when we notice that different senses of a given lexeme *l* may in fact share several quanta of sense, or semes, some being more salient than others. However, we feel it is possible to give some criteria from a generative perspective, which are never 100% true, but which may be good indicators.

Before examining examples, it is important to distinguish a polysemy internal to a verb (Martin 79) from an external polysemy. In the internal polysemy, the meaning of the verb changes for different reasons, independently of the impact of the arguments (but of course, subcategorized

arguments are then different). Changes can be metaphorical, or they can be motivated by the adjunction or the withdrawal of 'semes' or meaning quanta (Cruse 86). A good example in French is the verb *couler* (to flow), which has the following usages:
L'eau coule (water is flowing or running).
Le vase coule (the vase leaks).
Le bateau coule (the boat sinks).
In the external polysemy, the meaning of the verb is fixed, but the composition with different arguments produces different meanings, e.g.:
Le sable coule (metaphor, the sand flows).
There is no formal criterion to distinguish different senses, only a contrastive analysis of utterances permits their identification. Given a fixed sense for a verb, we now examine the different variations entailed by arguments.

A first point is that fixed forms or quasi-fixed forms should be considered as separate senses. By quasi-fixed forms, we mean constructions where the degree of variability of the arguments is very limited, making the expressions often weakly compositional. We have e.g. the case of *arrêter* (to stop) with quasi-fixed forms such as:
arrêter une décision/ une procedure (to finalize a decision/ a procedure) where arrêter = finaliser, décider, prendre.
In this construction, remains the idea of ending a process, the process related to the taking of a decision, but no object is physically stopped, as would be a car or a wheel. Similarly, a sentence such as *gagner sa vie* (to earn one's living) may be considered as a fixed form, meaning either that someone earns a sufficient amount of money to live or simply that someone works and does not depend on someone else. More strikingly is the case of *produire des témoins/ des documents* (literally to produce witnesses / documents, meaning to present witnesses to the court), where almost no features of *produire* are kept.

The meaning variations due to different syntactic forms (alternations) should also not be considered, since they are designed to describe syntactic operations of a given sense.

A second point is the arity, i.e. the number of required arguments. Although defining the arity of a lexeme is not a simple task, we may consider that two usages involving different arities correspond to different senses. Consider e.g. the verb *produire* (produce) which has the following usages:
produire une voiture, un film, un son (object of type physical-object) means fabriquer; it is of arity 2, and describes the physical construction of something
produire une impression sur quelqu'un (to produce an impression on someone) means provoquer une impression, its arity is 3 and the two senses are clearly distinct. We have the same phenomenon with *éprouver*, for which

we distinguish 3 senses:
(1) *éprouver une machine / quelqu'un* means to test a machine / someone, the arity is 2, and both types machine and human are kept at the same level since it is quite hard to say which one is the most primitive,
(2) *la maladie éprouve Jean* (the sickness affects/ afflicts John) means afflict, distress, arity is 2, and the subject is a cause, usually non-human,
(3) *éprouver quelque chose(+psychological state) pour quelqu'un* means to experience something for someone; it is of arity 3 and is clearly a distinct sense. Notice that this latter sense has derived usages, also of arity 3:
éprouver des difficultés à faire quelque chose
éprouver des doutes sur quelque chose
(to experience difficulties in doing something / doubts about something).

Another criterion for distinguishing senses is the belonging to two different WordNet ontological classes (possession, body care, communication, cognition, etc...). If two usages of a given lexeme *l* refer to two different WordNet classes, then we may consider that they are related to two different basic senses. This is not always so clear, of course, since there are metaphorical extensions over classes which are relatively close (e.g. cognition and communication verb semantic classes). For example *gagner* has a sense that belongs to the verbs of possession *gagner un objet/une récompense* (to win an object, a reward), and a different sense that belongs to movement verbs: *gagner la rive* (to reach the bank). Other examples are: *échouer* (= to fail = psychological verb, or to wreck / to run into = movement verb, or, alternatively, verb of creation/destruction), *inspirer, irriter, supporter*. However, each of these lexemes have a few features shared by their different senses, for example *inspirer* has the idea of something (air or ideas) coming into the body or brain, while *irriter* has the idea or irritation, either psychological or physical. One can consider that some of these senses are constructed under metaphor, but they nevertheless convey different meanings. This view is different from having a fixed sense for the verb and having metaphorical uses of arguments.

The next criterion for distinguishing senses is the difficulty of finding a generative operation to relate two usages. Difficulty may be felt not be a criterion, but it should at least be a kind of heuristics. When no generative operation can be clearly identified, we have usages corresponding to different basic senses.

Another important criterion for distinguishing senses, used in WordNet, are selectional restrictions. When selectional restrictions associated with a lexeme are very different, when they come from very different conceptual domains, and if there is no immediate metaphorical link, then we also may consider that we have different senses. This is for example a crucial criterion for the verb *supporter*. The basic sense, which is the most concrete, conveys

the idea of holding up / maintaining something in a certain position, as in: *La poutre supporte le toit.* (= soutenir).
The object of the verb is a physical-object, often an element of a structure. A second sense of *supporter* is accept, admit, endure, tolerate, as in the following usages:
supporter la violence, la paresse, les insultes, une personne = accepter, tolérer
supporter les épices, le froid, le gel, le feu = résister à.
In this case, the type of the argument is either 'psychological attitude' or 'aggressive physical phenomenon'. Notice also that this verb has also a sense related to its English counterpart to support:
supporter quelqu'un = soutenir, or *supporter un logiciel, une architecture* = accepter (to accept a certain software).
The type of the object is here quite specific, either 'computer element' or 'human'. Finally, there is a more restricted sense, meaning to take on, to take over or to assume, as in:
assumer des dépenses, une tâche, un rôle, un commandement = assumer, accepter.
In this latter case, the type of the arguments is something like 'administrative role'. As can be noted, a basic sense can be identified by one or more types. We have indeed the possibility of defining polymorphic types for the arguments of a basic sense of a lexeme. Finally, as advocated above, some types of arguments such as those expressing manner, cause or means may be felt to be less discriminatory than others (e.g. agentivity) for sense delimitation.

Finally, let us note two slightly marginal points which may help identify senses. The first point is the etymology and history of the word. Even if the sense attached to the word has changed a lot, the origin may help identifying basic senses. The second point is more technical and pragmatic: we believe that if we are unable to formally represent the semantics of a verb sense and its variations via a core representation and various instantiations, then it is preferable to split that sense into as many subsenses as necessary.

5.4. THE GENERATIVE LEXICON: MAIN CONCEPTS

The approach developed by J. Pustejovsky in a number of articles (including (Pustejovsky 95)) is based on the notion of type coercion, which is a specific form of inference.

5.4.1. *Argument structure in the GL*

In the GL, four types of 'arguments' are considered:

- *true arguments*, which must be present in the realization of the verb,

- *default arguments*, which may or may not be realized in the syntax, but which are nevertheless in the subcategorization frame,
- *shadow arguments*, which are usually semantically incorporated into the verb,
- *adjunct arguments* which are essentially the modifiers.

For example, for the verb *peindre* (to paint), there are two true arguments, the agent and the object being painted, and possibly a default argument representing the material used; an adjunct could e.g. characterize the instrument. For *butter*, there are also two true arguments (the agent and the object being buttered) and a shadow argument, the butter.

5.4.2. *The event structure*

In the GL, three types of elements are considered: states, events and transitions. This approach is justified in (Pustejovsky 88, 91) and compared to those of (Vendler 67) and (Verkuyl 89). The GL develops approximately the same typology for the organization of events, which may overlap, precede or follow each other. The event structure in the GL is used in some cases of type coercion, in particular in conjunction with *aspectual* verbs (i.e. verbs subcategorizing an object of type event) as in the example of *begin a book*, where we do not know what action (writing, reading, printing, copy-editing, etc.), related to the concept of book, is beginning.

5.4.3. *The Qualia structure*

The Qualia structure is the most innovative structure of the GL. It describes most of the essential attributes of an entity. Although any lexical item may be given a Qualia structure, it is mainly designed for nouns and nominal structures and allows a more powerful taking into account of the semantics of nouns in verb phrases and in propositions. The Qualia structure is a convenient way of implementing the compositional relations between a predicate and its arguments. It does not define the denotational space of the entity in the world, but rather, and broadly, its interpretative space.

The Qualia structure is subdivided into four roles:

- **The formal role**, that distinguishes the object from those which are more generic. Globally, this role includes descriptions close to the *isa* relation, but probably with more flexibility. In particular, we may find indications about orientation, size, form, shape, color, and position.
- **The constitutive role**, which, essentially, describes the parts of the object. Again, this role has a wide coverage and may require some subdivisions to be usable. The GL distinguishes material, weight, parts, etc.

- **The telic role**, which defines the purpose and the goal of the object. All the actions which can be realized on the object are listed at this level (e.g. for book: read, copy-edit, bind, etc.).
- **The agentive role** describes the actions which allow the creation of the entity. For example, we have: an agent/creator, a causal chain or the natural type. Book will then be associated with actions such as write. Publish or copy-edit may also be candidates. Similarly, under this role also fall actions that describe the destruction of the object.

We have in the Qualia on the one hand two roles which describe the structure of the entity considered: the formal and the constitutive roles, and on the other hand two roles of a very different nature which are closer to the event structure: the telic and the agentive roles. The elements in the roles are predicates or names of entities denoting sets, which point to other lexical entries. For example, the Qualia structure of *novel* is the following:
Novel: noun, novel(Y)
constitutive: paper, cover, chapters, references, ...
formal: book(Y),
telic: read(X,Y), print(X,Y), etc...
formal: write(X,Y).

5.4.4. *The generative operations*
The GL essentially considers 3 generative operations: selective binding, co-composition and type coercion. The two first operations have been presented above, let us now concentrate on type coercion, which is in fact the most important generative operation in the GL.

Type coercion is based on the notion of type shifting (Pustejovsky 91), which is a relatively well-known operation in formal linguistics and in computer science. Type shifting allows the creation of new types from subcategorization frames when the type expected for an argument does not correspond to the type found for the realized argument. Let us assume that there exists a certain set Σ of shifting operations defined on a given lexicon L, which can change the type of a lexical item. The *Function Application with Coercion* (FAC) procedure can then be defined as follows. Let α be of type c and β be of type $< a, b >$, then:

(i) if the type of c = a, then $\beta(\alpha)$ is of type b,

(ii) if there exist $\sigma \in \Sigma$ so that $\sigma(\alpha)$ results in an expression of type a, then $\beta(\sigma(\alpha))$ is of type b,

(iii) otherwise there is a type mismatch and the sentence is ill-formed.

For example, the type of a VP headed by *begin* is $<$ *event*, *event* $>$, which is for example realized in:
begin to read a novel.
In a sentence such as:

begin a novel,
there is a type mismatch since *a novel* is of type physical-object. At this stage, the Qualia of *novel* can be used to derive a new type which would meet the verb's requirement. Indeed, in the telic or in the agentive roles of the Qualia of novel, there are predicates such as write(X,Y), or read(X,Y) which refer respectively to the lexical entries write and read, both of type event. Therefore, the type of write and read, which is the type expected by the verb begin, can be 'pumped' and the VP *begin a novel* is well-formed. It is ambiguous and may mean begin reading or writing a book, but of course, the GL is not designed to resolve ambiguities. All the types which can in this manner be derived in one or more steps from the Qualia of novel define the *generative expansion* of the type of novel via type coercion and Qualia structure.

In the sentence:
Milano won the cup,
There is also a type mismatch at the level of the subject of *win* since the type of the expected subject is human, not a town or a location. Now, if, via the contitutive role of Milano, we know that it is a town (via the inheritance structure) composed of buildings, football teams, etc... and then, if we select football team, and consider again its constitutive role, we notice that a football team is composed of humans. Then the expected type is found. From the observation of a number of utterances, we believe that the depth of type pumping, to be psychologically realistic, should not exceed two levels.

The system is very large and powerful and certainly overgenerates. It is necessary to introduce additional constraints. One possible solution we think would be useful is to consider, for each verb semantic class, which roles of the Qualias, for each argument, are the most relevant. For example, for 'aspectual' verbs such as *begin*, coercion for the object is mainly concerned with agentive and telic roles. Formal and constitutive roles do not lead to any viable solution. This short example shows that verb semantic classes can be used to organize and constrain the power of type coercion associated with Qualia structures.

Note, finally, that type coercion can be applied to a large number of syntactic constructions, including adjectives and nominals (Busa 96). However, we think that a generative theory entirely based on types, which are relatively coarse-grained, and on a representation of the world such as the Qualia structure, may not be sufficient to account for all the case of sense variation presented above. Systems based on semantic representations rather than types may be more adequate. This remains to be shown and investigated in more depth.

In this volume, elements of the Gl are used and further elaborated in the article by F. Busa and in the panel paper.

6. Perspectives

Lexical semantics of predicative forms, as can be seen from this short introduction -which is far from comprehensive - is a very large field, with a number of theoretical and practical aspects and challenges. Lexical semantics seems to be now one of the main corner-stones of language processing and will certainly much evolve in the next few years.

Several national and international efforts are being carried out to better organize this field in general, or along a number of aspects. For example, several projects aim at developing a common set of base types that everyone could use to specify e.g. selectional restrictions. On a larger scale, let us mention EuroWordNet (documentation at university of Amsterdam), which has, in addition, a multi-lingual perspective that WordNet does not have. Let us also mention the European project EAGLES, whose aim is precisely to provide users of language technology with surveys and guidelines when looking for or developing resources or applications (http://www.ilc.pi.cnr.it/EAGLES96/rep2/).

Acknowledgements I thank Bonnie Dorr and an anonymous reviewer for their useful comments which greatly helped improving this document.

References

Baker, M.C., *Incorporation: A Theory of Grammatical Function Changing*, Chicago University Press, 1988.

Busa, F., *Compositionality and teh Semantics of Nominals*, PhD. Dissertation, Brandeis University, MA, 1996.

Chomsky, N. *Aspects of the Theory of Syntax*, MIT Press, Cambridge, MA, 1965.

Chomsky, N., *Lectures on Government and Binding*, Foris, Dordrecht, 1981.

Cruse, A., Some Thoughts on Agentivity, *Journal of Linguistics*, vol 9-1, 1973.

Cruse, A., *Lexical Semantics*, Cambridge university Press, 1986.

Davidson, D., The logical Form of Action Sentences, in *The Logic of Decision and Action*, N. Rescher (ed.), University of Pittsburgh Press, 1967.

Dorr, B., *Machine Translation, a view from the lexicon*, MIT Press, 1993.

Dorr, B. J., Garman, J., and Weinberg, A., From Syntactic Encodings to Thematic Roles: Building Lexical Entries for Interlingual MT, *Machine Translation*, 9:3-4, pp.71-100, 1995.

Dorr, B., Jones, D., Role of Word Sense Disambiguation in Lexical Acquisition: Predicting Semantics from Syntactic Cues, in proceedings of *Coling 96*, Copenhaguen, 1996.

Dowty, D., On the Semantic Content of the Notion of Thematic Role, in G. Cherchia, B. Partee, R. Turner (eds), *Properties, Types and meaning*, Kluwer, 1989.

Dowty, D., Thematic Proto-roles and Argument Selection, Language, vol. 67-3, 1991.

Fellbaum, C., *English Verbs as Semantic Net*, Journal of Lexicography, vol. 6, Oxford University Press, 1993. (see also paper in this volume)

Fellbaum, C., Co-occurence and Antonymy, *Journal of Lexicography*, vol. 8-2, Oxford University Press, 1995.

Fellbaum, C. (in press). A Semantic Network of English Verbs, in C. Fellbaum (ed.) *WordNet: An Electronic Lexical Database*. Cambridge, MA: MIT Press, 1997.
Fillmore, C., The Case for Case, in *Universals in Linguistic Theory*, E. Bach and R.T. Hams (eds.), Holt, Rinehart and Winston, New York, 1968.
Goldberg, A., *Constructions: A Construction Grammar Approach to Argument Structure*, University of Chicago Press, 1994.
Grimshaw, J., *Argument Structure*, Linguistic Inquiry monograph no. 18, MIT Press, 1991.
Gross, M., *Méthodes en syntaxe*, Masson, Paris, 1975.
Gruber, J., *Studies in Lexical Relations*, MIT doctoral dissertation and in *Lexical Structures in Syntrax and Semantics*, North Holland (1976), 1967.
Jackendoff, R., *Semantic Interpretation in Generative Grammar*, MIT Press, Cambridge, 1972.
Jackendoff, R., *Semantics and Cognition*, MIT Press, Cambridge, 1983.
Jackendoff, R., The Status of Thematic Relations in Linguistic Theory, Linguistic Inquiry, vol. 18, 1987.
Jackendoff, R., *Semantic Structures*, MIT Press, 1990.
Kaplan, R., Bresnan, J., Lexical Functional Grammar: A Formal System for Grammatical Representation, in *The Mantal Representtaion of Grammatical Relations*, J. Bresnan (ed.), MIT Press, Cambridge, 1982.
Lakoff, G., Johnson, M., *Metaphors we Live By*, Chicago University Press, 1980.
Levin, B., *English verb Classes and Alternations: A Preliminary Investigation*, Chicago Univ. Press, 1993.
Levin, B., Rappaport Hovav, M., *Unaccusativity: At the Syntax-Lexical Semantics Interface*, Linguistic Inquiry monograph no. 26, MIT Press, 1995.
Levin, B., Rappaport, M., *Unaccusativity, at the Syntax-Lexical Semantics Interface*, Linguistic Inquiry monograph nb. 26, MIT Press, 1995.
Marantz, A. P., *On the Nature of Grammatical Relations*, MIT Press, 1985.
Martin, R., La polysémie verbale, *Travaux de Linguistique et de Littérature*, vol. 17, 1979.
Pesetsky, D., *Paths and Categories*, MIT doctoral dissertation, 1982.
Pinker, S., *Learnability and Cognition: The acquisition of argument structure*, MIT Press, 1989.
Pugeault, F., Saint-Dizier, P., Monteil, M.G., *Knowledge Extraction from Texts: a method for extracting predicate-argument structures from texts*, in proc. Coling 94, Kyoto, 1994.
Pustejovsky, J., The Geometry of Events, in: *Studies in Generative Approaches to Aspect*, C. Tenny (ed.), MIT Press, 1988.
Pustejovsky, J., The Generative Lexicon, *Computational Linguistics*, vol. 17, MIT Press, 1991.
Pustejovsky, J., *The Generative Lexicon*, MIT Press, 1995.
Rappaport, M., Levin, B., What to do with θ-roles ?, in *Syntax and Semantics 21: Thematic Relations*, W. Wilkins (ed.), Academic Press, 1988.
Ravin, Y., *Lexical Semantics without Thematic Roles*, Oxford Univ. Press, 1990.
Reuland, E., Abraham, W., (eds) *Knowledge and Language, Vol II*, Kluwer Academic, 1993.
Roca, I.M. (ed.), *Thematic Structure: its Role in Grammar*, Mouton de Gruyter, Berlin, 1992.
Saint-Dizier, P., Verb Semantic Classes in French, IRIT research report, December 1995, also in proc. Coling96, poster-demo, Copenhaguen.
Stowell, T., *Origins of Phrase Structure*, MIT doctoral dissertation, 1981.
Talmy, L., Lexicalization Patterns: Semantic Structure in Lexical Forms, in *Language Typology and Syntactic Description 3: Grammatical Categories and the Lexicon*, T. Shopen (ed.), 57-149, Cambridge University Press, 1985.
Vendler, Z., Verbs and Times, *Philosophical Review 56*, 1967.
Verkuyl, H., Aspectual Classes and Aspectual Composition, *Linguistic and Philosophy*,

vol. 12, 1989.

Williams, E., *Thematic Structure in Syntax*, Linguistic Inquiry monograph no. 23, MIT Press, 1994.

Wiston, M.E., Chaffin, R., Hermann, D., A Taxonomy of Part-Whole Relations, *Cognitive Science*, 11, 417-444, 1987.

A COMPARISON OF DIFFERENT LEXICAL SEMANTICS APPROACHES FOR TRANSFER VERBS WITH A PARTICULAR EMPHASIS ON BUY/SELL

FEDERICA BUSA
Brandeis University

DANIELE DUBOIS
Ecole Nationale Supérieure - Centre national de la recherche scientifique

CHRISTIANE FELLBAUM
Ryder and Princeton universities

PATRICK SAINT-DIZIER
Institut de Recherche en Informatique de Toulouse - Centre national de la recherche scientifique

AND

EVELYNE VIEGAS
New Mexico State University

Abstract. In this paper, are discussed and compared different approaches to lexical semantics, reflecting presentations during the panel discussion at the Workshop on Predicative Forms in Natural Language. The panel was organized with the aim of evaluating different approaches to the study of word meaning, by comparing their advantages and disadvantages from both a theoretical and an applied perspective. The verb pair *buy* and *sell* was taken to support comparisons. The following areas are surveyed: WordNet and the organization of verbs of possession, the Lexical Conceptual Structure, The Generative Lexicon, an integrated view adopted e.g. in Mikrokosmos and a psycho-linguistics perspective.

1. Introduction

This paper is somewhat experimental: the authors discuss and compare different approaches to lexical semantics, reflecting their presentations dur-

P. Saint-Dizier (ed.), Predicative Forms in Natural Language and in Lexical Knowledge Bases, 53–91.

ing the panel discussion at the Workshop on Predicative Forms in Natural Language. The panel was organized with the aim of evaluating different approaches to the study of word meaning, by comparing their advantages and disadvantages from both a theoretical and an applied perspective. Because the discussion was quite successful in reflecting some various important concerns in lexical semantics, it seemed a useful exercise to gather and present some of the main perspectives currently held in the field.

Comparing and collecting notes into one single contribution produced an interestingly heterogeneous picture, reflecting the different concerns of the contributors and the lively debate in lexical semantics. In fact, this paper merges radically different approaches such as WordNet, Lexical Conceptual Structure (LCS), a Natural Language Processing (NLP) oriented approach for constructing verb semantic classes for French, Mikrokosmos, the Generative Lexicon theory (GL), and a psycholinguistic perspective. All these are evaluated around the topic of transfer verbs. Within this class of verbs, which is quite large (more than 200 verbs), we will, in particular, focus on the pair *buy/sell*, to make descriptions and statements more concrete; this allowed for concrete examples, descriptions, and conclusions, which could in many cases be compared and evaluated.

The analysis of the set of transfer verbs provided the common starting point for addressing how the different frameworks handle methodological issues, achieve coverage of the data, and produce suitable results for NLP applications. The particular choice of transfer verbs was guided by a number of reasons: transfer verbs have been broadly studied, they are rather stable cross-linguistically, and they do not display a very high degree of polysemy. The differences that emerge allow us to concentrate on representation issues and to raise the important issue of the nature of lexical semantic knowledge. In particular, as lexical semantics becomes a central concern of theoretical and computational linguists, the role of the lexicon and the questions surrounding a precise characterization of lexical knowledge are changing. The approaches presented reflect precisely this transition in the field and provide distinct answers to the problem of how the ever changing nature of word sense should be handled.

In the first section, Christiane Fellbaum presents the lexical database WordNet, where sets of lexical items are organized into a network by means of conceptual-semantic and lexical relations. WordNet provides a semantic characterization of English words in terms of their position in the net. A lexical resource, WordNet is gaining more popularity as a backbone for developing the lexicon of many NLP systems. While WordNet contains multiple senses of polysemous words, it lacks mechanisms to account for how meaning shifts take place under systematic and predictable conditions.

In section 2, Patrick Saint-Dizier presents a proposal for combining the lexical resources provided by WordNet and syntactic alternations (called contexts in his system) with a Lexical Conceptual Semantics framework for representing the semantics of verbs in an organized way and for reducing the number of senses for a given lexical item.

A similar concern is addressed in section 3, which presents the view held in GL theory. In particular, Federica Busa addresses the role of underspecified lexical representations not only as a useful alternative to sense enumeration but as linguistically motivated characterizations of the behavior of lexical items.

In section 4, Evelyne Viegas presents the computational lexical semantics methodology developed in Mikrokosmos, a knowledge-based Machine Translation system. The methodology reflects the tension between theoretical concerns and practical NLP issues. The questions which are asked are much broader in nature than those addressed by the previous authors, in a certain sense, and the result is a crucially different architecture where the distinctions between linguistic and commonsense knowledge may be hard to draw.

Finally, Danièle Dubois presents some experimental data showing that the semantic classification achieved by a traditional linguistic and language processing methodologies tend to overlap with the results of psycholinguistic experiments.

The reader will find out that the approaches presented below expose very strong biases either towards the theoretical side or towards the applied one. Although this may appear to be a weakness, it can also be a strength if, ultimately, researchers are interested in exchanging their expertise in a particular subarea of a field which is changing shape very rapidly. Our view is that lexical semantics is now a label which characterizes a domain of research that is growing larger and larger as the lexicon acquires a new and increasingly more prominent role in both theoretical and computational linguistics.

2. The representation of some transfer verbs in WordNet

2.1. CONCEPTS, SYNONYMS, AND POLYSEMES

WordNet (Miller, 1990; Miller and Fellbaum, 1991) is a database containing well over 120,000 English word-sense pairs. While conventional dictionaries are built around individual words, the smallest unit in WordNet is a synonym set (synset), which contains all the words that lexicalize a given concept. [1]

[1] "Concept" is notoriously hard to define; for the purposes of this discussion, "concept" and "meaning" are interchangeable.

For example, the verbs *buy* and *purchase* are interchangeable in most contexts, and can thus be considered synonyms. These two word forms are members of a synset representing a concept that could be expressed with the phrase *obtain or acquire by means of a financial transaction.* When a given word form is polysemous (or has homonyms), it is a member of several synsets. Thus, *buy* occurs in a separate synset to represent its meaning as a cognition verb, roughly equivalent to *believe* or *accept* (as in the phrase *I cannot buy his excuse*). Polysemes and homographs in WordNet are therefore not connected other than through their shared word form.[2]

Many languages share in common their conceptual inventories. For example, the two French verbs *dédommager* and *indemniser* denote the concept expressed by the English synset *indemnify, compensate, recompense.* But while equivalent French-English concept pairs can be identified in most cases, the words expressing the concepts in question do not always pattern the same way across languages. For example, a polysemous verb in one language may not be polysemous in another language. Thus, the French verb *gagner* means, under one reading, *earn* (as a salary, for example); its English equivalent, *earn*, is part of the synset that includes *make, clear, bring in, take in.* The second meaning of *gagner* is *win* (such as a game), which is represented by a synset that shares no members in common with the synset containing *earn.* These two senses of *gagner* are distinguished on the basis of their selectional restrictions (a salary or wage vs. a prize or something one covets); moreover, while the first sense implies that the earner returns work for what she or he earns under some kind of contractual agreement, the second sense may involve luck or skill and usually occurs in a less routine, formalized context. (Of course, these two senses of *gagner* are not entirely unrelated, but the distinction between homonymy and polysemy is hard to draw.) In English, the distinction between the two concepts expressed by French *gagner* is reflected in two distinct word forms (*earn* and *win*), each in its own synset. Because English lexicalizes the two concepts by means of two different word forms, what is arguably polysemy in French is not involved in English.

Other cases are more subtle. For example, the English verb *award* arguably has several meanings, although the differences are rather subtle. In WordNet, *award* occurs once in a synset with *present*, and once in a synset with *grant.* In the first case, the verbs refer to the act of giving an honor or prize to someone (*The Oscars were awarded last night*); the second meaning expresses the giving of something, such as a grant, on the basis of merit

[2]By contrast, the approach described in Busa (this volume) and Pustejovsky (1991, 1995) combines many of WordNet's distinct polysemes into one entry and accounts for the semantic differences with elaborate mechanisms that generate multiple, though related, meanings.

(*Research funds are awarded to qualified individuals*). The French word *accorder* seems to be a good equivalent of *award*, since it shares both of these senses.

In its second usage, *award* has as a synonym *grant*. *Grant* in turn is a member of several synsets in WordNet: *allow, grant* (*grant permission*), *accord, allot, grant* (*grant a privilege*), *grant, give* (*grant a degree*), and *grant, deed* (*grant land*). While these senses can all be distinguished on the basis of selectional restrictions, *grant* here clearly refers to closely related concepts. The same is true with its synonyms *accord*, *allot*, and *allow*. The meanings and uses of all these verbs overlap, although they can be distinguished in some specific contexts, where one but not the other verb is felicitous. WordNet captures the semantic overlap between these words through their shared synsets, which can be thought of as overlapping areas in semantic space. The overlap between words and concepts renders a word:word translation across languages impossible. In certain contexts, the French verb *allouer* can be rendered as *allocate*; in others, it must be translated as *grant*, *allot*, or *allow*. The reason is, of course, that in French, these words and concepts overlap just as they do in English, though in a different way.[3]

2.2. CONCEPTUAL-SEMANTIC AND LEXICAL RELATIONS

The semantics of verbs are represented in WordNet not only in terms of synonymy and polysemy, but also by means of conceptual-semantic relations that interconnect the synsets. For example, hierarchical (ISA) relations link more general concepts to elaborated ones. Thus, the synset *allow, grant* is a subordinate of *give*. The meaning of the ISA relation may vary somewhat for different verbs, but the relation between a verb A and its superordinate B can usually be expressed by the phrase *to A is to B in some way or manner*. As Fellbaum (this volume) points out, most of the lexicon can be interrelated by means of the ISA relation; moreover, the productivity of this relation is reflected in the fact that, when new words and concepts come into the language, they usually express semantic elaborations (hyponyms) of existing words and concepts.

The relational organization of WordNet results in a rather clear-cut classification of the transfer verbs that are the focus of this chapter. Most of them fall into one of several distinct groups, each headed by a superordinate synset that expresses a basic concept of transfer of possession.

[3]Whereas WordNet explicitly recognizes distinct meanings for verbs like *award, grant*, and *accord* by postulating different synsets of which the distinct meanings are members, other theories account for the polysemy in terms of senses derived by various generative devices from one core sense (Pustejovsky 1991, 1995; Saint-Dizier, this volume).

The most richly represented cluster is headed by the synset containing *give*, which itself ultimately goes up to *transfer*. *Give* (in the sense of passing on some material possession to someone) is linked in WordNet to a number of semantically more elaborate concepts. Among these are *lease, rent*; *bestow*; *accord, allot, grant*; *concede, yield*; *distribute, deal*; *offer, proffer*; *lease, let*; *render, return*; *leave*; and *provide, supply*.

Synsets are further interlinked by a relation of semantic opposition (Fellbaum, 1990; in press). Thus, the root concept lexicalized by *give* is linked to the concept expressed by *take* via a particular kind of semantic opposition, which Cruse (1986) has called "reversive." *Take* in turn is the head of a large cluster of transfer verbs, including *steal*, *seize*, *wrest*, and phrasal verbs like *take away*.

Besides *give* and *take*, English transfer verbs include a third central root concept, expressed by the synset *get, acquire*. Subordinate concepts are *earn, clear, make* (with its own more elaborate subordinate *eke out*), *accept* (heading in turn *borrow*), *buy, purchase*, and it lease, rent. *Buy* is linked via semantic opposition to *sell* (similarly to the pair *give* - *take*).

Moreover, *buy* is related to *pay* via entailment (see Fellbaum, this volume, for details). The semantic link labeled "entailment" here states that when someone buys something, he also necessarily pays for it. *Pay* in turn is linked to further elaborated concepts that are lexicalized by *compensate, recompense, indemnify* and *remunerate*, as well as *refund* and *spend*. The latter has its own subordinates, *squander, blow*, and *lavish, shower*.

Figure 1 shows some of transfer verbs discussed above; while not all related words and concepts are shown, the figure is intended to illustrate the net-like character of WordNet's lexical representations.

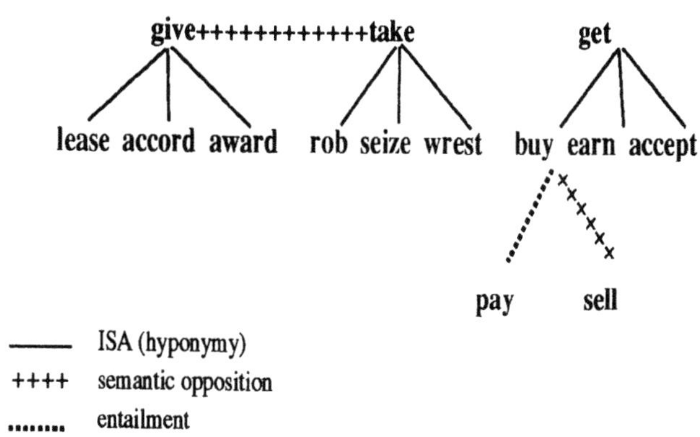

Figure 1. Illustration of WordNet's Lexical Representation.

Sell has a superordinate concept expressed by *exchange*, which is an important and central concept in the domain of transfer. *Sell* has a number of subordinates, including *auction*.[4] Finally, the concept expressed by *have, have got* is part of the semantic domain of transfer in that it denotes the state on which any kind of giving, taking, or getting is predicated. Most of the verbs we have considered here are thus linked to one of the basic concepts that constitute the "roots" of WordNet "trees." There are others that seem semantically related yet they are not attached to the same trees on WordNet. Among the verbs that Dubois (this volume) considers are *mail, post*, which is classified as a communication verb in WordNet; *subcontract* and *contract*, which are considered as verbs of social interaction; and the synset containing *levy, impose*, which is headed by the synset consisting of *charge, bill*; this concept is seen as a cognitive or communicative action resulting from a transfer, rather than a transfer act itself. Of course, communication can be seen as a kind of (abstract) transfer, so it is not surprising that verbs like *mail* can be classified as belonging to either domain.

In sum, the semantic classification of transfer verbs in WordNet results straightforwardly from WordNet's design. Basic, unelaborated concepts like *give*, *take*, and *have* are at the top of trees whose branches are semantically further elaborated concepts. Links run between semantically opposed concepts such as *buy* and *sell*, and verb pairs related by lexical entailment, such as *buy* and *pay*.

2.3. SEMANTIC AND SYNTACTIC RELATEDNESS

WordNet's semantic classification also correlates strongly with the syntactic behavior of the verbs. For example, verbs of giving participate in the so-called dative shift alternation, where the oblique argument that expresses the recipient can be expressed as the direct argument:

(1) a. Mary gave the bicycle to Max.
b. Mary gave Max the bicycle.

[4] Another way of representing the relation between *buy* and *sell* that captures the fact that they refer to the same event is to express the difference between these two verbs in terms of the mapping of their shared thematic roles onto the syntax. Thus Fillmore and Atkins (1992) discuss at length all the verbs that can be used to refer to an event of commercial transaction, along with the surface realizations of their semantic roles. In the case of *buy*, the BUYER is mapped onto the subject, the GOODS are expressed as the direct object, and the SELLER is optionally expressed as an oblique object. If the event is expressed with the verb *sell*, the argument referring to the SELLER is the subject and the BUYER is optionally expressed in a prepositional phrase; the GOODS are mapped as the direct object.

This syntactic alternation is the hallmark of transfer verbs (for references to the vast literature on the dative shift, see Levin 1993). Not surprisingly, many subordinates of *give* participate in this kind of syntactic behavior:

(2) a. Mary offered/awarded/granted/lent Max the money.
b. Mary offered/awarded/granted/lent the money to Max.

Other semantically related verbs cannot appear in the double object construction:

(3) a. Mary donated $1,000 to the library.
b. *Mary donated the library $1,000.

It is not clear whether this phenomenon can be given a semantic account. It turns out that *sell* exhibits the same syntactic behavior:

(4) a. Mary sold John a car.
b. Mary sold a car to John.

Here, too, the subordinate verbs do not necessarily share this alternation:

(5) a. They wholesale sweaters to the buyers.
b. *They wholesale the buyers sweaters.

While we have classified *sell* semantically as a verb belonging to the tree headed by *exchange*, the syntax groups it with *give*. This may indicate that the semantic aspect of giving contained in *sell* is more salient than the component of receiving (financial compensation) included in the concept associated with *sell*.

Many verbs of giving also participate in what Levin (1993) calls the Fulfilling alternation:

(6) a. Mary presented/entrusted/left the estate to Max.
b. Mary presented/entrusted/left Max with the estate.

Many verbs that can undergo the dative shift are excluded from the fulfilling alternation (Levin, 1993):

(7) a. *Mary gave/offered/granted/lent John with a car.

The syntactic behavior of verbs of giving is thus not homogeneous, and indicates that they fall into distinct semantic sub-classes (see Levin 1993 for a review of the relevant literature).

Many verbs that are subordinates of *get, obtain* in WordNet can also occur in the so-called benefactive construction (see Levin, 1993, for a discussion and an extensive bibliography):

(8) a. Mary bought/rented/got/steal a bicycle for Max.

b. Mary bought/rented/got/stole Max a bicycle.

This construction is quite similar to the Dative Shift, except that the recipient is a benefactor here. Thus, the different prepositions (*to* vs. *for*) appear to distinguish between semantically distinct roles for the arguments of verbs like *give* on the one hand, and *get* on the other hand.

Overall, we see that a syntactic grouping on the basis of diathetical alternations results in a very similar classification of verbs as the semantic one. Syntactic properties suggest smaller subclasses of verbs, but these may not go hand in hand with parallel semantic distinctions; in some cases, a the etymological or morphological make-up of a word may account for its particular syntactic behavior. See Saint-Dizier (this volume) for an analysis of the overlap between WordNet's semantic grouping and Levin's syntactic classification.

3. Verb Semantic Classes and Lexical Conceptual Structure

3.1. AIMS

One of the main practical goals of our investigations is to develop a relatively large-scale lexical knowledge base of predicative forms for French, with a priori language-independent techniques. It is well-known that a detailed and accurate semantic representation of a predicative form, and of a verb in particular, cannot be realized from a single representational system. In our approach, we want to investigate the integration of several a priori coherent paradigms. The different representational paradigms we consider are the following:

- **attribute-value pairs** of approaches like those of the Componential Analysis,
- **relations**, in particular those developed in lexical semantics (part-of, is-a, synonymy and antonymy) and, for example, used on a large scale in WordNet,
- **predicative and functional representations**, such as the Lexical Conceptual Structure or the Meaning Text Theory, or other representations based on graphs or semantic nets,
- **inferential systems**, using predicates and functions, and dealing with presuppositions, constraints of various forms, and describing the 'consequences' on the current world of the realization of the event described by the verb.

These different paradigms can be used to develop semantic descriptions in a coherent way. They describe different facets of knowledge, at different levels of granularity. We consider different classification criteria which induce different non-overlapping classifications. Each classification describes a certain *facet* of the predicate meaning.

3.2. METHODS, ADVANTAGES AND LIMITATIONS

Our first experiments show that the approach is feasible, with the different semantic approaches representing relatively independent, but complementary, semantic aspects. The feasibility is enhanced since we consider separately, on the basis of WordNet ontological categories, relatively modest sets of verbs, usually around 400 verbs. However, difficulties vary greatly depending on the family of verbs considered: while *possession verbs* aren't very difficult to treat, *cognition and psychological verbs* are much more delicate. Possession verbs indeed involve relatively simple and concrete notions of transfer (of objects and money) such as the trajectory, and additional properties, more abstract, which are relatively consensual, such as: moral judgements, and domains (family, law, education, etc.). Verbs are not very much polysemic and their arguments are usually standard (e.g. physical objects, humans). Finally, metaphorical extensions are also more regular and systematic. Cognition and psychological verbs refer to a number of properties which are non-consensual and hard to characterize: love, anxiousness, knowledge of something, etc.

It is also quite difficult to separate verb-senses, in order to make independent judgements for each sense. Our approach is to avoid defining too many senses for a verb, by limiting as much as possible the number of senses to very clearly delimited ones. Under a sense we include metonymic uses and different sense variations, allowing a certain *generative* dimension to our system. For example, for the French verb *ceder*, dictionaries and other sources (e.g. WordNet) identify many senses among which the following:

(9) a. *Céder un terrain*, (to sell a piece of land)
b. *Céder sa place*, (to give up one's seat)
c. *Céder sous la menace/ à la tentation*, (to yield to force/temptation)
d. *Céder sous le poids de quelque chose*, (to break under the weight of something)

We consider that all these occurences of *céder* are basically related to the same abstract sense of *céder*, namely to 'give up/give way to' something, with some kind of 'breaking off' (e.g. a transfer in the possessional field, or a break in the physical field), and possibly under some constraints. As can be seen, there are parameters and options in the definition of a sense. This leave space for various, more of less important, forms of metonymies and sense variations, whose meaning is compositionally elaborated upon from the verb's and the arguments semantic properties. We can, alternatively, postulate a single sense for cases (a) to (c) and a different one for case (d).

Criteria for sense delimitation are given in the introduction chapter of this book.

3.3. THE BUY / SELL PAIR

Since we work on French, the examples below are realized from *vendre* (buy) and *acheter* (sell). These two verbs are quite simple; the different levels above are advocated in an illustrative way. We give below different types of classifications, based on the different criteria stated above. We believe that they are sufficiently general to allow for generalization and the formation of classes. It is clear that we get different types of classes for each type of criteria, semantic similarities of a certain sort may exist despite syntactic differences and vice-versa. Overlaps as well as differences are then interesting to explain.

3.3.1. *Syntactic behaviors and classifications*

The general issue, addressed by Beth Levin (Levin 93) and others, of semantic similarity being mirrored in syntax, has shown that there is enough correlations to allow for generalizations, despite numerous exceptions. The verb *acheter* accepts the following 'alternations' (Saint-Dizier, this volume): middle reflexive, passive, nominalization with support verb 'faire', arity3-unspecified PP, arity3-object deleted, reflexive form of the indirect object. The verb *vendre* accepts the same alternations, except for the last one. The reflexive form associated with acheter characterizes the fact that you can buy something *for* yourself, but not *to* yourself. Consequently, the reflexive pronoun comes from a modifier PP, not from the indirect object (which must be absent if there is a reflexive pronoun):

(10) *X achète/buys Y pour/to Z à/for T.*

This construction cannot be used for the verb *vendre*. For that reason, those two verbs are not in the same verb class. It is however clear that these two verbs have a lot of syntactic behaviors in common, therefore, the 2 verb classes are quite close.

It should be noted that the syntactic characteristics of verbs of transfer of possession are quite diverse. For example, there are verbs which do not accept pronominal forms, because the agent cannot realize the action on himself, this is the case e.g. for verbs of exchange. Notice also that *vendre* and *acheter* have some similar metaphorical uses, but also a few others which differ: *vendre son âme, vendre une personne, acheter quelqu'un* (to sell one's souls, to sell someone, to buy someone), but **acheter des âmes* (to buy souls, a very unusual form, except may be in the Faust context). *Vendre son corps* (to sell one's body) is rather a 'semi-fixed' form, and *acheter son corps* (to buy one's body) is ill-formed.

3.3.2. *Thematic classifications*

Acheter and *vendre* have the following thematic grids (see definitions in Saint-Dizier, this volume), showing a symmetry between the two verbs, where each argument position may bear more than one thematic role:

(11) a. *acheter*: `[[agent, destination], [theme],[source]]`
b. *vendre*: `[[agent, source], [theme], [destination, incremental beneficiary theme]]`

As a comparison, here are a few other thematic grids of verbs which are more specific:

(12) a. *Céder*: `[[agent,source],[general theme],[destination]]`
b. *Echanger*: `[[agent],.[tg,acp],[tg,acp]]`
c. *Voler* (to rob): `[[agent,dest],[general theme],[source, incremental victim theme]]`

In (12b), échanger (to exchange) in the sense of exchanging something against something else (both object arguments are general themes + an accompaniement relation). From the thematic perspective, both verbs are classified in the same class of verbs of transfer of possession with a volitional agent, but they then appear in two different subclasses, since the trajectory of the second argument, the theme, is different, as indicated by the positions of the roles source and destination.

3.3.3. *A WordNet-like classification*

Our classification w.r.t. WordNet criteria remains modest so far. This classification for French was designed for testing pruposes, not for designing a real-size WordNet for French. We have a single sense for *acheter* and *vendre*. Both verbs are in the class of transfer of possession with a retro-transfer of money. This level of discrimination may equally be defined from the thematic grids. Both are also in the classes of 'commercial transaction' and 'legal transactions' (as opposed to robberies for example). These verbs are quite generic in the transfer of possession family and they do not show more distinctions in our description. We have considered here only the 'core' meaning of *buy* and *sell*. Let us now compare these classifications with those found in the English WordNet.

WordNet has 6 senses for *sell* which are the following:

1 'exchange for money,' is the sense we have considered as the 'core' sense in French.
2 expresses a manner,
3 expresses a kind of activity (commercial); it is related to the core meaning. Differences, we believe, are more syntactic than semantic.
4-6 describes 'metaphors': to sell one's principles, image or to sell someone. In our perspective, we view them as metaphorically derived from the 'core' sense, e.g. by generative devices.

WordNet has 4 senses for *buy*.

1 'purchase, get, acquire' corresponds to our core sense.
2 expresses the value of the object, it does not exist in French.
3-4 to buy someone and to buy an idea, and also a number of metaphors or sense-variations, in our terminology.

3.3.4. *Lexical semantic representation*

Let us now consider the LCS of these two verbs. The LCS has well-known formal and expressive limitations, but this approach is nevertheless one of the most interesting ones for representing meaning because it has a quite large validation on data and, modulo some minor extensions (for primitives and semantic fields), permits a quite abstract, but relatively exact and accurate, representation of a large number of verbs. The lexical semantics representations (LCS) of *acheter* and *vendre* share a lot of similarities. The trajectory of the thing being bought or sold is different, as seen also in the thematic grids. In the examples below, we show the progressive specification of the LCS associated with each of these verbs, from the generic class of transfer of possession verbs to the verbs themselves through their different classifications. The two basic operations on the LCS are:

1. the instantiation of an argument of a primitive by an LCS representation,
2. the adjunction of an LCS structure within a semantic field structure, represented between square brakets. This operation is realized by concatenation of LCS structures.

It is possible to show that the different elements of the LCS: conceptual categories, primitives, semantics fields, and also basic formula do correspond to elements of knowledge, at different levels of abstraction, acquired by humans. Language acquisition and generativity are therefore very close and probably based on the same elements of knowledge representation and lexicalization.

Let us now examine how complex representations, which originate semantic distinctions and semantic classes, are built up gradually. LCS templates can, in fact, be used as classification criteria to build verb semantic classes. Let us consider for that purpose the family of verbs of transfer of possession.

Family of transfer of possession:

Its representation is given by the primitive GO and by the specification of the possession semantic field: GO_{+poss}

This very abstract representation can be made a little bit more explicit by the introduction of an object J being transfered along a certain path Path, left underspecified:

(13) $[_{event}\ GO_{+poss}([_{thing}\ J\ )]$

The underspecified fields can be made more visible by the introduction of λ abstractions possibly typed with semantic types or fields:

(14) $\lambda\ Path : [_{path}\ ],\ \lambda J,\ [_{event}\ GO_{+poss}([_{thing}\ J\ ],\ Path)]$

where, for example, Path is typed to be of type *path.*

Class: transfer of possession with a volitive agent:
This class has been constructed from thematic criteria. It includes many verbs. Notice that the volition is here translated as a CAUSE, but in fact, depending on the verb, the degree of causality is subject to variations. This gradability cannot be represented by the LCS. From an LCS point of view, the Agent is often the cause of the transfer:

(15) $\lambda\ Path : [_{path}\ ],\ \lambda I, J,\ [_{event}\ CAUSE([_{thing}\ I\ ],$
$[_{event}\ GO_{+poss}([_{thing}\ J\ ],\ Path)]$

Classes: trajectory of the (main) transfer:
At this level, *acheter* and *vendre* are in different classes, because the trajectory isn't the same. The variable Path is reduced to two different structures. For *acheter*:

(16) Path $= [_{path}\ FROM_{+poss}([_{thing}\ K\ ]),\ TO_{+poss}([_{thing}\ I\ ])]$

where K is the indirect object and I the subject. For *vendre*, we get:

(17) Path $= [_{path}]$

Subclass: transfer with retro-transfer of money:
We can now adjoin to the event of change of possession the fact that there is a retro-transfer of money. This property is a priori not lexicalized in the argument structure of both verbs, but may be realized in the surface form by means of a modifier expressing the cost of the object being transfered. This property is represented for *acheter* by:

(18) $[_{event}\ GO_{+poss}([_{thing}\ money\ ],$
$[_{path}\ FROM_{+poss}([_{thing}\ I\ ]),\ TO_{+poss}([_{thing}\ K\ ])])]$

yielding the following comprehensive representation:

(19) $\lambda I, J, K,$
$[_{event}\ CAUSE([_{thing}\ I\ ],$
$[_{event}\ GO_{+poss}([_{thing}\ J\ ],$
$[_{path}\ FROM_{+poss}([_{thing}\ K\ ]),\ TO_{+poss}([_{thing}\ I\ ])]),$
$GO_{+poss}([_{thing}\ money\ ],$
$[_{path}\ FROM_{+poss}([_{thing}\ I\ ]),\ TO_{+poss}([_{thing}\ K\ ])])])]$

The retro-transfer of money is introduced by concatenation to the main transfer. Notice that the notion of money which was implicit (or incorporated) is made explicit here. It is not associated with any position in the argument structure, since it is not lexicalized. The fact that this retro-transfer appears after the main transfer in the LCS indicates that it is less prominent. To express the notion of transfer, the primitive EXCH may be introduced, having scope over the retro-transfer. Finally, the argument 'money' can slightly be made more informative by substituting it for the more 'pragmatic' function VALUE-OF(J), which produces the 'social' value of the object being transfered. *Vendre* has then the following LCS representation (with the function VALUE-OF):

(20) $\lambda I, J, K,$

$$[_{event}\ CAUSE([_{thing}\ I\ ],$$
$$[_{event}\ GO_{+poss}([_{thing}\ J\ ],$$
$$[_{path}\ FROM_{\hat{+}poss}([_{thing}\ I\ ]),\ TO_{+poss}([_{thing}\ K\ ])]),$$
$$GO_{+poss}([_{thing}\ VALUE-OF(J)\ ],$$
$$[_{path}\ FROM_{+poss}([_{thing}\ K\ ]),\ TO_{+poss}([_{thing}\ I\ ])])])])]$$

The representations of *acheter* and *vendre*, which are quite generic (or prototypical) verbs, are relatively comprehensive. Additional information such as the fact that it is a legal and commercial transfer could be added by means of an attribute-value system. The fact that, for *acheter*, 'I no longer possesses the object but I am richer' could be part of the inferential system. Similarly, presuppositions could be that I possesses J and that K has money to buy J.

3.3.5. *A few considerations on generativity*

Let us consider a metaphorical use of vendre:

(21) *vendre son âme,* (to sell one's soul).

The LCS allows us to say that the 'soul' has a certain value. If we assume that the function VALUE-OF(Y) above is quite general and reflects the 'social' value of the entity Y, then this function can be used also for souls as well. Now, there is clearly no retro-transfer of money, but of something else, quite undetermined, probably advantages of some sort. This can be still accounted for by not mentioning any value at all for Y, and then the representation becomes equivalent to that of verbs such as 'exchange'. In the case of *vendre quelqu'un* (to denounce someone), there is a retro-transfer of money, most probably, but less salient, and the value of Y is the value that that person represents in the society. As can be seen on this simple example, the LCS can also be used, with minor changes which can be parametrized, to account for metaphorical uses of some verbs.

Without going into many details here about our approach to generativity, let us simply mention that we do not think that the Qualia structure is a necessary structure. We have shown (Mari and Saint-Dizier 97) on a few motivational examples that we can develop generative devices that account for sense variation, metonymies and some metaphors by means of general knowledge rules, a set of inherent properties associated with lexical items (in particular nouns) and a set of compositional rules which allow for the construction of various semantic forms (in LCS) a corresponding to the various usages of a verb. Acceptance of a proposition is defined on the basis of the existence of a well-formed LCS representation of that proposition.

To conclude, let us also consider the case of the verb *payer* (to pay), which has core uses such as:
payer un salaire which is the basic sense/usage
payer une marchandise, un objet, which is a kind of metonymy,
but also less frequent uses, which can be accounted for by co-composition, a system where the meaning of the expression is defined from a complex interaction of the predicate and the argument, as in:
payer de sa vie (to pay with one's life)
and more crucially again:
payer un verre à quelqu'un (to pay a glass to someone)
se payer un zéro (to get a zero)
which may be considered as semi-fixed forms. In fact, on may consider that these forms were perfectly compositional ones in the past, but their use has progressively disappeared. These sentences are then just remnants of more compositional forms.

4. The Nature of Representations in Generative Lexicon

Lexical representations in Generative Lexicon, (henceforth, GL), are viewed as a model for compositional semantics completely divorced from semantic decomposition in the traditional sense (cf. Katz and Fodor, 1963, Jackendoff, 1983, Levin and Rappaport, 1995). The aim of developing a generative model of the lexicon requires a set of assumptions whereby the linguistic properties of a lexical item cannot be separated from the structuring of the semantic information that it carries. To illustrate this point, it is probably useful to make an analogy with syntax, where the explanatory power does not reside in the particular phrase- structure description of a given sentence, but rather in the relationship between components (e.g. c-command, binding, case- assignment, etc.) and the transformations that they allow. In GL, the focus is not simply on how much or what kind of semantic components are encoded in a lexical entry, but on how that information is structured into coherent relations. This is a necessary move if a semantic theory is to account for cases of systematic polysemy, meaning shifts and the contextual dependence of lexical items. In the discussion below, I will present the basic methodological issues involved in establishing the semantic properties that characterize the set of transfer verbs. In other words, I will sketch a GL answer to the following question: What is it that makes a transfer verb a transfer verb?

4.1. METHODOLOGICAL ISSUES

An essential methodological aspect for determining the semantic properties of a particular set of predicates is to recognize that semantically related lexical items display like grammatical behavior. This notion of "semantically related" items should not be equated with membership in a well-defined semantic class, a notion which is very hard to characterize precisely. Consider the following predicates:

(22) a. John gave a book to Mary.
b. John mailed a letter to Mary.
c. John handed the package to Mary.
d. John read a story to Mary.
d. John told a lie to his wife.

(23) a. John exchanged presents with Mary.
b. John shared a cookie with Mary.

The examples in (22) and (23) are just a small sample illustrating that lexical items are cross-classified. In particular, although *read* and *tell* share some of the grammatical properties of *give* and *mail*, they do not fall into the set of transfer verbs. Similarly, *exchange* and *share* involve different types of transfer, but they have different subcategorization properties from (22). As pointed out in Levin and Rappaport (1995), such a high degree of cross-classification requires that membership in a semantic class be established by syntactic tests as well as semantic properties. On this topic, GL takes a different stand. In particular, although the notion of semantic class is descriptively useful, is not adequate for stating semantic generalizations, without an appropriate abstraction reflecting the semantic properties that make a lexical item a member of that class. Thus, the notion of semantic relatedness is interpreted as equivalence or close equivalence of lexical representations, and it is arrived at by investigating the relatedness of the underlying semantic type of a given set of lexical items. I will turn now to discuss these issues in relation to verbs of transfer.

4.2. IDENTIFYING SEMANTIC RELATIONS AMONG TRANSFER VERBS

Although the expressibility of arguments reflects important distinctions among the set of transfer verbs, I will not mention these here, since their subcategorization behavior has already been presented above. Rather, I will focus on other aspects concerning the semantic properties that these predicates have in common. The predicates *give*, *take*, *buy* and *sell* are semantically related in that they denote a relation between individuals which can be characterized as a transfer of possession. Within this set of predicates a number of distinctions can be made reflecting different degrees of specificity. Consider, for instance, the behavior of *get* and *rent*, in (24) and (25):

(24) a. John got a book from Mary. (i.e. *take*)
b. John got a book for Mary. (i.e. *give*)

(25) a. John rented the apartment from Mary. (i.e. *take*)
b. John rented the apartment to Mary. (i.e. *give*)

With predicates such as *get* and *rent* the subcategorization determines the direction of the transfer (cf. Bierwisch, 1983, Jackendoff, 1983, Pustejovsky, 1997), an option which is not available with *give*, *take*, *buy* and *sell*, in (26) and (27), where directionality is lexically specified:

(26) a. John gave a book to/*from Mary.
b. John took a book *to/from Mary.

(27) a. John sold a book to/*from Mary.
b. John bought a book *to/from Mary.

Note that (26d) is grammatical under a derived reading, i.e. *bring*, which is not part of the core semantics of *take*. Consider, first, the lexical representations of *give* and *take*, which are presented in a simplified form:

$$\text{(28) a.}\quad \begin{bmatrix} \textbf{give} \\ \ldots \\ \text{EVENSTR} = \begin{bmatrix} \text{E1} = e_1 : \tau_1 \\ \text{E2} = e_2 : \tau_2 \\ \text{REST} = e_1 <_\alpha e_2 \end{bmatrix} \\ \text{QUALIA} = \begin{bmatrix} \text{FORMAL} = \textbf{have}(e_2,\textbf{z},\textbf{y}) \\ \text{AGENTIVE} = \textbf{give\_act}(e_1,\textbf{x:human},\textbf{y:obj},\textbf{z:human}) \end{bmatrix} \end{bmatrix}$$

$$\text{b.}\quad \begin{bmatrix} \textbf{take} \\ \ldots \\ \text{EVENSTR} = \begin{bmatrix} .. \end{bmatrix} \\ \text{QUALIA} = \begin{bmatrix} \text{FORMAL} = \textbf{have}(e_2,\textbf{x},\textbf{y}) \\ \text{AGENTIVE} = \textbf{take\_act}(e_1,\textbf{x:human},\textbf{y:obj},\textbf{z:human}) \end{bmatrix} \end{bmatrix}$$

The subevents that characterize the transfer are encoded in the AGENTIVE and FORMAL role respectively, which are taken to express an underlying causal relation between events, whose ordering is given in terms of precedence, e.g. $<_{\propto}$. Note that, as a verb of transfer of possession, *give* weakly presupposes that there is an event of taking involved, however this does not seem to have any effect in the lexicalization. The same holds for *take*. Things are different with *buy* and *sell*. These predicates incorporate simultaneously the semantics of *give* and *take*, in that for every event of *selling* there are events of giving and taking, in addition to a (possible) transfer of money, (cf. St.Dizier, above). Partial evidence for this distinction is provided by the following contrast:

(29) a. John gave the book to Mary for a week.
b. John took the book from Mary for a week.
c. *John sold the book to Mary for a week.
d. *John bought the book to Mary for a week.

In (29a-b), the adverb *for a week*, specifies the length of time during which Mary and John had the book. One possible explanation for the unavailability of the adverb with *sell* and *buy* is that there is no unique final state which is modifiable. In the next section, I will discuss what it means for a predicate to have more than one resulting event in its representation.

4.3. DEGREES OF SPECIFICITY

Since *buy* and *sell*, as well as *rent* or *lend*, incorporate both the relations **give** and **take**, which are taken to be semantic types, we need a way to make explicit how a semantic type may express simultaneously more than one relation. Pustejovsky (1995, 1997) introduces the notion of a complex type or *dot object*, which is a Cartesian type product of *n* distinct types characterizing the interpretation of lexical items that carry multiple senses simultaneously. By themselves, the types that constitute a *dot object* would be contradictory without a relation that expresses a coherent structuring. In the case of predicates and event nominals, this relation is a partial ordering between events. I will not detail here the specific aspects of the proposal, which is outlined in Pustejovsky (1995, 1997). Suffice it to say, for the current purposes, that this treatment captures interesting generalizations about the properties of nominals and predicates which cannot be given a satisfactory account either by enumeration of senses or by meaning postulates. In order to illustrate the nature of complex relational types, consider, first, the predicate *rent*, which expresses the complex relation **give.take**, where the directionality of the transfer is unspecified. This can be represented as in (30) below:

(30) $\mathcal{R}_{give}.\mathcal{R}_{take}$ = rent the apartment

With predicates of this type, the directionality is specified by different PPs acting as filters, (cf., Pustejovsky, 1997 for details). As mentioned above, *sell* and *buy* are different in that the directionality of the transfer is expressed in the lexicon. In GL these distinctions are taken to be the result of focusing or heading one of the two relations, which for *buy* is the relation **take** and for *sell* is the relation **give**. Since the transfer of money is common to both *buy* and *sell*, for simplicity I will consider it a constant, notated as $\mathcal{T}$. This is shown below, where * marks the head and acts as a filter over the two relations:

(31) buy $= \mathcal{R}_{give}.\mathcal{R}^*_{take}.\mathcal{T}$
$*[\mathcal{R}_{give}, \mathcal{R}_{take}, \mathcal{T}] : \mathcal{R}_{take}$
$\exists \mathcal{R}_{give} \exists \mathcal{T} [\mathcal{R}^*_{take} \wedge \mathcal{R}_{give} \wedge \mathcal{T}]$

(32) sell $= \mathcal{R}^*_{give}.\mathcal{R}_{take}.\mathcal{T}$
$*[\mathcal{R}_{give}, \mathcal{R}_{take}, \mathcal{T}] : \mathcal{R}_{give}$
$\exists \mathcal{R}_{take} \exists \mathcal{T} [\mathcal{R}_{take} \wedge \mathcal{R}^*_{give} \wedge \mathcal{T}]$

Note the headedness triggers existential quantification of the non headed relation, which is crucially part of the semantics. The qualia structure representations of types such as (31) and (32) is shown below, where I have ignored the constant $\mathcal{T}$ for simplicity, and where the subscripts F and A refer to the relations that are encoded in the AGENTIVE and FORMAL roles of *give* and *take*, respectively (cf., (28) above):

(33) a.
$$\left[\begin{array}{l} \textbf{buy} \\ \ldots \\ \text{EVENSTR} = \left[\begin{array}{lcl} \text{E1} & = & e_1 : \tau_1 \\ \text{E2} & = & e_2 : \tau_2 \\ \text{E2} & = & e_3 : \tau_3 \\ \text{E2} & = & e_4 : \tau_4 \\ \text{REST} & = & (e_1 <_\alpha e_3) \circ_\alpha (e_2 <_\alpha e_4) \end{array}\right] \\ \text{QUALIA} = \left[\begin{array}{lcl} \text{FORMAL} & = & \mathcal{R}_{give^F}(e_3, z, y).\mathcal{R}^*_{take^F}(e_4, z, y) \\ \text{AGENTIVE} & = & \mathcal{R}_{give^A}(e_1, x, y, z).\mathcal{R}^*_{take^A}(e_2, z, y, x) \end{array}\right] \end{array}\right]$$

b.
$$\left[\begin{array}{l} \textbf{sell} \\ \ldots \\ \text{QUALIA} = \left[\begin{array}{lcl} \text{FORMAL} & = & \mathcal{R}^*_{give^F}(e_3, z, y).\mathcal{R}_{take^F}(e_4, z, y) \\ \text{AGENTIVE} & = & \mathcal{R}^*_{give^A}(e_1, x, y, z).\mathcal{R}_{take^A}(e_2, z, y, x) \end{array}\right] \end{array}\right]$$

Note that in (33) above, the events of giving and taking are related by temporal overlap. In addition, given the way arguments are specified in the AGENTIVE role, the complex relations in the FORMAL role make reference to the same arguments. Thus, formally *buy* and *sell* are distinguished by which relation in the AGENTIVE role is headed, such that we can represent the predicates as follows:

(34) a.
$$\begin{bmatrix} \textbf{buy} \\ \ldots \\ \text{QUALIA} = \begin{bmatrix} \text{FORMAL} = \mathcal{P}_{give^F.take^F}(y,z) \\ \text{AGENTIVE} = \mathcal{R}_{give^A}(x,y,z).\mathcal{R}^*_{take^A}(z,y,x) \end{bmatrix} \end{bmatrix}$$

b.
$$\begin{bmatrix} \textbf{sell} \\ \ldots \\ \text{QUALIA} = \begin{bmatrix} \text{FORMAL} = \mathcal{P}_{give^F.take^F}(y,z) \\ \text{AGENTIVE} = \mathcal{R}^*_{give^A}(x,y,z).\mathcal{R}_{take^A}(z,y,x) \end{bmatrix} \end{bmatrix}$$

The representations in (34) make an interesting prediction. Note that with *sell* the relation which is headed in the AGENTIVE role refers to the argument x, i.e. the "giver", which is not expressed in the FORMAL role. This event- argument configuration can be seen as similar to the distinction between *kill* and *die*, for instance, or alternatively as the causative / unaccusative alternation with certain predicates. We can thus expect that, in certain languages, *sell* could be treated as a causative version of *buy*, which is the case for German, where the distinction is marked via a causative morpheme. In turn, nothing would rule out that a language which does not lexically mark *buy* and *sell* with headedness specifications, would have a unique logically polysemous item similar to *rent*.

4.4. COMPARISON WITH OTHER APPROACHES

The analysis sketched above, for the set of transfer verbs, is meant to address a number of methodological issues illustrating how the generative properties of the lexicon emerge from the combinatorics of semantic types. This approach differs substantially from the ones presented above. As discussed in Section 2 above, the classes obtained in WORDNET distinguish *give* and *take*, or *buy* and *sell*, or *rent* in terms of a fixed set of semantic relations, among which are hyponymy and semantic opposition. In addition, a predicate such as *rent* is classified as a more elaborate concept than *sell*. The distinctions that are drawn by adopting the methodology of GL are quite different. First of all, there is no proper hierarchical organization between *sell*, *buy* and *rent*, which are related by focusing different components of the representation, which simultaneously encode the relation `give` and `take`. I have argued above that this treatment has the potential of making interesting cross-linguistic predictions concerning the lexicalization patterns of these predicates. Hence, *sell*, *buy* and *rent* are not semantically opposed, but their underlying type is specified differently, minimally by heading one of the relations.

In addition to headedness specification, transfer verbs are expected to differ in the specification of other parameters. Consider, for instance how *buy* and *sell* may differ from *exchange*. The complex relational type of *exchange* most likely involves the relation **give.give**, with additional information concerning the symmetry of arguments. Under this view, the definition of a semantic class is achieved by identifying the crucial components of a lexical item, which are expressed as semantic types. This allows for a more precise statement of semantic class, in that transfer verbs can be defined as complex relational types encoding the semantics of **give** and **take**. This certainly does not exhaust all the issues, although it provides a promising method for avoiding a proliferation of semantic classes while maintaining fine- grained distinctions among different predicates within the same class. Spelling out the nature of the generative devices deriving complex relations out of simple types and specifying the parameters for those types is a matter of ongoing research.

5. Views of the Lexicon: Contemplative and Inquisitive

5.1. THEORETICAL FRAMEWORK - GOALS

The ultimate goal of studying semantics within the framework of Mikrokosmos (Nirenburg et al., 1996), is to investigate the mechanisms of obtaining a meaning representation given a text and vice versa. Work on the lexicon is a component of this program of study: the knowledge which is acquired is, from the outset, intended to help extract and represent meaning given a text or realizing a text given its meaning.

Our overall goal is to use our lexicons in NLP non-toy applications. We are interested in the trade-offs between computational linguistics concerns (including empirical evidence from cross-linguistic examinations) and NLP concerns. As a consequence, our view of the lexicon is both contemplative (how good is it from a computational linguistic perspective) and inquisitive (how useful is it for particular NLP applications such as Machine Translation (MT) and Multilingual Generation).

Our lexicons are composed of superentries (Meyer et al., 1990), where each entry consists of a list of word senses, stored independently of their part of speech (the verb and noun form of *walk* are under the same superentry). Each word meaning is identified by a unique identificator, or lexeme (Onyshkevych and Nirenburg, 1994).

The information contained inside a lexeme is minimally described via **10 zones** corresponding to various levels of lexical information, relevant to phonology, orthography, morphology, syntax, semantics, syntax-semantic linking, stylistics, lexical relations (paradigmatic and syntagmatic information, lexical rules, and database management information.

The lexicon primarily connects with an ontology and an onomasticon (a special-purpose lexicon of named entities such as cities, corporations, or products). The set of symbols used in the lexicon, and possible relationships between them, are mostly grounded in a knowledge source called the *ontology*. The symbols are defined as *concepts* in the ontology. As described, e.g., in (Mahesh and Nirenburg, 1995; Mahesh, 1996); the ontology is a large collection of information about EVENTs, OBJECTs and PROPERTYs in the world, displayed in Figure 2. In addition to the taxonomic multi-hierarchical organization, each concept has a number (currently averaging 14) of other local or inherited links to other concepts in the ontology, via relations (themselves defined in the PROPERTY sublattice). For instance, the concept BUY is linked to the concept MONETARY-UNIT through the relation COST (see Figure 3). These links include case-role-like relations linking EVENTs to semantic constraints on the allowable fillers of those case-roles (i.e., selectional restrictions), properties (such as MANUFACTURER-OF) of things like COMPANYs, etc. A concept definition consists of this collection of links, each of which is itself defined in the ontology.

It is important to note that there need not be any correlation between syntactic category and semantic or ontological class. For example, the verb *destroy* may be represented by an EVENT, as will the noun *destruction* (with a different linking in the syntax-semantics interface, of course).

5.2. A CROSS-LINGUISTIC INVESTIGATION ON THE ONTOLOGICAL TYPE BUY: REPRESENTATIONAL ISSUES

A cross-linguistic study on the literal uses[5] of the pairs "buy/sell" in English, "acheter/vendre" in French and "comprar/vender" in Spanish, shows that, in absence of preferences such as focus, topic or theme/rheme opposition, they could be used interchangeably (assuming the arguments were changed too), thus sharing the same semantics. Therefore both elements of the pair can be mapped into the same concept which we have labeled BUY, illustrated in Figure 3, as a corpus study showed that both verbs of the pairs were about equally used.[6]

Figure 3 shows the concept BUY which is defined by features (such as ISA, SUBCLASSES, AGENT, COST, ...) along with their values (such as EVERYDAY-FINANCIAL-EVENT, MONETARY-UNIT ...)[7].

In the following, we focus on the SYN and SEM zones of the lexical entries for the pair "buy/sell" in Figure 4 and Figure 5 respectively, using

[5]By literal uses we mean where the pairs have their literal meaning as opposed to, say, a metaphorical use.

[6]The corpora search we did on the Wall Street Journal 1987, 1988, 1989 for English; Agence française de Presse for French and Infosel for Spanish indicates that there is a preference for using the nominalizations rather than the verbs.

[7]The label Sem is omitted throughout Figure 3 apart from HAS-PARTS for room purposes. Defaults however only appear here for HAS-PARTS.

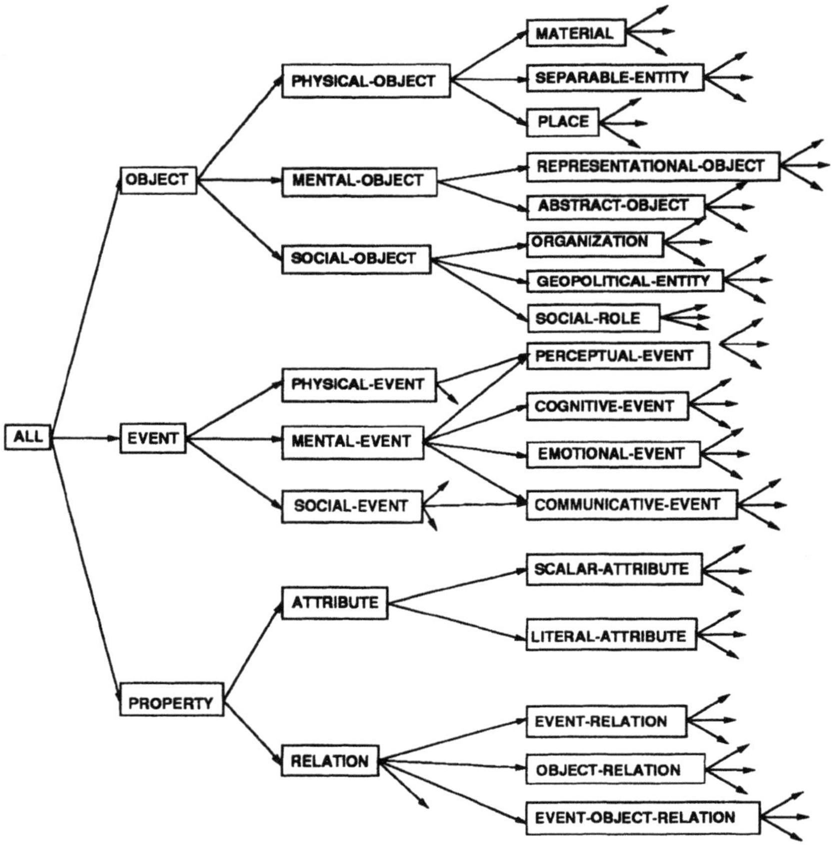

Figure 2. Ontology Top Level.

typed feature structures (tfs) as described in (Pollard and Sag, 1987).

These entries provide the interface between the syntax and the semantics via the indexes. The only obligatory functional arguments for the lexical items *buy* and *sell* are the subject and object, whereas the obliques are optional. What constitutes a BUY event is the presence of five semantic arguments, namely, who is doing the financial transfer, who is the origin or source of the financial transfer, what is being transferred, who is the beneficiary of the financial transfer, and what is the price or cost of the transfered object. This semantic frame is syntactically realized by requiring a subject and an object. The COST, BENEFICIARY and SOURCE are not

Concept Name: BUY	
Definition:	a transfer of possession event (buying or selling) that involves the exchange of money for merchandise
ISA:	EVERYDAY-FINANCIAL-EVENT TRANSFER-POSSESSION
SUBCLASSES:	AUCTION BUYOUT
AGENT:	HUMAN
THEME:	OBJECT (NOT HUMAN)
COST	MONETARY-UNIT
DESTINATION:	HUMAN
SOURCE:	HUMAN
HAS-PARTS:	Default: transfer-object Sem: event
BENEFICIARY:	HUMAN

Figure 3. Illustration of the concept BUY.

necessarily lexicalized at the surface level.[8]

Briefly, other zones in the lexicon (such as lexical relations) will provide additional information on the paradigmatic and syntagmatic uses of *buy* and *sell*, whereas lexical rules will flag the lexicon entry for the different rules which apply locally. For instance, in the case of EVERYDAY-FINANCIAL-EVENTs, the lexicalization - grammaticalization of this semantic class involves rules such as ditransitive, passive, as in (Levin, 1993). Therefore, *buy* can accepts the following (made-up) realizations:
Joan bought a doll from Sears for Marie for $50.

[8]However, in the domain of stock market, one can easily elide the object as in *Traders sold/bought the whole morning*, this is due to "redundant" semantic information in *stock trader* and the object of their trade, namely *stocks*. This case is comparable to examples such as *I ate (the meal)*, except that in the stock example, it is semantic information coming from both the verb and the noun which enable such elision. However, such examples do not require to create another entry for *buy/sell*, it can be calculated at processing time.

$$\begin{bmatrix} \text{buy-V1} & \\ \text{syn:} & \begin{bmatrix} \text{root:} & \boxed{0}\begin{bmatrix}\text{cat:} & \mathbf{V}\\ \text{sem:} & \boxed{00}\end{bmatrix} \\ \text{subj:} & \boxed{1}\begin{bmatrix}\text{cat:} & \mathbf{NP}\\ \text{sem:} & \boxed{11}\end{bmatrix} \\ \text{obj:} & \boxed{2}\begin{bmatrix}\text{cat:} & \mathbf{NP}\\ \text{sem:} & \boxed{21}\end{bmatrix} \\ \text{oblique:} & \boxed{3}\begin{bmatrix}\text{root: } \mathbf{for} & \\ \text{obj:} & \boxed{3}\begin{bmatrix}\text{cat:} & \mathbf{NP,opt+}\\ \text{sem:} & \boxed{31}\end{bmatrix}\end{bmatrix} \\ \text{oblique:} & \boxed{4}\begin{bmatrix}\text{root: } \mathbf{from} & \\ \text{obj:} & \boxed{4}\begin{bmatrix}\text{cat:} & \mathbf{NP,opt+}\\ \text{sem:} & \boxed{41}\end{bmatrix}\end{bmatrix} \\ \text{oblique:} & \boxed{5}\begin{bmatrix}\text{root: } \mathbf{for} & \\ \text{obj:} & \boxed{5}\begin{bmatrix}\text{cat:} & \mathbf{NP,opt+}\\ \text{sem:} & \boxed{51}\end{bmatrix}\end{bmatrix} \end{bmatrix} \\ \text{sem:} & \begin{bmatrix} \boxed{00}\ \mathbf{buy} & \\ \text{agent:} & \boxed{11}\ \mathbf{human} \\ \text{theme:} & \boxed{21}\ \mathbf{object} \\ \text{beneficiary:} & \boxed{31}\ \mathbf{human} \\ \text{source:} & \boxed{41}\ \mathbf{human} \\ \text{cost:} & \boxed{51}\ \mathbf{monetary\text{-}unit} \end{bmatrix} \end{bmatrix}$$

Figure 4. Partial Lexicon Entry for the English lexical item *buy*.

Marie was bought a $50 doll by Joan from Sears.
Joan bought Marie a $50 doll from Sears.
...

If it happened that a lexical item belonging to this semantic class accepts different lexicalization - grammaticalizations, these latter will be encoded in the lexical rule zone of this particular entry.

Each entry will also include triggers for generation, specifying how to lexicalize the concept BUY into *sell* or *buy*. The difference between the entries for *buy* and *sell* consists of a different syntactic-semantic mapping, the alternations are common to both.

5.3. PROS AND CONS OF SUCH AN APPROACH

We pursue coverage along the dimensions of both depth (the amount of information on each lexical item) and breadth (the number of lexical items) in order to use our lexicons in non-toy applications.

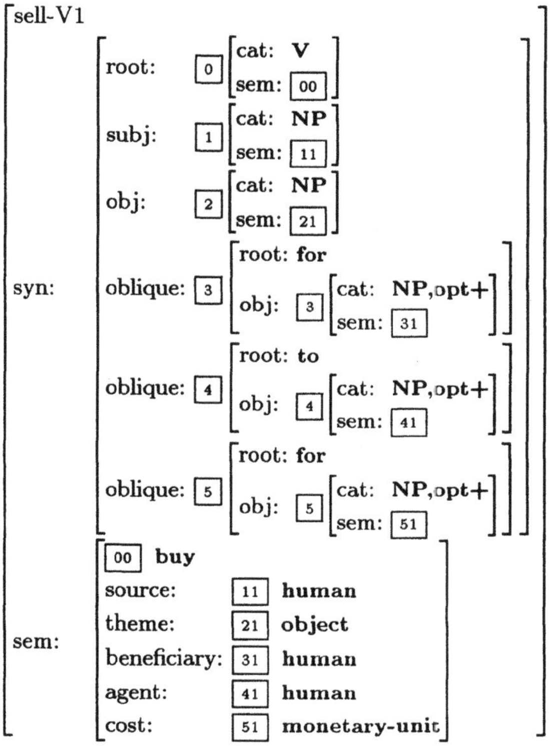

Figure 5. Partial Lexicon Entry for the English lexical item *sell.*

We have acquired for the Mikrokosmos project[9] an ontology of about 5000 concepts and have acquired semi-automatically a core Spanish lexicon of about 7000 word senses. This semi-automatic acquisition of the Spanish core lexicon, has been extended to the automatic acquisition of about 32000 new entries, using morpho-semantic lexical rules (derivational morphology), as reported in (Viegas et al., 1996).

Our approach has obvious limitations in that it is expensive to evaluate comprehensively on a large corpus of inputs, due to high start-up costs in knowledge acquisition and the inevitable complexity of any realistic application-oriented evaluation scheme. However, our experience shows that after a year of semi-automatic acquisition for the ontology (1.5 person/year - 5,000 concepts) and a Spanish core lexicon (4 person/year - 7,000 word sense entries), we could use this same lexicon to accelerate the acqui-

[9]Mikrokosmos, an MT project, can be seen at the following URL: http://crl.nmsu.edu/Research/Projects/mikro/Home.html

sition in the same language (in our case, Spanish) multiplying the numbers of root entries per 27 (Viegas et al., 1996) by using derivational morpho-semantic rules applied to verbs (the final lexicon contains about 40,000 word sense entries). Moreover, having the lexicon of a language, mapped into a conceptual world, rendered easier the production of the generation lexicon (Viegas and Beale, 1996) which thus mainly consisted of reindexing the analysis lexicon on concepts. Also, we produced an English lexicon using the Spanish core lexicon as a start. Compared to the acquisition from scratch, the acquisition of the core English lexicon was 12 times faster than the acquisition of our first core lexicon (Spanish).

The Spanish lexicon has been used to analyze unrestricted Spanish texts which had, on an average, 17 sentences with over 21 words per sentence. We correctly disambiguated 91% of open class words, counting only ambiguous words, and 97% counting all open class words. In Mikrokosmos, we consider as ambiguous words, homographs, homomorphs and polysemes, whereas results on word sense disambiguation reported by other researchers are usually results on disambiguation of homographs only (Yarowski,1992), and mainly homomorphs (Wilks and Stevenson, 1997), with correct disambiguations of 96% and 92% respectively, counting all open class words.

However, one should mention that the development of knowledge sources, such as the Mikrokosmos ones, can be considered an overkill for some NLP applications such as Information Retrieval: in this case a statistical approach might indeed be more cost-effective than a knowledge base approach.

The English lexicon is being tested for lexical choice on the generation of English texts; results are forthcoming.

5.4. COMPARISONS WITH OTHER APPROACHES

In this section we compare our approach to the other computational (lexical) semantics approaches presented in this chapter. We aim at showing the main differences, which should not been taken as criticisms of the other approaches per se. It is indeed sometimes difficult to render justice to all the work involved in the different approaches which have involved years of research, all the more so when the research goals are different.

The Mikrokosmos ontology differs from the primitives in the LCS à la Jackendoff in that we do not have a frozen set of primitives.[10] Although Wordnet can be said to subsume the Mikrokosmos ontology, along the ISA links, the latter differs in two ways: it can be used for any language with minimum enhancement, as the symbols in the ontology are just labels, English labels for convenience or could be Greek symbols or whatever, and are not English words as is the case in Wordnet. For instance, we added

[10]See (Viegas et al., this volume) for a discussion on primitives.

only about a dozen concepts to create the English lexicon from the Spanish core lexicon (which contains 7000 word senses). Second, the concepts in the ontology encode more information than Wordnet, with relations (such as case roles) and attributes. The Mikrokosmos lexicon is different from the Generative Lexicon (GL) approach in that it maps the lexical items mostly to concepts in a rich, well structured and organized conceptual world (ontology), whereas at this stage, the Lexical Conceptual Paradigms (LCPs) of GL entries seem to be undefined labels. Our semantic classification for EVENTs is different from the classification of verb classes such as the one developed in (Levin, 1993). Our experience working on aspectuals shows that Levin's classes hardly helped us to characterize them. For instance, *cease* appears in Beth Levin's classification under the following sections:

- Aspectuals (Begin Verbs)

Another aspectual, *begin*, is also attributed the same classification:

- Aspectuals (Begin Verbs)

This classification does not allow a system to differentiate between *The book chapter began with a citation* and **The book chapter ceased with a citation.*

This could illustrate the fact that semantics does not parallel syntax, and as such, Levin's verb classes seem to present a lexical syntactic theory of verb classes rather than a lexical semantic organization of verbs. In the Mikrokosmos approach, a semantic class is built as an event hierarchy where every event, along with its associated required links to other concepts, constitutes a semantic class, such as the EVERYDAY-FINANCIAL-EVENT which constitutes the semantic class for the lexical items which map to either EVERYDAY-FINANCIAL-EVENTs or BUY or AUCTION. The lexical items which belong to a semantic class minimally accept the lexicalization - grammaticalizations of the semantic class, and other lexicalization - grammaticalizations are local to the lexical items.

Another main difference between Mikrokosmos and the other approaches discussed here is our concern for an inquisitive as well as contemplative view on the lexicon: try to reconcile theoretical concerns with NLP concerns.

Finally, our experience showed that it is possible to use a knowledge base approach to NLP, and that we performed better than statistic based approach in one of the most difficult NLP tasks: word sense disambiguation. This demonstrates that a (lexical)-semantic approach to NLP is feasible, and as such, whatever the semantic approach adopted (Wordnet, LCSs, GL, Mikrokosmos, etc...), it is worth considering it for NLP applications, in particular when statistic-based approaches fail.

Further research in Mikrokosmos involves applying the whole methodology of acquisition and processing to the analysis of Chinese and the generation of English; including, as part of the generation process, a summarization component; and, finally, incorporating the system within a hybrid architecture, allowing the use of different approaches, in order to lessen the costs of knowledge-based acquisitions, which are expensive but cannot on the other hand be avoided. More basic research involves working on a set of microtheories such as: reference, anaphors, aspect, language gaps, planning, lexical choice, etc., from the viewpoints of representations and processing.

6. A psychological investigation of cognitive representations for transfer verbs

The present section is concerned with a psychological investigation of semantic structures for transfer verbs, with a specific focus on the pair *vendre* (sell) and *acheter* (buy). It investigates how native speakers stucture the semantic space for those verbs, and along what dimensions and principles of categorization. In contrast with the previous approaches within this chapter, it relies on empirical data collected on different groups of subjects, in order to identify the psychological structures and processes for semantic representations and interpretations for verbs. It is thus concerned with *language in use* by speakers and not with language *per se*, and with the relevance the abstract entities that linguistic and formal analyses manipulate in language processing.

Besides the intrinsic psychological elaboration, more developped elsewhere, the results presented here propose a comparative analysis of (1) the psycholinguistics data and interpretations with (2) the formal descriptions previously given in the this chapter. We don't suppose any *a priori* necessity of mapping between the structural properties of the psychological data and those of the formal representations. We neither want to evaluate the "psychological reality" of whatever formalism, nor to directly ground linguistic properties in any psychological reality.

In this first descriptive step, we restrict ourselves to outline the isomorphism, overlap or discrepencies between these different and autonomous descriptions (formal and psychological) of semantic verb classes. From a psychological point of view, the formal descriptions could be considered as powerful descriptive tools that allow us to make explicit the relevance of different structural properties of language, regarding how native speakers process transfer verbs in a classification task. Specific attention will be devoted to the analysis of the differential relevance of syntactic alternations, thematic roles or WordNet psychological descriptions, diversely enhanced in the different formalisms presented in this chapter.

In turn, from a formal point of view, the psycholinguistic data, being related to general psychological processes, could just suggest cognitive hypothesis of grounding the formal properties of language in knowledge resources on language. Through the classification paradigm used by native speakers, we may in some way evaluate how experts (linguists) make explicit formal properties of semantic verb classes from their linguistic competence in comparison to common sense (i.e. folk) metalinguistic judgments that "naive" subjects did. As such, the overlap of criteria from both populations may bring argument regarding the structural properties of the linguistic competence. However, we will prevent ourselves from drawing causal inferences from the results from such specific metalinguistic and reflexive activities of classification - from experts or from naive subjects - to language processing in general (i.e. spontaneous production or comprehension) as well as in discursive contexts.

6.1. TWO EXPERIMENTS

The first purpose of the two experiments reported below was to identify along what principles native speakers were able to explicitly structure a set of transfer verbs. In both experiments, the subjects were involved in a free sorting task of the set of verbs. Each verb of the set was printed on a chip of paper, and subjects were instructed to "sort all the verbs within as many piles they wish, along the semantic resemblances or differences among them, and to give the reasons why they did so."

In the first experiment, more detailed in Dubois & Saint-Dizier (1996) *acheter* (buy) and *vendre* were included within a set of 50 transfer verbs and were processed by a group of 20 adults. They have been selected as members of an *a priori* formal class identified along syntactic criteria (Saint Dizier, this volume). We obtain the following results:

- **vendre** belongs to a 4 element class, together with *louer, revendre, soustraiter* (approximately translated as: rent, resell, subcontract)
- **acheter** belongs to an 8 element class, together with *accorder, allouer, assigner, attribuer, choisir, destiner, racheter* (grant, allot, assign, chose, re-buy)

They both share the same core of syntactic properties, or alternations: middle reflexive, passive, nominalization with support verb *'faire'*, arity 3- unspecified PP, arity3-object deleted, but, whereas *acheter* accepts the reflexive form of the object, *vendre* does not (see Saint-Dizier, supra).

In the second experiment, the two verbs *acheter* and *vendre* were presented within a set of 46 transfer verb selected from the same *a priori* formal syntactic classes. These classes were in that latter case formed from syntactic criteria and a few semantic criteria, in particular thematic roles (see Saint-Dizier's paper in this volume). The results obtained are then the following:

- *acheter* belongs to a larger class of 13 verbs which includes *accorder,* **acheter**, *adjuger, allouer, arracher, assigner, attribuer, décerner, enlever, octroyer, offrir, prodiguer, retirer,* (assign, withdraw, offer, etc.),
- *vendre* belongs to a class of 14 verbs which includes *accepter, acquérir, attendre, dérober, emprunter, louer, obtenir, poster, prêter, recevoir, revendre, soutirer,* **vendre**, *voler* (accept, acquire, expect, steal, borrow, rent, obtain, mail, receive, rob).

The two classes contrast on the "trajectory" of the second argument, the theme, that is "symmetrical" either as source or as destination.

Within this second free sorting task, 6 groups of 20 subjects were solicited (from teenagers of 11 years old to teachers from public high schools), in order to follow the late psycholinguistic evolution in semantic organization.

6.2. ANALYSIS OF THE RESULTS

Within each experiment, the subjects' partitions were processed along a classification algorithm that computes distances from between-subjects' agreements (Barthélemy and Guénoche 1988) : the more subjects have sorted two verbs within a class, the smaller the distance between these two verbs is and the closest (the more semantically similar) the verbs can be considered. Let us consider the results of the two experiments we have carried out:

From **Experiment 1** we notice that *vendre* is closely associated to *revendre*. This first grouping can be interpreted as the anchor for a larger class. It progressively aggregates:
"louer, payer, acheter, racheter, facturer, rembourser, rémunérer, dépenser"
then *"sous-traiter"*,
and further, *"rétribuer, dédommager, indemniser"*.

The comparison between this psychological pattern of classification and the *a priori* formal - syntactic - structure shows both major differences and a significant overlap. First, it can be noticed that the psychological classes include more items than the syntactic ones, and *acheter* and *vendre* are closely related within the same psychological class whereas they belong to two differents formal classes. Only 2 out of 8 of the items from *acheter*-syntactic class are present in the psychological classification (only *acheter* and *racheter* are at a certain stage in the *vendre*-class). The remaining items of the *acheter*-syntactic class *(accorder, allouer, assigner, attribuer, choisir, destiner)* are psychologically related to another class that includes *adjuger, décerner, prodiguer, procurer* and that is finally linked to *promettre.*

This result suggests that the syntactic criteria are not the only ones of psychological relevance for semantic classifications carried out by human subjects. Two additional verbs, *dépenser, facturer*, also belong to the psychological class even if their syntactic behaviors differ from the other verbs.

However, it is worth noting that the psychological *acheter/vendre*-class exhaustively gathers the 6 items of another syntactic class:
dédommager, indemniser, payer, rembourser, rémunérer, rétribuer
a set that includes, besides the syntactic criterion of a quasi-incorporated argument, a semantic one (that can be named as "kind of money exchanged"), or even a textual or domain specific one, such as technical vocabulary for administrative purposes.

In other words, the two structures obtained from formal and psychological criteria are similar but with interesting recompositions: The formal description categorizes the verbs along all or none (i.e. boolean) criteria whereas the psychological one clearly allows one to identify a "core" subclass including *vendre, revendre* and *louer* whereas *soustraiter* is much more distant. Symetrically, technical verbs (i.e. more specialized verbs of transfer of possession: *adjuger, décerner*, etc.) related to a syntactic quasi-incorporated argument remain grouped in the psychological representation, but at different distances from *acheter, racheter*. These latter verbs can be considered as a second semantic core of the psychological classification.

In short, the data suggest that the formal (and here mainly syntactic) criteria may contribute to the description of the psychological organization or, that the observed psychological organization can rely on syntactic regularities but that it also and more heavily relies on subjects' semantic evaluations. Psychological structures are more heterogeneous than what the formal descriptions may account for: the psychological data suggest to include other structural principles in the description of the semantic space, such as prototypicality, genericity/specificity and domain (or textual) specificity, that may, or may not, overlap with formal linguistic criteria. Besides the identification of these subtle within-class criteria, the main criterion that subsumes the psychological class across other classes is that all the verbs gathered with the two target verbs *acheter* and *vendre* are actually **verbs that imply money**.

A confirmation of such an interpretation can be found in the verbalisations given by the subjects after their sorts :

- *"renvoie à des actions mercantiles"*
- *"concerne l'argent, contrat social"*
- *"symbolise l'argent"*
- *"verbes qui se conjugent de la même manière"*
- *"en rapport avec une monnaie d'échange"*

Examples of subjects' comments for the classes that included *acheter* and *vendre*

From **Experiment 2**, the main result reported here is that the pair *acheter/vendre* regularly occurs within the same cluster for most groups of subjects, in spite of differences on formal grounds, regarding both their syntactical behavior and their thematic roles. The following table summarizes the clusters within which *acheter/vendre* occurs: the distances as given by the aggregations within the tree representation are represented by levels of embedding within brackets, for the "core" ones, and by (+ ...), for the ones which are more distant.

6th grade (11 years old)	((**vendre**/*revendre*) **acheter**/*payer/rembourser*)
3rd grade (14 years old)	((**vendre**/*revendre*/**acheter**) (*payer*) + ... (*louer*))
2nd grade (15 years old)	(((**vendre**/*revendre*)(**acheter**/*payer*)) + ... (*dédommager/indemniser*) + ... (*rémunérer/rembourser*))
"Terminal" grade (18 y.)	((**vendre**/*revendre*) *louer*)
Technical school (20 y.)	((**vendre**/*revendre*) + ... (**acheter**/*emprunter/louer*) + ... ((*dédommager/indemniser*) *rémunérer/rembourser/payer*)))
P (Teachers)	((**vendre**/*revendre*) (*soustraiter/louer*)) ((**acheter**/*gagner*) (*acquerir/obtenir*) (*recevoir/accepter*))

Across all groups, *vendre* is regularly associated to *revendre* and aggregated with *acheter/payer* and *louer*. There are two exceptions: The 18 years old children did not regularly classify *acheter* which did not aggregate with any other verb, and in Teachers, *vendre* and *acheter* agregate two different classes in a more fine grained classification than the children.

To summarize, it can be stated that :

- A tight relation between *vendre* and *revendre* is regularly observed, explaining the fact that, in human memory, lexical representations include morphological properties that structure the mental lexicon.

- *vendre* can be considered as a "basic" level term and even a prototype inasmuch it is commonly used as a core from which other terms aggregate to form a class that can be named "transfer of possession involving money."
- *acheter* plays a similar role in another closely related class including an added semantic property of a "**positive** transfer involving money," positive meaning "directed towards the subject who gets something."
- *payer* structures a third node either closely related or, in the older subjects, remoted to the set of **technical** terms for "transfer of possession thru money" (as if, within this context, *payer* was the prototype of the technical meaning for "transfer of possession involving money"

In short, the results suggest that **transfer involving money** is the salient criteria motivating the grouping of *acheter/vendre* within the same class for most of the subjects. Whereas *vendre* seems "neutral" regarding the orientation of the transfer, *acheter*, as connected to *gagner, acquerir, obtenir* is a rather **positive transfer** than a generic one.

In this second experiment also, subjects' verbal comments on their sorting further confirm the previous interpretations. For example, 6 grade children (11 years old) mainly commented the *vendre/ acheter* class as follows:

- *tout ce qui est argent (all what is money)*
- *parce qu'il veut dire qu'il donne rembourse et offre (because it meant that he gives or refunds)*
- *parce qu'on peut acheter, offrir un cadeau (because we can buy and offer a present)*
- *donner un cadeau*
- *payer*
- *a rapport avec l'argent*
- *concerne l'argent (has to do with money)*
- ...

In contrast, it can be inferred from teachers'comments that they clearly sort out *vendre* as a generic transfer, lightly weighted with a negative orientation:

- *la perte (the loss)*
- *donner par rapport à une chose (give w.r.t. something)*
- *vente*
- *dessaisir*
- *donner quelque chose de matériel*
- *argent*
- *vendre*
- *du don, l'argent*

whereas *acheter* is explicitly as a positively-egocentered transfer:

- *gain*
- *recevoir (recieve)*
- *vente (sell)*
- *obtenir quelque chose (get something)*
- *obtenir avec de l'argent*
- *réussite (succeeds)*
- *cadeau*
- *achat*
- *de l'extérieur à moi*
- ...

In comparison to the formal descriptions of *acheter* and *vendre* given above, we can conclude that the psychological organization of semantic classes for verbs do include criteria from different knowledge sources on language: morphological, syntactic as well as multidimensionnal semantics (inclusion, entailement, synonymy, as found e.e. in WordNet). To be more precise, especially regarding the previous formal descriptions, most of the criteria taken into account in the various analyses do also seem relevant in the subjects' psychological classifications : Subjects do organize the semantic classes that involves *acheter* and *vendre* taking account of:

- morphological criteria (*racheter* and *revendre* respectively clusered with *acheter* and *vendre*),
- regularities in syntactic behavior such as alternations as described by Levin and discussed above by Saint-Dizier,
- thematic roles,
- lexical properties such as directionality of the transfer (Saint-Dizier and Busa) and of course
- ingredience, inasmuch the psychological classification of the class of transfer verbs including *acheter* and *vendre* is mainly grounded on the presence of money in the transfer. The semantic dimension itself includes different specific semantic properties such as
- antonymy (as taken into account in Wordnet, see above) or more pragmatic considerations such as
- the context of use : subclasses of terms terms that may behave the same inasmuch they represent a paradigm within the same type of (specialized) text, a dimension that WordNet does account for as well as suggested by Saint-Dizier or Viegas, this chapter).

However, on a formal and structural basis, one major difference between formal and psychological descriptions which has to be stressed is that whereas any formal description is required to be at the same time partial and systematic in giving clear-cut categories, psychological descriptions have to account for the now well-known properties of psychological categories (Rosch, 1978; Dubois, 1991): in order to be accurate and efficient (rather than true and exhaustive). Human cognitive systems do not compute the multimentionality of potential linguistic criteria of classifications. Human cognitive processing rather reduces processing costs by evaluating global distances of any item from a prototype or from a prototypical representation that summarize the (verb) class. This evaluation is not done by computing necessary and sufficient conditions or all or none criteria (as required - and given - by a strict formal analysis) but through global evaluation along the main representative properties of a category, according to family ressemblance rather than according to strict membership decisions.

Furthermore, besides linguistic properties (i.e. properties specific to langage *per se*) that can be identified in subjects' partitions (and that can be seen as a result of language contraints on psychological activities), psychological classifications also involve subjective specific criteria : as such the "from the subject/to the subject" dimension may be considered as a psychological version of the more abstract "orientation of the transfer" elicited in formal descriptions. This observation can be taken into account by linguistic systems that have issues relevant to casual users (Wordnet for example). Symetrically, we can consider that the principles of organization for verb semantic classes elaborated by linguists do also belong to naive subjects' linguistic knowledge, and knowledge use. On this latter point, it is not surprising that our results converge with Wordnet structural principles that indeed take into account psychological motivation of conceptual organization for semantic classification.

To conclude on a more theoretical level regarding the interactions between linguistic (formal descriptions) and psychological ones the comparisons of the descriptions presented here lead us to be very cautious (and barely dogmatic) regarding the subtle play of constraints between language as an autonomous system and its psychological processing : any rapid conclusion that directly ground language into an abstract cognition or, symetrically that consider psychological structures as a mere reflexion of linguistic structures is too simple and counterproductive regarding the developpement of any interesting theory of langage and language processing.

7. Conclusion

This paper presented a quite large number of perspectives on the analysis and the representation of transfer of possession verbs, with a special emphasis on the pair buy/sell. It is clear that we can't draw any simple and easy conclusion from the confrontation of these perspectives, nor should we be satisfied with simply stating that they present different facets of verb meanings and different manners of meaning representation. The preparation of this text presented a challenge. To make it more than a simple enumeration of analysis and representations and to contrast these representations in a positive way, the authors have focussed on the main features of their system/approach, highlighting many current problems in lexical semantics. The result leaves us with a richness of questions, arguments, and counterarguments, centered around a small set of verbs at the core of the discussion, that need to be answered, debated, tested, and refined. Furthermore, the psycho-linguistics results and the ensuing discussion avoid the trap of indicating which perspective is more 'realistic' than another one. It also avoids to attach preferences to perspectives, but rather highlights the differences between the semantics viewed by 'linguistic engineering specialists' and the notion of meaning as manipulated and perceived by human language users.

One of the many lessons is that humans do not make (at least so explicitely) all of the quasi-ideological distinctions computational linguists make between syntax, semantics, pragmatics, lexicon, etc... It is clear that one could argue that these distinctions are just for the purpose of separating out difficulties and making systems more modular. This is probably true, and an efficient way to work on so complex a domain, but this may also entail new difficulties and more or less artificial problems, as, e.g. the cooperation between these modules (syntax-semantics, semantics- pragmatics) that humans, speakers of a language, do not percieve as such. On the other hand, humans have a more or less concious idea of polysemy and of meaning shifts; they play with words, with their contexts of enunciation and interpretation as well as with morphology and morphological resemblances. The panel as well as the contributions by the individual authors in this book demonstrate that the study of lexical semantics, in particular when focused on verbs, is complex and difficult; it is also rewarding in reminding us of the colorfulness and creativity with which speakers manipulate word meanings.

References

Barthélemy, J.P. and Guénoche, A. (1988) *Les arbres et les représentations des proximités*, Paris: Masson.

Bierwisch, M. (1983). "Semantische und konzeptuelle Reprasentationen lexikalischer Einheiten," in R. Ruzicka and W. Motsch (eds.), *Untersuchungen zur Semantik*, Berlin, Akademische-Verlag.

Cruse, D. A. (1986) *Lexical Semantics*. New York: Cambridge University Press.

Dubois, D. (1991) *Sémantique et cognition*, Paris: CNRS.

Dubois, D. and Berleux (1997) Cognitive categories for verbs of transfer in French (in preparation).

Dubois, D. and Saint-Dizier, P. (1996) Construction et représentation de classes sémantiques de verbes: une coopération entre syntaxe et cognition. *Actes du colloque RFIA '96* Rennes, 16-18 jan.1996.

Dowty, D. (1989) On the Semantic Content of the Notion of Thematic Role, in G. Cherchia, B. Partee, R. Turner (eds), *Properties, Types and meaning*, Kluwer.

Dowty, D. (1991) Thematic Proto-roles and Argument Selection, Language, vol. 67-3.

Fellbaum, C. (1990) *English Verbs as a Semantic Net*, Journal of Lexicography. (see also paper in this volume)

Fellbaum, C. (in press). A Semantic Network of English Verbs. In Fellbaum, C. (ed.) *WordNet: An Electronic Lexical Database and Some of its Applications.* Cambridge, MA: MIT Press.

Jackendoff, R. (1983) *Semantics and Cognition*, MIT Press.

Jackendoff, R. (1987) The Status of Thematic Relations in Linguistic Theory, Linguistic Inquiry, vol. 18.

Jackendoff, R. (1990) *Semantic Structures*, MIT Press.

Katz, J. and J. Fodor (1963) "The Structure of a Semantic Theory," *Language*, 39.

Levin, B. (1993) *English Verb Classes and Alternations.* The University of Chicago Press.

Levin B. and M. Rappaport, (1995) *Unaccusativity: At the Syntax-Lexical Semantics Interface*, MIT Press.

Mahesh, K. and S. Nirenburg (1995) A situated ontology for practical NLP, In Proceedings of the Workshop on Basic Ontological Issues in Knowledge Sharing, *International Joint Conference on Artificial Intelligence* (IJCAI-95), Montréal, Canada, August 1995.

Mahesh, K. (1996) *Ontology Development: Ideology and Methodology.* Technical Report MCCS-96-292, Computing Research Laboratory, New Mexico State University.

Mari, A., Saint-Dizier, P. (1997), Générativité: au-delà d'une théorie des types, in proc. *TALN'97*, Grenoble.

Meyer, I., Onyshkevych, B. and L. Carlson (1990). *Lexicographic Principles and Design for Knowledge-based Machine Translation.* Technical Report CMU-CMT-90-118, Carnegie Mellon University.

Miller, G. A. (1990). Nouns in Wordnet: A Lexical Inheritance System. International Journal of Lexicography 3.4, 245-264.

Miller, G. A., and Fellbaum, C. 1991. Semantic Networks of English. *Cognition*, special issue, eds. Levin, B., and Pinker, S., 197-229. Reprinted in: *Lexical and Conceptual Semantics*, eds. Levin, B. and Pinker, S., Cambridge, MA: Blackwell. 197-229.

Nirenburg, S., Beale, S., Mahesh, K., Onyshkevych, B., Raskin, V., Viegas, E., Wilks, Y., and Zajac, R. (1996) Lexicons in the MikroKosmos Project. In Proceedings of *Artificial Intelligence and System Behaviour*, University of Sussex, Brighton, England.

Onyshkevych, B. and Nirenburg, S. (1994) *The lexicon in the scheme of KBMT things.* Memoranda in Computer and Cognitive Science, MCCS-94-277. Las Cruces, N.M.: New Mexico State University. Reprinted as: A lexicon for knowledge-based machine translation, in: Dorr and Klavans 1995 (eds).

Pollard, C. and I., Sag (1987) *An Information-based Approach to Syntax and Semantics: Volume 1 Fundamentals.* CSLI Lecture Notes 13, Stanford CA.

Pustejovsky, J. (1991) The Generative Lexicon. *Computational Linguistics*, 17:409-441.

Pustejovsky, J. (1995) *The Generative Lexicon*, MIT Press.

Pustejovsky, J. (1997) The Semantics of Lexical Underspecification, in *Folia Linguistica.*

Rosch, E. (1978) *Categorization and Cogntion*, Hillsdale, N.J.: Erlbaum.

Viegas, E. and S. Beale (1996) Multilinguality and Reversibility in Computational Semantic Lexicons. In Proceedings of the *International Natural Language Generation 96*, Sussex, UK.

Viegas, E., Onyshkevych, B., Raskin, V. and S. Nirenburg (1996) From *Submit* to *Submitted* via *Submission*: on Lexical Rules in Large-scale Lexicon Acquisition, in *Actes de Association for Computational Linguistics'96*, Santa Cruz, Juin 23-28.

Yarowsky, D., (1992) Word sense disambiguation using statistical models of Roget's categories trained on large corpora. In *Proceedings of the 14th International Conference on Computational Linguistics.* pp. 454-460, Nantes, France.

Wilks, Y. and Stevenson M. (1997) Sense Tagging: Semantic Tagging with a Lexicon. In Proceedings of the SIGLEX workshop *Tagging Text with Lexical Semantics: Why, What, and How?*, Washington DC.

THE ORGANIZATION OF VERBS AND VERB CONCEPTS IN A SEMANTIC NET

CHRISTIANE FELLBAUM
Cognitive Science Laboratory, Princeton University
Princeton, New Jersey, USA

Abstract. The focus of this paper will be the representation of verbs forms and verb concpets in WordNet, a semantic network of English (Miller, 1990; Fellbaum, in press). A brief comparison between WordNet and decompositional models of the lexicon as well as Pustejovsky's (1991, 1995) Generative Lexicon is followed by an examination of synonym sets, the units of the network. We consider the semantic and syntactic constraints that operate on non-literal extensions of verb senses. The paper shows how the particular semantic classification of verbs in WordNet reflects the regular syntactic behavior of verbs in terms of transitivity alternations. [1]

1. Verbs in WordNet

WordNet is an electronic lexical database of English containing over 120,000 concepts expressed by nouns, verbs, adjectives, and adverbs (Miller, 1990; Miller and Fellbaum, 1991; Fellbaum, in press). In this paper, the discussion will be limited to verbs.

The design of WordNet is that of a network of interlinked nodes representing word meanings. The nodes are sets of unordered, roughly synonymous word forms (synsets), all of which express one given concept. For example, the synset containing the verb forms *put, place, set, pose, position, lay* stands for the concept, which can be referred to by any one of its members. The most current version (1.6) of WordNet contains some 10,300

[1]Preparation of this paper was supported in part by contract N66001-94-C-6045 with the Advanced Research Project Agency, contract N6601-95-C-8605 with the Advanced Research Project Agency, Computer Aided Education and Training Initiative and grant IRI-9528983 from the National Science Foundation.

P. Saint-Dizier (ed.), Predicative Forms in Natural Language and in Lexical Knowledge Bases, 93–109.

distinct verb forms (strings) organized into more than 12,000 verb synsets, representing an equal number of distinct verb meanings. Most verb forms are single words, but phrasal verbs and some collocations are included as well. The fact that the number of senses or synsets is higher than that of distinct verb forms reflects the polysemous nature of verbs; it is interesting to note here that verbs are more polysemous than nouns. WordNet also includes many idiomatic verb phrases, such as *kick the bucket* and *come around.* These are treated as single lexical items, or "long words," although they do not always represent syntactic categories.

The synsets are linked to one another by means of conceptual-semantic relations, and WordNet's weblike structure results from the connections via these relations among synsets[2]. Most verb synsets are interrelated by a pointer standing for a manner relation, dubbed TROPONYMY, or MANNER, RELATION (Fellbaum, 1990; Fellbaum, in press). Troponymy is fairly analogous to the hyponymy, or ISA relation among noun synsets, which builds hierarchical structures linking generic (superordinate) concepts to more specific (subordinate) ones. For example, the synset *snooze, doze* is represented as a troponym (a kind of subordinate or hyponym) of the synset *sleep, slumber.* The troponymy relation here states that *snooze* (or *doze*) means "to sleep (or slumber) in a particular manner." Other examples of verb concept pairs linked by troponymy are *whisper-talk*, *sneak-walk*, and *guzzle-drink*, where the former member of each pair refers to a particular, semantically more specified, manner of performing the action referred to by the latter verb.

Similarly to the ISA-relation among nouns, the manner relation builds hierarchical tree structures several levels deep (however, while the "trees" in the noun lexicon can have a dozen or more levels, verb hierarchies tend to be rather "bushy" and more shallow). For example, *move* is the root node of a tree that includes the two increasingly specific levels represented by the nodes *run* and *sprint.*

Other conceptual-semantic relations linking both entire synsets and individual verb forms are CAUSE, ENTAILMENT, and SEMANTIC OPPOSITION (for details see Fellbaum, 1990; in press).

WordNet distinguishes itself from other current representations of lexical meaning in several ways. First, its units are whole lexical items. By contrast, much of lexical semantics focuses on components of word meanings. Semantic Decomposition, popular in the 1960s and 1970s (Katz and Fodor, 1963; Katz, 1972; Gruber, 1976; Lakoff, 1970; Jackendoff, 1972; Schank,

[2]Synsets are further linked by lexical relations, which hold among specific lexical items only rather than among all the members of the connected synsets. The principal lexical relations in the verb component of WordNet is that of semantic opposition. In this paper the discussion will be limited to semantic relation holding among synsets only

1972; Miller and Johnson-Laird, 1976) has more recently made a comeback in the form of Lexical Conceptual Structures (Hale and Laughren, 1983; Jackendoff, 1983; Talmy, 1985; and others). An analysis of the verb lexicon in terms of semantic components may account in an economic fashion not only for the structure of the verb lexicon, but also for much of the verbs' syntactic behavior, but the exact number and nature of the components has yet to be determined.

WordNet's focus on the lexical, rather than the sublexical, level results from its original aspiration to model human lexical organization and lexical memory. Its design may appear to be inconsistent with compositional analyses. But as Fellbaum (1990; in press) notes, shared semantic features are in fact implicit in WordNet and function to structure verb classes, hierarchies and semantic relations that link verbs and verb synonym sets. For example, the generally accepted semantic primitives CHANGE and MOVE, which are crucial in accounts such as Jackendoff (1983), are represented as full-fledged verbs in WordNet that head large hierarchies of semantically more elaborated verbs all sharing significant semantic and syntactic properties. CAUSE, another important semantic primitive, has the status of a semantic relation in WordNet, where it connects causative-inchoative verb pairs like *fell-fall* and *raise-rise*.

The Generative Lexicon constitutes a second major framework of current work in lexical semantics (Pustejovsky 1991, 1995; Busa, this volume). The generative approach departs radically from previous models in abandoning the enumerative lexicon in favor of underspecified entries and a number of devices that generate specific, context-dependent readings. WordNet by contrast is very much an enumerative lexicon in that it attempts to represent explicitly the different senses of polysemous strings. This may seem uneconomical in comparison with a generative lexicon. On the other hand, WordNet contains the kinds of generalizations that generative mechanisms rely on to yield new senses. For example, the generative lexicon accounts for the polysemy of the verb *enjoy* in terms of its direct object nouns. Thus, *enjoy* in (1a) means *enjoy smoking*, because the object here is is a kind of tobacco product; in (1b), *enjoy* has the sense of *enjoy eating*, because the object is a kind of edible food; and in (1c) it means *enjoy drinking*.

(1) a. John enjoyed the cigar.
    b. Mary enjoyed the lunch.
    c. Sam enjoyed the French vintage Bordeaux.

The qualia structure that derives the appropriate meaning of *enjoy* in each instance must rely on the kind of semantic noun hierarchies that WordNet provides. To generate the sense of *enjoy* that means *enjoy smoking*, the noun must in each instance come from the noun hierarchy headed by the node *tobacco product*; in sentences like (1b), *enjoy* co-occurs with nouns from the tree headed by *food, nutrient*, and the verb in (1c) gets its particular meaning because it selects for a noun from the hierarchy that has *beverage* as its highest node.

A viable generative lexicon, therefore, appears to require lexical information that is organized in accordance with hierarchical structures, such as the ones that are explicitly represented in WordNet. The number of entries in the generative lexicon can be kept small because of the power of inheritance structures in the lexicon; the work that is necessary to coerce the appropriate sense in each context is done by the generative devices, which exploit these lexical-semantic structures.

Hierarchical structures such as the ones headed by *tobacco product* and *beverage* cannot account for cases of metonymy as in (2), where *dinner* refers not to the food but to the company.

(2) a. John enjoyed the fundraising dinner.

In WordNet, metonymy is handled like other kinds of polysemy. A string is polysemous when it occupies more than one position in the network, that is, when it is a member or more than one synset. But polysemous words necessarily have different superordinates, subordinates (troponyms), opposites, etc. Thus, the string *play* in one of its meanings is a troponym of *perform*; another sense is a subordinate of *compete*, while a third sense has *gamble* as a superordinate:

(3) a. The New York Philharmonic played Brahms.
b. The Knicks played the Nets.
c. My aunt plays the slot machines.

There is no distinctions in WordNet between homonymy, polysemy, and metonymy. However, in many cases, senses that are closely related semantically are grouped together. For example, the two senses of *play* in the phrases "play a melody" and "play an instrument" are grouped together so that their semantic similarity is indicated; by contrast, the meaning of *play* as in "The spotlight played on the politician" is not closely related in meaning and is not part of the same sense group. The related sense groupings often reflect those made in traditional dictionaries, where major sense distinctions are differentiated from minor ones and metonymous uses of a word. Of course, the borderline between polysemy and homonymy is far from clear-cut, and WordNet avoids drawing a categorical distinction.

In the case of sentences like (2), the string *dinner* has two WordNet senses, one referring to the meal (food) and the other to the group of people assembled for the meal. In such cases, a modifier would have to do the work of disambiguating the head noun in favor of the *company* sense over the *food* sense.

The kind of regular polysemy illustrated by *dinner* is not found systematically in WordNet. And it is not clear whether separate entries (implying distinct senses) are desirable, since many contexts require an underspecified interpretation, as in (4), where both *meal* and *company* are felicitous readings:

(4) a. John enjoyed the dinner.

While WordNet explicitly marks many cases of regular polysemy (Apresjan, 1973), (Nunberg, 1978), (Ostler and Atkins, 1991), metonymy is not accounted for in a systematic way; in fact, it is not clear that this would even be possible, given the many ways in which speakers extend the meanings of words in creative ways. Thus, WordNet has some shortcomings typical of an enumerative lexicon, as discussed in Pustejovsky (1995), but it contains much of the information required by a generative lexicon.

While its relational design and its deliberate limitation to whole lexical items distinguish WordNet from most current models of the lexicon, it contains much implicit information about verb classes and their semantic and syntactic properties that can be uncovered within its web structures. Our goal in this paper is to show that the WordNet perspective and its particular way of word sense representation can yield interesting generalizations about the semantic and syntactic properties of verbs.

2. Synsets

The building block of the net is the synset, which consists of all the words that can express a given concept. For example, *behead* and *decapitate* both lexicalize the same concept of cutting off the head of (a person or an animal). The synset containing these two verbs tells us that the meaning of *behead* is the same as that of *decapitate*, and vice versa. Synonymy therefore is one important principle of meaning representation in WordNet (in addition, word meaning is represented by the links between synsets). Like all semantic and lexical relations, synonymy is a polysemous relation between words and can be further differentiated into a number of subtle sub-relations (see Cruse (1986) for an enlightening discussion).

Absolute synonymy, whereby two synonyms are substituable for each other in every context, is rare and would be uneconomical for a language.[3] Synonyms in WordNet are words that can be substituted for each other in most, but not necessarily all, contexts. The members of a synset correspond roughly to what Cruse (1986) calls cognitive synonyms, in that their sub-

[3] See Cruse (1986) for a discussion of the connotational differences between synonyms like *behead* and *decapitate*.

stitution for one another does not change the truth value of the sentence in which they occur.

The combination of words into sets of synonyms, even in the loose sense of WordNet, can reveal interesting properties of semantically similar words. Synsets allow one to examine the question as to whether words that refer to the same concept do so accidentally and arbitrarily, or whether synonymy is determined by rules that constrain what kind of words can denote the same concepts. Another way of framing the same question is from to ask, from the vantage point of polysemy, given that a string can have several meanings, how are these meanings constrained? As (Lakoff and Johnson 1980) showed, the spatial meanings of words, for example, systematically take on temporal meanings. The assumption is that a word cannot just take on any meaning in addition to its core meaning, but must be constrained in the kind of concepts it can refer to by its core meaning.

2.1. EXTENDED MEANINGS AND SEMANTIC SIMILARITY

A good place to begin investigating the constraints on the mapping of forms and meanings is in the realm of non-literal language. Metaphors and idioms refer to the same concepts as their non-literal counterparts, from which they differ commonly in register or selectional restrictions (Cruse, 1986), but not usually in their denotation. In WordNet, metaphoric and idiomatic verbs and verb phrases senses are treated in the same way as literally referring expressions and are included in the appropriate synsets. [4]

For example, the string *die* is found in several synsets, one of which it shares with *break down*; the sense of these words is the one found in sentences like *My car died/broke down last night. Die* in the synset *die, break down* is a metaphoric extension of the *die* found in the synset *die, decease, perish, go, exit, pass away, expire*, where it has a literal meaning.

We do not know how non-literal meanings arise. But the meaning of *die* to signify "break down" or "stop functioning or operating" indicates that there exist semantic constraints that allow the metaphoric extension of verb meanings in ways that relate non-arbitrarily to the literal meanings. In the case of *die*, the sense extension is predicated on the parallelism between the cessation of an animate being's vital functions and the termination of a vehicle's or a machine's functioning. The first sense is doubtless older and

[4] We call a verb "idiomatic" when it is part of an idiomatic phrase, such as ***hit*** in the idiom verb phrase ***hit the nail on the head***. What we call "metaphoric" verbs are simple verbs (not part of a larger phrase) whose meanings have been derived from a core, central meaning of the same verb form. Of course, in some cases it can be a matter of debate whether a given meaning of a polysemous word is the primary, literary meaning, or whether it is an extended meaning.

existed before cars or machines were invented; therefore, the direction of the sense extension is uncontroversial.

Consider also the example of *kill* meaning not "take life," but "consume" or "eliminate" or "destroy" as in (5). The extended sense captures the meaning of destroying or removing or taking away that *kill* shares with the primary sense of "taking life." This parallelism allows the easy interpretation and acceptance of the non-literal synonym. The semantic element of erasure, overpowering, or elimination that is common to both literal and extended senses of *kill* is expressed in shared links to other synsets that express related concepts.

(5) a. He killed an entire bottle of vodka. (*kill, down, finish off,..*)
b. The editor-in-chief killed my story. (*kill, rule out, wipe out,..*)
c. Artichokes kill the taste of other foods. (*kill, overwhelm,..*)

Semantic similarity plays a role in constraining the non-literal extensions of word meanings. This hypothesis is testable in a semantic network like WordNet, where semantic similarity is defined as shared membership in a synset and shared links to superordinate, subordinate, and other related concepts. Synonymity is the closest form of semantic similarity, so examining the members of a synset turns up cases of literally and non-literally (primary and extended) referring word forms that denote the same concept. Another interesting area for investigating the nature of sense extensions would be to examine hierarchically related words, and semantically opposed words.

2.2. EXTENDED MEANINGS AND SYNTACTIC SIMILARITY

Perlmutter (1978) first drew the important distinction between two kinds of intransitives, unergatives and unaccusatives. The derivation of the latter involves movement of the internal object in d-structure to subject position in s-structure, whereas no such movement is involved in the derivation of unergatives. Unergatives often are activity verbs, whereas unaccusatives frequently denote a change of some kind. Examples (6) and (7) are unergatives and unaccusatives, respectively:

(6) a. Peter read for an hour.
b. Mary is dancing.

(7) a. Peter arrived last week.
b. Mary fell in the dark.

The syntactic differences between transitive, unaccusative, and unergative verbs have been associated with the differences in the semantics of verbs belonging to these syntactic classes. Some verbs of change, such as *rot*, *mold*, *rust*, and *die* are typically unaccusative. These verbs are inchoatives, i.e., they denote the beginning of a state that is brought about by the change denote by the verb; in the cases of the verbs listed here, this state cannot be induced by an AGENT but only by an inanimate CAUSE. Many verbs denoting a change of state or motion have both an unaccusative and a transitive (causative) use, which includes the expression of an argument referring to an AGENT or CAUSE; examples are *grow, blacken* , *turn*, and *bounce* (see Levin, 1993, and references therein). Examples of causative-inchoative (unaccusative) pairs are given in (8) - (9):

(8) a. John bounced the ball.
b. The ball bounced.

(9) a. Mary bent the wire.
b. The wire bent.

In many cases, an examination of the synonyms that co-occur in the same synset with an unaccusative verb in WordNet shows that these synonymous verbs have extended meanings, which refer to events denoting changes of state or certain kinds of motion. Interestingly, the semantic similarity extends to a syntactic similarity, in that both the literal and the extended meaning are expressed by unaccusative verbs. Thus, the verbs *break, break down* and their metaphoric synonym *die* are all unaccusatives. Similarly *fall* (*sick, in love*) is an unaccusative verb, just like its literal synonym *become.*

Many unaccusative verbs with extended meanings do not have simple synonyms, but are paraphrased within the same synsets by passives, a syntactically related construction; both unaccusative and passives have only one argument that is generated in d-structure to the right of the verb and is moved to the external position in s-structure.

For example, *fall* (as in *This task fell to me*) lacks a good synonym, but its meaning can also be expressed by the passive verb *be assigned to.* Similarly, *go* in *The building went up* means *was erected*; in *This prize went to Mary*, *go* means *was awarded.* When a book *comes out* or *appears*, it *is issued*, and when it *goes to press* or *goes on the block*, it is *printed* or *auctioned. Come* in *This shoe comes in three colors* means *is found* or *is available.* When something *comes* or *springs* to mind, it *is associated or remembered.* If a word *escapes* someone's lips, it inadvertently *is articulated.* In all these examples, a simple unaccusative verb has a paraphrase that is expressed by the passive construction.

Such synsets provide evidence for the thesis that the semantics of verbs are reflected to some extent in their syntactic behavior. Verbs in the same synset that refer to a change of state or a certain kind of motion are all unaccusatives; this is true as well for the extended senses. Semantically similar change and motion verbs that do not have synonyms can be paraphrased only by passives; the passive construction is syntactically very close to the unaccusative; these cases are therefore parallel to those involving an unaccusative.

In sum, WordNet's structuring of the verb lexicon into groups of synonyms and synonymous verb phrases reveals that sense extensions appear to be constrained by the syntactic properties of the literally referring verbs. Synonyms and synonymous paraphrases share their syntax with their literal counterparts. These observations further strengthen the common assumption that the syntactic and semantic properties of verbs are interrelated.

3. Verb Hierarchies and Syntactic Classes

We have looked at some cases where the semantic organization of WordNet in terms of cognitive synonymy is paralleled by syntactic similarity. In this section I will examine the extent to which semantic class membership, as represented in WordNet's hierarchical structures, parallels syntactic class membership. Levin (1993) has amassed evidence showing that verbs that share certain kinds of syntactic behavior, such as transitivity alternations, or diatheses, also tend to form semantically coherent classes. By contrast, WordNet organizes words on a purely semantic basis without regard to their syntactic properties. But we would expect that the kinds of syntactic regularities studied by Levin are reflected in semantic relatedness as it is represented by WordNet's particular structure. That is to say, if the theory that syntactic similarity goes hand in hand with semantic similarity, syntactic regularities should provide a means to test one's semantic classification, and verbs that are semantically related in WordNet should generally exhibit syntactically similar behavior. We already saw an example of semantic and syntactic parallelism in the case of the members of the synsets. Synonyms are by definition semantically equivalent, and we noted that metaphoric extensions of senses are of the same syntactic class as their more literal synonyms or paraphrases.

3.1. CAUSATIVES

A well-studied alternation is the so-called causative alternation, exemplified in (10) - (11), in which a large number of English verbs participate (see references in Levin, 1993):

(10) a. John dropped the glasses.
     b. Mary bent the wire.

(11) a. The glasses dropped.
     b. The wire bent.

Smith (1970) and Levin (1993) have pointed out that this alternation is generally available to verbs expressing a change of state or a change of position. Thus, the verbs undergoing the causative-inchoative alternation are semantically characterizable and form a homogeneous class. In a hierarchical organization like WordNet, verbs like *bend* and *drop* can be grouped within a few (large) trees, whose roots are *move* and *change*; these top node verbs themselves participate in the causative-inchoative alternation. The particular syntactic properties of these verbs are therefore reflected in their membership in semantic classes (hierarchies) that form the core of the semantic network, although WordNet's construction did not take the shared syntactic behavior of these verbs into consideration.

3.2. THE MIDDLE

Another common transitivity alternation in English is the Middle, exemplified below by the verb *sell*:

(12) a. Honda sells thousands of cars each year in the U.S.
b. This model sells easily.

The semantic relatedness as well as the differences (e.g., the requirement of a modifier, such as an adverb, for the middle) between the verbs in (12a) and (12b) has been extensively studied (see references in Levin, 1993). Unlike in the case of the causative-inchoative alternation, the relatedness between the two senses, as in the many other transitive-middle pairs, is not reflected in WordNet. This is due to the fact that the transitive and the middle each have a different superordinate, whereas the causative and inchoative readings of a verb often share the same superordinate, *change* or *move*. While transitive *sell*, as in (12a), is grouped with other possession verbs whose common superordinate is *exchange (for money)*, the middle verb, meaning "be sold at a certain price or in a certain way," (as in (12b)) is a stative verb linked to its superordinate *be*. The troponyms of the transitive verb (*scalp, auction, bootleg, remainder, dump*, etc.) are not all available for the middle alternation; thus, the stative *sell* is not connected in WordNet to verbs like *scalp*, which is part of the transitive hierarchy only. On the other hand, stative *sell* has many co-troponyms, namely other middles like *wash, clean, test, close, open, compare*, etc. These verbs all come from very different domains of the lexicon and are semantically heterogeneous. But due to the stative nature of the middle senses of these verbs, they are all daughters of (one sense of) the verb *be* in WordNet. One might object that *be* is not a felicitous superordinate concept for these verbs. However, *be* seems to be the only candidate for the superordinate here, and it appears regularly as the genus term in dictionary definitions of middle senses.

The semantic representation of WordNet therefore does not yield a semantically uniform class of verbs that have a middle alternation. The corresponding transitive verbs all belong to different semantic hierarchies, while the middles are grouped, together with many other non-middle verbs, under the common superordinate *be*. This is interesting in light of the fact that for speakers of English, transitive verbs like *sell* do constitute a semantic class: Speakers know which transitive verbs have a middle alternation and which ones do not. But WordNet does not account for this knowledge in semantic terms.

In sum, WordNet's semantic organization does not always go hand in hand with a syntactic organization. English has no verb that can serve as a root node to all (or many of) those verbs that exhibit the transitive-middle alternation; as a result, the transitive and the middle senses of verbs like *sell* are found in very different parts of the lexicon. By contrast, English does have verbs like *move* and *change* that serve as superordinates to many verbs both in their transitive (causative) and in their inchoative senses.

4. Lexical Gaps as Common Superordinates

In constructing a semantic net by connecting synsets via semantic relations, one is occasionally confronted with a gap, i.e., a seemingly missing node. For example, English has no verb(s) expressing the concept BE PERCEIVED, of which particular manners are *smell* (as in "The soup smells"), *look* ("The house looks attractive"), *sound* ("the orchestra sounded terrible"), *taste* ("the soup tastes delicious"), and *feel* ("The massage felt relaxing"). The concept BE PERCEIVED arguably exists, because it ties together the class of verbs including the ones listed. Moreover, a corresponding "active" concept, PERCEIVE, is lexicalized. *Perceive* has among its troponyms *smell*, *sound*, *taste*, and *feel*, in their transitive uses ("I smell the soup"; "I taste the garlic", etc.). We consider lexical gaps in the English verb lexicon like BE PERCEIVED to be accidental; such concepts may well be lexicalized in other languages.

The reason for positing a non-lexicalized node like *be perceived* is the need for a superordinate synset that is shared in common by such co-troponyms as *smell*, *look*, *sound*, *taste*, and *feel*, which constitute a semantic class. Co-troponyms also tend to be syntactically similar (the class including *smell*, for example, subcategorizes for an adjective). The syntactic properties of co-troponyms can provide the basis for making further, more subtle, distinctions of verb subclasses.

An example is discussed by Fellbaum and Kegl (1988), who consider creation verbs like *fabricate* and *compose*. Many creation verbs participate in the so-called Material/Product alternation (Levin, 1993):

(13) a. They wove a carpet from the sheep's wool.
b. They wove the sheep's wool into a carpet.

(14) a. He molded a head from the plaster.
b. He molded the plaster into a head.

A number of other verbs, including *fabricate* and *compose*, which are also creation verbs, do not participate in this syntactic alternation, despite their semantic similarity to verbs like *weave* and *mold*:

(15) a. The tabloid fabricated a story out of the witness's report.
b. *The paper fabricated the witness's report into a story.

(16) a. He composed a song out of the ancient melody.
b. *He composed the ancient melody into a song.

Fellbaum and Kegl (1988) suggest, on the basis of such sentence pairs, a more fine-grained sub-classification of creation verbs that distinguishes a class of verbs referring to acts of mental creation (such as *fabricate* and *compose*) from verbs denoting the creation from raw materials (such as *weave* and *mold*). This distinction can account for the systematic difference among the two sets of verbs in most cases.

Levin (1993) distinguishes these verbs in terms of membership in one of two classes: the BUILD class, which comprises verbs like *mold* and *weave*, and the CREATE class, which includes such verbs as *compose* and *fabricate*. In WordNet, we have drawn the same distinction by means of two generic verb concepts, *create from raw material* and *create mentally*; these concepts in turn are co-troponyms of the common superordinate *create*. Although they are not lexicalized in English, the two nodes heading distinct trees appear justified on the basis of the syntactic facts. Mapping the verb lexicon as a semantic network allows for a satisfactory way of accounting for semantic and syntact distinctions among verb classes.

5. Classifications based on Verb and Noun Hierarchies

Levin (1993) lists verbs like *clap* and *nod*, which participate in what she terms the Understood Body-Part Object Alternation:

(17) a. Peter clapped his hands.
b. Mary nodded her head.

(18) a. Peter clapped.
b. Mary nodded.

These verbs are all subordinates of the synset *gesticulate, gesture, motion*. So once again, the syntactic behavior that is common to *clap*, *nod*, *wink*, *shrug*, and others is paralleled by a semantic similarity, which is reflected in WordNet by the verbs' shared membership in the same hierarchy.

The understood object here is usually a particular body part; other body parts or nouns referring not to body parts cannot be the understood objects:

(19) a. *Peter clapped his feet/his shoes.

In addition to the semantic and syntactic homogeneity of the verbs that participate in the Understood Body Part Object Alternation, the similarity here extends to the class of nouns that these verbs take (or that are the understood objects in the intransitive alternations). The object in each case is a particular body part, i.e., a member from the noun hierarchy headed by *body part*.

In the verb hierarchy, as well as in the noun hierarchy that contains the class of possible objects for the verbs, WordNet reflects the lexical knowledge that speakers of English must possess when encountering or using these verbs of gesturing. It is very likely that in their mental lexicons, speakers store both the verbs and the nouns in an efficient way, of which WordNet's semantic hierarchies constitute one particular model.

A similar case is presented by the so-called psych (*anger, crush, enrage*, etc.) and cognition verbs (including *advise, convince*, and *enlighten*). These verbs participate in what Rizzi (1986) has termed the detransivitized pro-Arb object alternation, in which the object of the verb receives a non-specific, generic interpretation:

(20) a. The movie angered the audience.
b. The editorial convinced many readers.

(21) a. The movie angered.
b. The editorial convinced.

In WordNet, the verbs here are homogeneous classes (specifically, they are emotion and cognition verbs, each with several distinct subtrees that are ultimately joined, showing their semantic relatedness).

Levin (1993) notes that the objects of these verbs are affected. While this is clearly a correct generalization, it is not useful for Wordnet, where characterizing these verbs in terms of a common superordinate *affect* would yield too large a class; moreover, it would include such verbs as *hit* and *sell* as well, which denote actions affecting their object arguments, yet they are not similar to psychological or emotion verbs like *anger* and *convince*, nor do they participate in this pro-Arb object alternation.

But the pro-Arb verbs can be characterized as a class in terms of WordNet's structure by the fact that their object arguments all come from the noun hierarchy headed by *people* in the *group* file.

5.1. VERB CLASSES DEFINED IN TERMS OF NOUN HIERARCHIES

These psych and cognition verbs could be regarded as a particular sub-class of a larger class of verbs that can appear both transitively and intransitively. The so-called unspecified object deletion is a transitivity alternation that is shared by a variety of verbs. As Mittwoch (1982) noted, unspecified object deletion is accompanied by an aspectual change that is manifested in the (in-)compatibility with certain time adverbials. While verbs with an overt object are telic (and denote accomplishments), the intransitive counterparts are atelic (and denote activities):

(22) a. John ate a pizza.
b. Mary cleaned the house.

(23) a. John ate.
b. Mary cleaned.

In addition to receiving an activity reading, the intransitive variant is characterized by the fact that the understood object must be a typical instance of an object of the verb. Thus, (23a) can never be interpreted to mean that John ate his hat; (23b) can not mean that Mary cleaned her teeth.

Resnik (1993) searched text corpora to determine the class of nouns that can appear as objects of transitive verbs, and classified the nouns in terms of WordNet noun hierarchies. Resnik found that verbs like *eat* and *drink* have objects that belong to relatively few hierarchies and can thus be said to be semantically homogeneous. These verbs tend to participate in the unspecified object deletion alternation. The reason seems to be that the deleted objects are relatively predictable and carry little information value, since they come from only a few semantic hierarchies. By contrast, verbs like *put*, which are "promiscuous" and select a wide variety of semantically heterogeneous objects cannot appear in the intransitive alternation. In these cases, the semantic class of a deleted object is not predictable and carries real information.

Verbs such as *eat* and *clean* are not related in WordNet. WordNet's hierarchical organization therefore does not rely on any semantic generalization to characterize the class of object-dropping verbs. However, Resnik's results show that these verbs are characterizable in WordNet in terms of the semantic classification of their object nouns. [5] Fellbaum and Kegl (1989) also

[5]However, there are apparent counterexamples to Resnik's explanation that the semantic coherence and thus the predictability of the objects makes them potentially deletable. The class of denominal verbs termed "braid" verbs by Levin (1993) cannot drop their object even though these objects are highly constrained and only a very specific object noun, referring to a body part, is usually selected. These verbs include: coldcream, henna, manicure, rouge, crimp, condition, etc.

studied verbs that can undergo Indefinite Object Deletion and focused on the verb *eat*:

(24) a. Paul ate a hamburger.
b. Paul ate.

When one examines the troponyms of *eat* one finds that they fall into two syntactic classes: those that must always be used transitively, and those that are always intransitive. The first class includes the verbs *gobble*, *guzzle*, *gulp*, and *devour*; the second class includes denominal verbs like *dine*, *lunch*, *brunch*, and *snack*. Fellbaum and Kegl propose two distinct verb senses of *eat*; each sense occupies a different position in the network and is part of a different hierarchy. Intransitive *eat* expresses the concept "eat a meal'.' In many troponyms of this verb, such as the denominals *dine*, *breakfast*, *picnic*, and *feast*, the verb *eat* has become conflated with hyponyms of the noun *meal*: for example, *dine* simply means "to eat dinner." Because their semantic structure incorporates a noun that corresponds to a direct argument, these verbs are intransitive. Other intransitive troponyms of *eat* are *munch*, *nosh*, and *graze*. Although these verbs are not conflations of *eat* and a noun, they are semantically related in that they refer to eating informal kinds of meals or repasts. By contrast, the transitive verb *eat* has the sense of *ingest*, and its troponyms denote specific manners of eating: *gobble*, *gulp*, *devour*, etc.

Cases like *eat* show that differentiating among semantic hierarchies can account for the syntactic behavior of distinct groups of verbs.

6. Conclusion

We considered the representation of the verb lexicon in terms of a semantic network, in which synsets that represent verb concepts are interlinked via semantic and lexical relations. We examined the question whether organizing verbs in this manner sheds light on the organization of the verb lexicon, and found that the notion of synsets, which include non-literal verb meanings, provides for the possibility of examining the constraints on the extended uses of words. Second, we showed that a hierarchical organization of verbs can provide semantically and syntactially motivated classification criteria. WordNet's semantic organization reflects many syntactic regularities, mirroring Levin's (1993) work where syntactic classifications yielded semantic coherence. We expect that additional useful insights can be derived by further examining the representation of verbs in a semantic network.

References

Apresjan, Y. 1973. Regular Polysemy. *Linguistics* 142. 5-32.

Cruse, D. A. 1986. *Lexical Semantics.* New York: Cambridge University Press.

Fellbaum, C., and Kegl, J. 1988. Taxonomic Hierarchies in the Verb Lexicon. Presented at the EURALEX Third International Congress, Budapest, Hungary.

Fellbaum, C., and Kegl, J. 1989. Taxonomic Structures and Cross-Category Linking in the Lexicon.In: de Jong, K., and No, Y. (eds.). *Proceedings of the Sixth Eastern State Conference on Linguistics.* Columbus, Ohio: Ohio State University

Fellbaum, C. 1990. The English Verb Lexicon as a Semantic Net. *International Journal of Lexicography,* 3/4, 278-301.

Fellbaum, C. 1995. Co-Occurrence and Antonymy. *International Journal of Lexicography,* 8.4:281-303.

Fellbaum, C. (in press). A Semantic Network of English Verbs. In Fellbaum, C. (ed.) *WordNet: An Electronic Lexical Database and Some of its Applications.* Cambridge, MA: MIT Press.

Fillmore, C. J. 1986. Pragmatically Controlled Zero Anaphora. In: *Proceedings of the Twelfth Annual Meeting of the Berkeley Linguistics Society.*

Fodor, J. 1970. Three Reasons for Not Deriving "Kill" ¿From "Cause To Die." *Linguistic Inquiry* 1: 429-438.

Gruber, J. 1976. *Lexical Structures in Syntax and Semantics.* New York: North-Holland.

Hale, K., and Laughren, M. 1983. The Structure of Verbal Entries: Preface to Dictionary Entries of Verbs. Warlpiri Lexicon Project, MIT, Cambridge, MA.

Jackendoff, R. S. 1972. *Semantic Interpretation in Generative Grammar.* Cambridge, Mass.: MIT Press.

Jackendoff, R. S. 1983. *Semantics and Cognition.* Cambridge, Mass.: MIT Press.

Katz, J. J. 1972. *Semantic Theory.* New York: Harper and Row.

Katz, J. J., and Fodor, J. 1963. The Structure of a Semantic Theory. *Language* 39, 170-210.

Lakoff, G. 1970. *Irregularity in Syntax.* New York: Holt, Reinhart, and Winston.

Lakoff, G., and Johnson, M. 1980. *Metaphors we live by.* Chicago : University of Chicago Press.

Levin, B. 1993. *English Verb Classes and Alternations.* Chicago: University of Chicago Press.

Miller, G. A. 1990. (ed.). WordNet: An On-Line Lexical Database. Special Issue of *International Journal of Lexicography* Vol 3, No. 4.

Miller, G. A. and Johnson-Laird, P. N. 1976. *Language and Perception.* Cambridge, MA: Harvard University Press.

Miller, G. A., and Fellbaum, C. 1991. Semantic Networks of English. *Cognition,* special issue, eds. Levin, B., and Pinker, S., 197-229. Reprinted in: *Lexical and Conceptual Semantics,* eds. Levin, B. and Pinker, S., Cambridge, MA: Blackwell. 197-229.

Mittwoch, A. 1982. On the Difference Between *Eating* and *Eating Something*: Activities vs. Accomplishments. *Linguistic Inquiry* 13: 113-122.

Nunberg, G. 1978. *The Pragmatics of Reference.* Bloomington, IN: Indiana University Linguistics Club.

Ostler, N., and Atkins, B. T. 1991. Predictable Meaning Shift: Some Linguistic Properties of Lexical Implication Rules. In: Pustejovsky, J., and Bergler, S. (eds.), *Lexical Semantics and Knowledge Representation,* Proceedings of the SIGLEX Workshop.

Perlmutter, D. M. 1978. Impersonal Passives and the Unaccusative Hypothesis. Proceedings of the Berkeley Linguistics Society 4, 157-189.

Pustejovsky, J. 1991. The Generative Lexicon. *Computational Linguistics,* 17:409-441.

Pustejovsky, J. 1995. *The Generative Lexicon.* Cambridge, MA: MIT Press.

Resnik, P. 1993. *Selection and Information.* Unpublished Ph.D. Dissertation, University of Pennsylvania, PA.

Rizzi, L. (1986). Null Objects in Italian and the Theory of *pro. Linguistic Inquiry,* 17:501-

557.

Schank, R. C. 1972. Conceptual Dependency: A Theory of Natural Language Understanding. *Cognitive Psychology* 3: 552-631.

Schank, R. C., and Abelson, R. 1977. *Scripts, Goals, Plans, and Understanding.* Hillsdale, NJ: Erlbaum.

Smith, C. 1970. On Causative Verbs and Derived Nominals in English. *Linguistic Inquiry,* 3:136-138.

Talmy, L. 1985. Lexicalization Patterns: Semantic Structure in Lexical Forms. In: Shopen, T. (ed.). *Language Typology and Syntactic Description,* vol. 3. Cambridge: Cambridge University Press.

DESCRIBING VERB SEMANTICS IN A TYPE HIERARCHY

Disambiguation of Italian verbs

ACHIM STEIN
Institut für Linguistik/Romanistik, Universität Stuttgart
Keplerstraße 17, D-70174 Stuttgart, Germany

1. Introduction

This contribution presents aspects of ongoing research on the semantic representation of the Italian verb vocabulary which aims to build lexical resources for linguistic research and NLP.

The basis of the research is a detailed analysis of the Italian verb vocabulary: the 1643 most frequent Italian verbs (according to the IBM frequency list VELI 1989) have been described syntactically: for each verb, every construction retrieved either from one of the currently existing monolingual Italian dictionaries or from a 50 Million word corpus has been entered as a sub-entry to the dictionary, along with its German contextual equivalents. Sub-entries have been further split up according to semantic criteria, so that the dictionary contains about 12,000 sub-entries. The result of this analysis is a syntactic database of Italian verbs. It is represented on the left of figure 1, labelled "project 1".[1] This database does not contain any explicit semantic information, except for some argument classes (human subject etc.).

Current research, which will be presented in this contribution, focuses on verb semantics and is represented on the right side of the figure ("project 2"). We briefly outline the major tasks of this project:

1. Every construction in the database is assigned to a semantic concept. Concepts are types in a hierarchically ordered system. Each verb in the database, via its constructions and senses, is linked to the nodes in the hierarchy that define it. A verb which is linked to more than one node is polysemous. The set of nodes to which it is linked represents

[1]This work has been carried out in the DFG Project II A 2 - Bl 178/4-1 "Italienische Verben".

P. Saint-Dizier (ed.), Predicative Forms in Natural Language and in Lexical Knowledge Bases, 111–137.

its potential polysemy, i.e. the set of senses the verb can have in appropriate contexts. Section 2 deals with aspects of type hierarchies, and some theoretical views of verbal polysemy are presented in section 3.

2. The second step of semantic description is concerned with generalizations over the types attached to the constructions. Two questions will be discussed: in section 4.1 we are going to verify the hypothesis that members of the same semantic verb class behave similarly with respect to their potential polysemy, and section 4.2 presents an attempt to generalise over the types within a verb, i.e. to elaborate a single semantic representation for each verb which is not its core meaning, but an underspecified structure from which the contextual interpretations may be derived.

2. Ontology

In a general sense, an ontology is a framework for the representation of conceptual information. In many current NLP theories, the more abstract levels of information are organised in systems which exhibit ontological characteristics. Work in the AI domain has produced a de facto standard for the representation of ontologies which normally consists of defining (a) a subsumption lattice over types, (b) the information which is attributed to the types, and (c) methods of information inheritance. Nevertheless, many questions concerning ontologies and their use in NLP remain unanswered, which has led to a number of diverging approaches and models. With respect to their content, it is not clear what ontologies should organise: the world itself, our knowledge of the world, or the meanings of linguistic expressions? With respect to their design, ontologies differ in the extent to which they adopt formal criteria, i.e. if they subscribe to mildly formalised common-sense knowledge criteria in the tradition of Husserl's formal theory of objects (Husserl, 1900-01) or to strict logical methods introduced by Russell, Frege, Leśniewski and others (see Cocchiarella 1991). In his classification of ontologies for NLP, Bateman (1992) distinguishes three types, situated on different levels of the continuum between linguistic and world knowledge representation.

On the first level, i.e. close to lexical description, ontologies combine principles of ontology and lexicon design on a low level of abstraction, e.g. Jackendoff's (1990) *lexical conceptual structures* (LCS) or LCS-based applications like Dorr's ontology of lexicon entries for machine translation (Dorr et al., 1994). On an intermediate level, interface ontologies are characterised by a level of description which is clearly distinct from lexicon and grammar, but accepts only categories which are of lexical or grammatical relevance:

> "[they] provide a classification of any particular instances of facts, states of affairs, situations, etc. that occur in terms of a set of general objects and relations of specified types that behave systematically with respect to their possible linguistic realizations. This classification has itself many of the properties of an ontology [...] although [...] it must be seen as a strictly linguistically motivated ontology." (Bateman, 1992, 59)

An example of this type is the *(Generalised) Upper Model* (Bateman, Fabris, and Magnini, 1995). Finally, conceptual ontologies strive for language independent knowledge representation, although they may be closely related to NLP Systems. An example is the use of the Mikrokosmos ontology (Mahesh and Nirenburg, 1995) for knowledge-based machine translation, with explicit links between the lexicon and the ontology. Other systems, like Cyc (Lenat and Guha, 1988), lack a clearly defined application and usually face the problem of arbitrary acquisition and description of concepts, which makes re-usability in NLP systems problematic (Mahesh et al., 1996).

Since our point of departure is the syntactic database, it is obvious that our type hierarchy does not reach the level of abstraction of conceptual ontologies: it is meant to be a tool to pre-structure large amounts of linguistic data and make it accessible more easily for further linguistic analysis. According to this practical approach, our types are motivated grammatically and semantically, and not by theoretical or extra-linguistic aspects of ontology design.

The type hierarchy underlying our classification is partly inspired by systemic-functional process ontologies (Halliday, 1985; Matthiessen, 1992), which differentiate between major process types like material (doing and happening), relational (being and having), verbal (saying), and mental (thinking). These categories have also been adopted in the German component of the EUROTRA project, in which explicit syntactic tests have been postulated for the assignment of the various process types (Steiner et al., 1988). This method, however, produced semantically counter-intuitive categorizations and proved that strict syntactic criteria are not adequate for a taxonomy of verbal processes. Bateman (1992), for example, criticises the fact that in EUROTRA-D complement phrases in subject and object position automatically select for mental processes, regardless of the verb meaning: *That ... means that ...* is in fact a relational process, but could not be classified as such.

Although syntactic considerations have led to the introduction of types like `affect` and `happen` to set off causative from non causative readings, type assignment has been principally based on meanings and kept apart from syntactic considerations. Another property which is inspired by the systemic approach and concerns the more specific levels in the hierarchy is the "dimension" of processes: as described in the Generalised Upper Model (Bateman, Fabris, and Magnini, 1995), the "orientation" of processes is taken into account. Speech act verbs, for example (cf. section 4.1.2), can be message oriented (*say*) or addressee oriented (*accuse*). Implicitly, this type of distinction is also adopted in the more practical approaches, as in Levin's (1993) differentiation between *amuse* verbs (experiencer oriented) and *admire* verbs (phenomenon oriented). Unlike Levin's semantic classes for English verbs, our types (a) have not been assigned to verbs as such, but to each of its sub-entries (senses, constructions, as explained in section 1), (b) they are more detailed than Levin's and (c) hierarchically structured.

For the time being, the type hierarchy contains some 200 terminal nodes. In the hierarchy, features are attached to nodes, and children nodes inherit or specify the features of parent nodes. In the discussion of examples (section 4), we will use the path through the tree (or an abbreviation of the path) to name nodes and the verb classes associated with them, for example `material-affect-location` for a change-of-location verb, which inherits, for example, the features aspect=dynamic from the `material` node, subject=causer or object=affected from the `affect` node, affected_mode=moved from the `location` node etc. We cannot provide a detailed account of types and features here, but some of the categories will be mentioned in the examples below.

3. Polysemy I: theories

A widely discussed point in the polysemy discussion is the question of whether words have a core or literal meaning from which other meanings are derived. The assumption of a core meaning is widely accepted and spans from 19th century historical linguistics over structural and psychomechanic linguistics (e.g. Guillaume's 1964 *signifié de puissance* vs. *signifié d'effet*) to prototype theories (e.g. Kleiber 1990) and current theories in computational linguistics (e.g. Raskin and Nirenburg 1995), while other works demonstrate the implausibility of the core meaning view (e.g. Fillmore 1982). An intermediate position is taken by theories which consider polysemous acceptions as selections from a set of meanings assigned to lexemes (e.g. Rastier's 1987 actualised and virtualised meaning components or Bartsch's 1994 criticism of Searle's 1992 literal meanings). We think that for practical purposes like building lexical resources for NLP, there is no need to resolve the question on a theoretical level. All the meanings of a form must somehow be accessible from the lexicon, and questions of homonymy vs. polysemy or derived meanings vs. enumerated meanings are relevant for acquisition and implementation strategies only. In our analysis, for example, assuming a core meaning has facilitated generalizations about typical meaning shifts (section 4.1), but this core meaning does not appear as a preferred reading in the lexical representation of a verb (section 4.2). Since the notions of core meaning and polysemy are closely interlinked, we will briefly discuss some definitions of polysemy before we turn to the linguistic analysis proper, considering some recent proposals in NLP as well as traditional linguistic classifications.

Copestake and Briscoe (1995) distinguish between constructional polysemy and sense extension. Constructional polysemy is sense modulation induced by the context (or "co-composition", according to Pustejovsky 1995), whereas sense extension is a lexical, rule-based process which creates derived meanings from basic ones. A third term, "verbal" polysemy, is introduced by Pustejovsky and Busa (1995) for relations between different grammatical realizations of the same verb, for example causative and unaccusative constructions. Since "verbal polysemy" may also refer to polysemy of verbs, we prefer to call this type "grammatical polysemy". (Note that most instances of regular polysemy cited by Apresjan (1974) are also results of the adjunction of causal, potential, negative and other operators, and are therefore closer to the cases of grammatical polysemy.) The opposition between constructional polysemy and sense modulation is probably gradual, since Copestake and Briscoe (1995) define the former as "change of meaning" and the latter as "a question of selecting the appropriate aspect of the meaning".

An earlier classification in the perspective of structural semantics is Martin's (1972) distinction between *polysémie d'acceptions* and *polysémie de sens*. In (1), we paraphrase the examples of this classification, as represented in figure 2.

(1) case 1: *femme* 'woman' → 'wife'
case 2: *minute* '60th part of a minute' → 'short lapse of time'
case 3: *blaireau* 'badger' → 'brush made of the hair of a badger'
case 4: *rayon* 'line emitted by a luminous source' → 'line between the center and its surrounding circle'
case 5: *plateau* 'tray' → 'geological form'

Martin assumes a more literal meaning A, listed in the upper line, and a derived sense B. He distinguishes two degrees of polysemy:

(a) At a lower degree (*acceptions*), polysemy results from addition (case 1) or deletion (case 2) of semantic components, where both senses can either be co-hyponyms as above, or not, as in case 3.
(b) At a higher degree, polysemy results from component substitution, with co-hyponymy (case 4) or hyperonym shift (case 5).

Although Martin's classification has some weak points, like the definition of hyponymy, and, consequently, the distinction between cases 4 and 5, we consider it to be useful to retain some of these cases and to apply them to polysemy of verbs.

(a) The cases 1 and 2 are based on implication, i.e. B is a kind of A (case 1: specification) and A is a kind of B (case 2: abstraction): the verb *abbandonare* may have the more general sense 'move away from' as in *abbandonare la città* 'to leave the town', but it can also mean 'to move away from & not care for' as in *abbandonare la moglie* 'to leave one's wife'. The motion component in the latter interpretation can be deleted (case 2) as in *abbandonare il suo giardino* 'to neglect one's garden', which does not imply motion away from the garden.
(b) In case 3, A is related to B by some unspecified, but often causal or temporal relation: *confidare* means 'to trust somebody', but can also refer to the verbal process which is a result of the mental attitude, as in *confidò al suo amico che era malatto* 'he confided to his friend that he was ill' (see example 13 in 4.1.2).
(c) In cases 4/5, A and B share some semantic components. A verbal example would be *apprendere la morte del presidente* 'to hear about the president's death' and *apprendere una lingua* 'to learn a language' which both share the perceptional components and refer to a cognitive process.

Martin (1979) adds the distinction between *polysémie externe* and *polysémie interne* with respect to predicates, i.e. verbs and adjectives. He considers external polysemy as semantic ambiguity of the argument of a predicate while the meaning of the predicate remains unchanged, whereas internal polysemy implies modifications of the structure of the predicate itself. His example *cambrioler quelqu'un* 'to burgle somebody' is a case of argument reinterpretation which means in fact 'to burgle the place where somebody lives'. External polysemy, then, covers instances of argument metonymy, which can be accounted for either by adding argument specifications to the verb entry in a sense enumerative lexicon or by relating concepts (e.g. 'house', 'apartment') to an argument ('person') of its telic function value ('live in') in a generative lexicon, where such type coercions are triggered whenever 'normal' interpretation fails.

Martin's distinction seems convincing, but in everyday lexicographic analysis, it turns out that it his hard to make a clear-cut distinction between meaning changes of the predicate and meaning changes of its arguments: it is virtually impossible to prove that the meaning of a verb does not change at all when it appears in a different context. We checked some of Martin's analysis against the types we assigned to the corresponding Italian constructions in our hierarchy: in cases of external polysemy, the type should be the same for both contexts, in cases of internal polysemy, the types should differ. The results were contradictory: in some of Martin's examples, e.g. *le liquide coule* 'the liquid flows' vs *le sable coule* 'the sand flows' (an example which he does not find entirely convincing himself), both senses belong to the same subclass of motion verbs. On the other hand, the senses of the verb *apprendre*, which according to Martin illustrates external polysemy quite well (*apprendre la danse à qn* 'teach somebody to dance' vs *apprendre la nouvelle à qn* 'tell somebody the news'), belong to rather distant types in the **material-social-interaction** and the **verbal** domain, respectively. Although we do not agree with all of Martin's examples, we consider his work to be of relevance in any practical endeavour in the lexicographic domain:[2] systematic lexical descriptions require precise distinctions between phenomena which modify the internal structure of a word, i.e. its semantic representation(s), and phenomena which control the interpretation of the context of a word.

In current theories, polysemy of verbs is often limited to some of the types cited above, whereas others are excluded from the discussion. For example, Pustejovsky's treatment of polysemy of verbs focuses either on grammatical alternation (1995, Chap. 9) or on semantic specification. With respect to examples like *open the letter, open the trail* and *open the door*, he claims:

> what is distinctly different about verbs such as *open, close, break, repair*, and *fix*, is the functional dependence that the verb has on the argument denotation for arriving at the appropriate interpretation of the verb in context. (Pustejovsky, 1995, 220)

According to this view, the contextual interpretation of a verb contains parts of the argument representation (in the cases cited above the activity associated with the argument), and can be paraphrased as 'produce a state in which it is possible to read the letter/walk through the door' etc. Co-composition successfully circumvents the distinction external *vs* internal polysemy, but captures only variations where the complement adds components to the predicate (see Martin's case 1 above), and often distinguishes senses which in published dictionaries are not disjoint. The Italian equivalent of *open*, *aprire*, for example, has several senses which cannot be described as specifications of a change-of-state predicate:

(2) a. aprire una nuova era
introduce a new era
b. aprire un fossato
dig a trench

The complements in (2) cannot be associated with any activity in particular, and they are not affected, but produced by the main process expressed

[2]See also Martin (1994) for a discussion of polysemy with respect to machine readable dictionaries.

by the verb. These are cases where the change of meaning modifies the entire semantic structure and cannot be reduced to simple addition of components (see Martin's cases 4 and 5 above).

We think that in-depth linguistic analysis is a prerequisite for the construction of systems which interpret or disambiguate verbs in context. This analysis must be systematic, i.e. consider all the senses a form can have and avoid the arbitrary sense distinctions of traditional dictionaries. In our approach, the database of Italian verbs is the reference for completeness, and the type hierarchy discriminates word senses. A verb is polysemous if its senses are associated with more than one type, and disambiguation is successful if types can be distinguished in context.

Our account of polysemy is therefore two-fold: The first part concerns definition of and generalizations about potential polysemy. In our approach, the potential polysemy is expressed by the set of senses a verb is associated with, i.e. by all the meanings a verb may have in an appropriate context. Generalization on this level means to describe regular co-occurrences of senses within the sets, and to identify the systematic relationships among them (see Atkins 1991). This procedure usually results in a fine-grained semantic sub-classification. Examples will be presented in section 4.1, where we consider for example verbs which a change-of-location reading and a change-of-mental-state reading, in order to delimit the meaning components which typically participate in this kind of meaning shift. This method resembles the work of Levin (1993) in that it yields verb classes, except that our approach strives for semantic instead of syntactic regularities. Since this part of the work abstracts from context and considers only relations between senses, it gives insight in systematic linguistic properties which, in Saussure's terms, are situated on the *langue* level. The second part is complementary and concerns the concrete use of a form, i.e. the *parole* level: when an entity of the *langue* is used in a speech act, this entity is actualised. Actualization is context-dependent selection of the appropriate sense, and a sense disambiguation system should be able to imitate this selection process at least approximately.

4. Polysemy II: examples

4.1. THE CONTEXT-INDEPENDENT APPROACH

In a context-independent view, polysemy is considered as paradigmatic relation between senses of a form, independent from its context. Our research on context-independent polysemy consisted in investigating the relations between the semantic classes shared by a given verb, and was carried out in two steps: at a first stage, we calculated the degree of affinity between pairs of semantic types based on their co-occurrence in the set of types associated with each verb (as explained in section 3). For example, the coocurrence of the types **material-affect-location** and **material-affect-mental** (see the discussion below) is significantly more frequent than the coocurrence of **material-affect-attach** (*attach* verbs) and **psych-perception-directed** (directed perception verbs). The second stage consists in detailed analysis of the verb groups which share the more frequent type co-occurrences in order to delimit the meaning components which trigger or inhibit these meaning shifts. Sections 4.1.1 and 4.1.2 present two example analysis.

4.1.1. *Regularities between* location *and* mental *types*

Both **location** and **mental** are subtypes of the **material-affect** branch, which subsumes externally caused processes. The types which are held responsible for meaning shifts are situated on the lower levels of the hierarchy. We distinguish between licensing and blocking types. Types which license the meaning shift from **location** to **mental** are situated under **directed** (3) and **manner-of-motion** (4).

(3)
a. **down**:
abbassare 'put down' → 'disparage'
affondare 'sink' → 'plunge'
deprimere 'press' → 'humiliate'
b. **away/towards**:
allontanare 'remove' → 'alienate'
attrarre 'attract' → 'entice'
divertire 'divert' → 'entertain'
orientare 'position' → 'direct (somebody's attention)'
rimuovere 'remove' → 'suppress (a feeling)'
trascinare 'draw' → 'carry somebody away'
c. **around**:
avvolgere 'wrap' → 'embroil'
circondare 'surround' → 'overwhelm'
d. **into**:
affondare/immergere 'immerse' → 'plunge'
introdurre 'introduce' → 'instruct'

(4)
a. **by forcing**: *forzare* 'force', *premere* 'press', *stringere* 'contract'
b. **by throwing**: *gettare* 'throw'
c. **by carrying**: *caricare* 'carry'

The members of these verb classes have access to the path from **location** to **mental**, and the rules associated with this path transform the features in a regular way: **down** changes into **negative**, and **manner** loses the concrete local features which distinguish for example *forzare, premere* and *stringere* by the physical position of the agent with regard to the object. Figure 3 visualises that *abbassare* is attached to the directional subtype down, from where it may access the path between location and mental.

The parallelism between local and mental interpretations is obvious and seems to be typical for many of the most frequent meaning shifts we encountered in our statistical analysis: the shift from one interpretation to another concerns only few and clearly delimitable types in the representation of the verb, as the level below directed in the present case. Other properties, no matter if they are attached to upper level types, like causation, or to lower level types, like the ambiguity between telic and untelic processes, remain unchanged. If we ignored these regularities we would miss important generalizations, and therefore abstract types like generic-down have been introduced as context-dependent, underspecified primitives in our representation (see section 4.2).

Types which block the mental interpretation of affect-location verbs are goal and configuration: they contain explicit information about where the object is moved to (*alloggiare* 'into a house', *imboccare* 'into the mouth', *incassare* 'into a box'), or about the configurational relation of the moved object to other objects or to the goal (*adattare* 'adapt', *affiancare* 'side by side', *contrapporre* 'facing each other' etc.). Interestingly, the type configuration blocks the meaning shift to affect-mental, but allows the shift to psych-action (mental activities without an external causer) as in *adattarsi, allinearsi* 'adopt another opinion', *confinarsi* 'limit oneself', *riposare* 'rely on'.

One might of course argue that these meaning shifts are induced by the arguments, especially by the respective direct objects of the verb. But context does not help to define exactly which verbs can be affected by this kind of induction, and, what is more, in many cases of polysemy object classes are not relevant for disambiguation. Human objects, for example, can be affected by both locational and mental processes, as in (5), where the ambiguity of *divertire*, basically a psych verb, cannot be resolved contextually. Another example is *schierare*, cited in (6) below, with a local core meaning ('position').

(5) a. Le sue barzellette mi divertono.
His jokes amuse me
b. divertire le forze nemiche
distract the enemy forces

The relation between the various types and the behaviour of the verb is fairly straightforward when the type acts as a specifier for the arguments of the verb: directed specifies the path, goal specifies the place. Hence, it is obvious that types which would produce a conflict with the general specifications of the target type inhibit meaning shifts: a lexically specified object or object class as in *pilotare* 'to pilot' or a lexically specified concrete goal as in *archiviare* 'to archive' are incompatible with types under **affect-mental**, where the object must denote an entity with a brain and the prepositional phrase a resulting mental state. The scope of other types is more ambiguous: **manner** can refer either to arguments (agent or instrument) or to the event itself, **configuration** can imply instrument or object characteristics. In these cases, further sub-classification may be necessary to derive the potential polysemy.

Apart from specific features like incorporated arguments, there are also more general principles which could be used to account for licensing and blocking of meaning shifts, for example those of agency and telicity: Whereas directed movement can, but need not be telic, verbs denoting movement into a configuration are telic by definition, since the configuration is the result of the action. It is not surprising, that actions on human mental states can be expressed by the directed movement verb class which does not presuppose the resulting state, just as it is impossible to anticipate the result of a mental influence.

The type **configuration** seems to be less restrictive than the type **goal**: a few verbs may in fact shift to the mental domain (*legare* 'embed' → 'occupy', *schierare* 'position' → 'convince'), but in these cases the configurational aspect either tends to disappear or is less prominent from the start: *legare* 'bind' loses its obligatory prepositional phrase (*a*+NP 'bind sth to sth'), which, in the local reading, denotes the goal of the attachment, and *schierare* has only a weak specification of the configuration and means 'position in some useful way' (a typical object of *schierare* would be *soldiers*). In contexts where the configurational features prevail, such as 'line' in *allineare*, however, the verb behaves as predicted and mental interpretation is ruled out:

(6) a. Il suo discorso ha {schierato/*allineato} la nazione dietro la Casa Bianca.
His speech has aligned the nation behind the White House
'...made them adopt the opinion of...'
b. È un lavoro che mi lega {troppo / in città}.
It is a job which {occupies me too much / keeps me in the city}.

Even the few examples discussed so far show clearly that a very detailed semantic description is necessary before generalizations about meaning shifts of verbs can be made. Systematic queries for counter-examples often provide further hints to subclassification: the query for **affect-location** verbs which do specify a direction, but do not have any **affect-mental** construction produces a rather homogeneous list of verbs with the directional features **out** (*confinare* 'ban', *fugare* 'chase', *prelevare* 'pick up', *togliere* 'pull out' etc.) and **back** (*respingere* 'push back', *rilanciare* 'throw back', *rimandare* 'send back', *ritirare* 'pull back', etc.).

4.1.2. *Regularities between* **psych** *and* **verbal** *types*

The **psych** and **verbal** types (thinking and saying) have been chosen in order to point out some aspects of polysemy which are typical of meaning shifts between the more abstract process types. In a relation between a concrete and an abstract type, such as `location-mental` in 4.1.1, the core meaning is usually concrete and the meaning shift is directed towards the abstract type, and the rules for the transformation of subtypes, such as **down** → **negative** are relatively homogeneous across members of a verb class. Abstract types, on the other hand, often refer to more complex realities with temporal, causal or presuppositional features which have to be taken into account by the transformation rules. The following examples have been selected to demonstrate some aspects of this complexity.

Rules which link **psych** and **verbal** types can be additive or transformational. Since additions are ordered temporally and logically, we can distinguish several kinds of meaning shifts. Let us consider three of them:

(1.) Verbs with a psych core meaning shift to the verbal domain. The rules specify that the verbal features are added without overwriting the psych features. The mental activity precedes the speech act and is implied by it. Verbs of this type are listed in (7).

(7) *analizzare* 'analyse'
auspicare 'hope'
azzardare 'dare'
immaginare 'imagine'
inventare 'invent'
ipotizzare 'imagine'

Typical verbal features are roles such as addressee, message or content. In Italian, the addressee is typically realized as a dative argument (*a*+NP) which differs from other prepositional phrases because it can be replaced with the dative pronoun (*Parlo a Gianni* → *Gli parlo* 'I speak to him').

In (8) and (9), both *inventare* and *analizzare* obviously denote speech acts which refer to the result of the mental process or state. On the syntactic level, however, both verbs maintain the properties of psych verbs in that they do not accept the dative complement, although it is semantically unproblematic to interpret *alla commissione Affari...* as the addressee of the speech act.

(8) Ora ha anche inventato che la colpa è tutta mia.
Now he also claims that it is all my fault
'He invented it and said it'

(9) Il ministro dell'Interno Scotti, analizzando la situazione di Gela alla commissione Affari costituzionali del Senato, aveva affermato che...
..., presenting the situation of Gela to the Senate committee for constitutional matters...
'He analysed the situation and communicated the result of his analysis'

Another example of meaning shifts which are not necessarily reflected on the level of argument structure is *immaginare* in (10). What triggers the contextual meaning here is not the addition of a typically verbal argument: there is not even a potential candidate for the addressee role, and the complement phrase alone cannot distinguish psych from verbal readings. The fact that knowledge about the world — a painter cannot dream *in* one of his paintings (10a) — or knowledge about textual dimensions — the global context of (10b) is not Ariost's mental state, but the content of his work — is required for the interpretation is a strong argument in favour of rule-based lexical representation of polysemy:

(10) a. In quell'affresco il pittore ha immaginato che il deserto si animi improvvisamente di vita, di uomini, piante e animali.
In this fresco the painter has represented how the desert is suddenly animated by life, humans, plants and animals

b. L'Ariosto immagina che Astolfo salga con l'ippogrifo sulla luna.
Ariost describes that Astolfo mounts with the hippogryph to the moon

Note that the possibility of these meaning shifts does not tell anything about style or frequency, but the root of the verb might be a useful indicator for the commonness of meaning shifts: *immaginare* is semantically close to *ipotizzare*, but its root 'image' is less closely related to speech acts than *ipotesi* 'hypothesis': *immaginare* is used verbally in artistic and literate contexts only.

The sample of verbs produced by co-occurrence queries indicates the common features of this verb class: they refer to a volitional, experiencer controlled act of mental representation (`psych-action-representation-volitional`). A look at the semantically closest counter-examples (see Figure 4) yields the adjacent classes for which the path to verbal is not accessible: on the one side, there are the *dream* verbs which are `non-volitional` (*sognare*), on the other side there are the *plan* verbs which are volitional, but not representational (*pianificare, ideare*). Consider also *augurare* and *auspicare* 'hope, wish'.

(2.) The second case is also an example of additive meaning shift. Again, polysemy rules operate on psych verbs and add verbal features. However, compared to the verbs of group (1.), the temporal order is inversed, since the speech act is presented as preceding and causing the mental activity. Logically, there is no implication between both types: the mental activity is not a necessary consequence of the speech act and need not be caused it. The speech act meaning is neither message oriented nor content oriented, but addressee oriented, i.e. the addressee appears as direct object in the speech act readings (11b, 12b), although native speakers sometimes hesitate between the dative and the accusative interpretation of clitics such as *mi* in (11b):

(11) a. Questo rassicura la mia coscienza. (GA)
This calms my conscience
b. Mi rassicurò che non m'avrebbe abbandonato. (LUI)
He assured me that he would not leave me alone

(12) a. L'avevano prevenuto contro di noi.
They had prepossessed him against us
b. Prevengo i miei ascoltatori che si tratterà di un discorso breve e alla buona.
I draw the attention of my public to the fact that it will be a short and good speech

(3.) The third group contains verbs with no additive relation between psych and verbal features: the interpretations exclude each other. The example of *confidare* demonstrates that the verbal features either erase the psych features or re-assign them to other arguments:

(13) a. Ma io confido che ciò non avvenga.
But I believe that this won't happen
b. Il giovane Heseltine confidò a un amico che intendeva diventare primo ministro.
The young Heseltine confided to a friend that he wanted to become Prime Minister
c. "Difficile elaborare una qualsiasi strategia quando dalla mattina alla sera cambiano le carte in tavola" , confidava qualche giorno fa Frans Andriessen.
"...", confided F.A. some days ago.

In (13a) the verb refers to the subject's mental attitude (confidence). In (13b) it is combined with a dative argument and thus acquires a verbal meaning. The type `mental attitude` does not apply to the content of the complement phrase, but is transferred to the addressee of the speech act, *un amico*. In this case, there definitely is a meaning shift, whereas in the cases (1.) and (2.) both related meanings remain actualised. Meaning shift here is indeed a verbal property (internal polysemy, as defined in section 3): it is not triggered by the syntactic realization, but can apply even in contexts where the addressee is omitted, as in (13c).

4.2. THE CONTEXT-DEPENDENT APPROACH

The results gained from the research about systematic meaning shifts exemplified in section 4.1 are currently being integrated in the semantic lexicon. The design of the lexicon entries is guided by the following principles:

1. Each verb should have a single representation which subsumes all its contextual interpretations.
2. The representation should be broad enough to capture all the senses of a verb, i.e. enclose the entire set of types the verb is associated with in the hierarchy.

3. The representation should be precise enough to set a verb off from its semantically closest neighbours.
4. The representation should take advantage of the regularities of meaning shifts as illustrated in the previous section.
5. Linking of the semantic representation with syntactic structure should be possible.

The meaning of verbs is represented in structures similar to LCS: the meaning is decomposed into primitives which bind the arguments of the verb. Compared to the static meaning representation à la Jackendoff, some modifications are introduced into the representational framework in order to cope with contextual ambiguity: polysemy is transferred from the lexical level to the sub-lexical level of primitives, which leads to a generalization of polysemy rules: they are functions which take primitives and argument features as values and return a more specific interpretation of the primitive. We call this process primitive refinement by sub-lexical rules. With respect to the problem of core meaning discussed in section 3, the representations do not stand for the predominant meaning of the word, but for an abstraction of the totality of its meanings.

In section 4.1.1, we considered cases where the assumption of generalised primitives (like `generic-down` for *abbassare*) can help explain typical shifts of meaning. Instead of this rather straightforward example, we present in what follows the analysis of the relatively polysemous verb *abbandonare* 'abandon, leave'. The example will illustrate how the the principles cited above have been applied, how lexical representations are constructed and how they can be used in a system which disambiguates polysemous verbs. *Abbandonare* is more polysemous than its English counterpart *abandon*: figure 5 lists examples for the readings distinguished in the syntactic database along with their English paraphrases:

According our principles, the representation of *abbandonare* has to be different from its semantic neighbours. From its hyperonym, which according to most monolingual dictionaries (Devoto and Oli, 1992) is *lasciare* 'leave' (lit. 'let'), and from its closest synonyms (*allontanarsi* 'move away', *partire* 'leave', *separarsi* 'separate'), *abbandonare* is differentiated by a concession introduced by the primitive DESPITE. Note that adjunction of meaning components is only one of several advantages of LCS-based structures. Theoretical aspects of the representational formalism are not our point here, but see Dorr et al. (1994) for other examples, e.g. the representation of inherent objects and the handling of translation mismatches. See also Pause et al. (1995), who suggest a similarly general representation for the French verbs *quitter* 'leave' and *partir* 'go away'.

Abbandonare is represented in (14) as a process where the agent a causes his relation with b to end, although the relation is expected to continue.

Note that for the sake of readability, we use a simplified notation which is more readable than our internal Prolog notation.

(14) *abbandonare*: [CAUSE [a [[END [RELATION [a, b]]]]]]
[DESPITE [EXPECTED [NOT [END [RELATION [a, b]]]]]]

Since we have transferred verbal disambiguation to the level of primitives, contextual information refines the general primitives and thus yields the different readings of *abbandonare*. Refinement should produce the following interpretations of RELATION (the numbers refer to the readings in figure 5):

1. *a* executes *b*. *b* denotes events or actions. (1,4,5)
2. *a* executes an action on/with *b*. *b* denotes objects. (2,6)
3. *a* is at *b*. *b* denotes places. (3)
4. *a* is together with *b*. *b* denotes objects. (7)
5. *a* keeps *b* in an expected state. *b* denotes *a* (via the reflexive pronoun *si*), *a*'s body or bodyparts. (8,12-16)

Refinement of RELATION requires, of course, knowledge about the arguments of the primitive: when the arguments of RELATION are looked up in the knowledge base, the structure is refined with all the more specific relations which match either classes of the arguments or the arguments proper. This knowledge base is built up gradually to meet the requirements of verbal disambiguation, and it contains only information which is relevant for this purpose. Its design is inspired by the approach described in Kayser and Abir (1995); the attribution of some fundamental object classes to Italian nouns follows the CISLEX convention described in Langer (1995). Before discussing any more details concerning contextual disambiguation, we will briefly introduce one practical application of the verb representations.

The semantic representations, the knowledge base and the rules for primitive refinement are modular parts of a system for linguistic analysis. They are kept in a Prolog database, designed for easy acquisition and maintenance of the semantic descriptions, with utilities which convert this system-independent format into modules for application systems. For the time being, our efforts concentrate on syntactic parsing and disambiguation of verbs in Italian text. We use Schmid's YAP, a parser for a simple constraint-based grammar formalism with two components: a context-free parser generates a parse forest, which is then evaluated by iterative recalculation of features until the result converges (Schmid, 1996). To parsing methods have been tested:

At a first stage, the syntactic verb descriptions of the DFG project cited in section 1, enhanced by semantic features taken from the type hierarchy, had been included in the feature structures of the verb entries in the parser lexicon. However, due to the fine granularity of the verb lexicon with an average of over seven (and up to over sixty) meanings per verb, disambiguation based on bottom-up parsing with extensive feature structures in the verb entries of the parser lexicon has turned out to be problematic.

We therefore decided to abandon the static meaning descriptions in the lexicon at a second stage. The function of the parser was restricted to pure syntactic analysis, with feature structures containing only morphological information, and semantics was dealt with in separate modules independent of the parser lexicon. In this design, syntactic and semantic analysis are almost completely independent from each other. Deprived of restrictions over feature structures, the parser digests relatively unrestricted input, e.g. sentences from newspaper corpora and the example sentences from the syntactic database, and produces a highly ambiguous output. This parse forest is then post-processed by a program which extracts all the potential argument structures and matches them against the semantic representations (some modules of the system are presented in figure 1, section 1).

(15) Il capitano abbandonò la nave a trenta minuti dal porto
the captain abandons his ship thirty minutes away from the port

A sample analysis of sentence (15), which represents one particular sense of *abbandonare*, will illustrate the steps of the disambiguation:

1. highly ambiguous syntactic parsing
2. extraction of possible argument structures from the parsing result
3. matching of argument structures with semantic information about the arguments
4. refinement of the underspecified verb representation

For (15), the parser produces a parse forest with 80 solutions (due to the various possibilities of PP-attachment), out of which post-processing extracts only two argument structures, since the *da*-PP is not considered to be a potential verbal argument:

(16) **Potential argument structures:**

```
1 capitano/np, abbandonare, nave/np, minuto/pp-a
2 capitano/np, abbandonare, nave/np,
```

The first argument structure is rejected, because the three-place predicate of *abbandonare* is incompatible with time or place PPs as indirect objects. The second argument structure is matched with the two-place predicate, but the internal polysemy of the verb cannot be completely resolved on the sentence level: the refinement of the relation between the *capitano* and *nave* produces a local relation `be-at`, which is defined as a default for static relations between humans and objects, as well as the more specific `maintain` relation which holds between humans and some artifact classes:

(17)
```
REJECTED: capitano abbandonare nave minuto

act :cause(capitano,end)
end :be-at <-- relation(capitano,nave)
RESOLVED: capitano abbandonare nave

act :cause(capitano,end)
end :maintain <-- relation(capitano,nave)
RESOLVED: capitano abbandonare nave
```

This analysis illustrates the types of information required for disambiguation: nouns are assigned to a hierarchy of object classes, where *capitano* and *nave*, respectively, are attached to `human-with-function` and `vehicle`, and refinement procedures specify primitives with respect to these classes. In the present case, disambiguation was not based on these lower level classes, but on the relations between `human` and `thing` vs `artifact`. According to our principles, object classes are refined only when necessary: the `human-with-function` class, for instance, is used to refine the `exist` primitive in *abbattere*, which distinguishes between the 'kill' and the 'force to resign' reading. The artifact domain is a good example for relatively fine-grained distinctions which are linguistically motivated: material artifacts default to the local `be-at` relation, *yen* as a (non-material) monetary unit in context 2 (figure 5) selects the `use` relation, whereas artifacts with temporal dimension (requiring human action) like *nave, casa* or *giardino* 'garden' select `maintain`.

The discussion of these few cases is sufficient to show that our concept of a knowledge base is not comparable with approaches striving for a more or less exhaustive representation of knowledge (see section 2): in the Mikrokosmos ontology, *captain* is associated not only with `human`, but also with concepts like `protection-role`. Even if this information might not be needed for verb disambiguation, it could be used for a more detailed meaning representation, i.e. for inferences like 'if the captain leaves his ship and if the ship is not in the port, the ship is sinking and the passengers have already left the ship'. Future investigations will have to evaluate to what extent the information required for sense disambiguation coincides with the information contained in general purpose ontologies.

Since the lexical representation of polysemy has only just started in our project, the shortcomings of the system still outnumber its advantages. For instance, our present analysis of *abbandonare* is unable to resolve the ambiguity between the 'leave' and the 'neglect' reading with respect to the arguments of RELATION, although human interpreters normally consider the second reading to be much more implausible: the captain might of course be the owner of a ship which he has not repainted for years, but this interpretation is blocked by the PP *a trenta minuti dal porto*. It is nevertheless desirable to preserve this ambiguity for subsequent levels of disambiguation. Further analysis could make use of information contained either in previous sentences or in adjuncts. In fact, since the parser does not prematurely foreclose any interpretations, adjuncts can easily be extracted and integrated into the argument structures. The restrictions on adjuncts, however, are often too vague for them to be captured in the same way as selectional restrictions on arguments. In (15), for instance, it would be insufficient to associate place adjuncts with the 'leave' reading and time adjuncts with the 'neglect' reading, since the latter is also compatible with place specifications which are attributes of the direct object:

(18) il capitano abbandona [$_{NP}$ [$_{NP}$ la sua nave] [$_{PP}$ a Lavagna]]
'the captain neglects his ship, which is at Lavagna'

Another question is how far the refinement should go. It is obvious that the resulting interpretations stop well below the level which would be necessary for selecting the contextually appropriate word in a target language: the example sentences for the sixteen contexts of *abbandonare* taken from the syntactic database are translated with over forty different German equivalents, and the difference between the number of distinguished senses and the number of German equivalents is much higher with other verbs. If, in Italian, it does not matter whether husbands, wives, cats or dogs are abandoned, the distinction between the object classes `human` and `animal` is relevant in German (*verlassen* vs *aussetzen*).

However, it would be inefficient to confuse principles of disambiguation in the source language with principles of lexical choice in the target language. For the time being, we consider the disambiguation process to be accomplished if the output, i.e. the refined lexical representation, is sufficiently specific to distinguish the senses associated with one type in the hierarchy from the others. The performance of the system is controlled by the granularity of the type hierarchy: our system does not distinguish between *aprire la porta* 'open the door' and *aprire la lettera* 'open the letter' (see Pustejovsky's example cited in section 3), since both senses belong to the same type. On the other hand, it is not even clear that more subtle sense distinctions produce better results in NLP: Dolan (1994) cites cases where unnecessary specification of word meanings misses useful information and has negative consequences for NLP systems.

5. Conclusion

We have presented a model for the representation of verb semantics which accounts for the potential polysemy of verbs as a result of systematic semantic analysis of Italian verbs with particular attention to generalizations about meaning shifts. The application of the model for the purpose of sense disambiguation in context has been demonstrated. The novelty of the approach is that contextual disambiguation happens on a sub-level of the lexical representation, i.e. on the level of primitives. We also argued that in both domains, theoretical description and the practical application, linguistic principles of structuring the necessary knowledge should be preferred to principles of general purpose knowledge representation.

References

Apresjan, Jurij D. 1974. Regular Polysemy. *Linguistics*, 142:5–32.

Atkins, Beryl. 1991. Building a Lexicon: the contribution of lexicography. *International Journal of Lexicography*, 3:167–204.

Bartsch, Renate. 1994. *The Myth of Literal Meaning*. Technical report, ILLC, Department of Philosophie, University of Amsterdam.

Bateman, John. 1992. The theoretical status of ontologies in natural language processing. In Susanne Preuß and Birte Schmitz, editors, *Text Representation and Domain Modelling — ideas from linguistics and AI*, KIT-Report 97, p. 50–99, Berlin. Technische Universität Berlin.

Bateman, John, Giovanni Fabris, and Bernardo Magnini. 1995. The Generalized Upper Model Knowledge Base. In Michael Wooldridge and Nick Jennings, editors, *Intelligent agents: Proceedings of the ECAI-94 Workshop on Agent Theories, Architectures, and Languages, Amsterdam, the Netherlands, August 8-9, 1994*, Berlin, New York. Springer.

Blumenthal, Peter and Giovanni Rovere. 1992. Valenz, Polysemie und Übersetzung. Eine Fallstudie. *Romanische Forschungen*, 104:1–26.

Cadiot, Pierre. 1994. Représentations d'objets et sémantique lexicale: Qu'est-ce qu'une boîte? *French Language Studies*, p. 1–23.

Cocchiarella, Nino B. 1991. Ontology II: formal ontology. In Hans Burkhardt and B. Smith, editors, *Handbook of metaphysics and ontology*, volume 2 Bde. Philosophia, München, p. 640–647.

Copestake, Ann and Ted Briscoe. 1995. Semi-productive Polysemy and Sense Extension. *Journal of Semantics*, 12:15–67.

Cowper, E.A. 1990. Apparent Polysemy and Thematic Underspecification. *Toronto Working Papers*, 12.

Devoto, Giacomo and Gian Carlo Oli. 1992. *Nuovo vocabolario illustrato della lingua italiana*. Felice Le Monnier, Milano.

Dolan, William B. 1994. *Word Sense Ambiguation: Clustering Related Senses*. Technical report, Microsoft Research, Advanced Technology Division, Microsoft Corporation.

Dorr, Bonnie et al. 1994. Concept-Based Lexical Selection. In *Proceedings of the AAAI-94 fall symposium on Knowledge Representation for Natural Language Processing in Implemented Systems*, New Orleans, LA. AAAI.

Egg, Markus. 1994. *Hochgradige Polysemie und ihre Repräsentation*. Arbeitspapier 63. Konstanz.

Fillmore, Charles. 1982. Frame Semantics. In Linguistic Society of Korea, editor, *Linguistics in the morning calm. Seoul International Conference on Linguistics, SICOL, Han'guk-Ono-hakhoe*. Hanshin, Seoul, p. 111–138.

Fuchs, Catherine et al. 1991. Polysémie, glissements de sens et calcul des types de procès. *Travaux de Linguistique et de Philologie*, 29:137–169.

Guillaume, Gustave. 1964. *Langage et science du langage*. Nizet, Paris, 2 edition.

Halliday, Michael A. K. 1985. *An Introduction to Functional Grammar*. Edward Arnold, London etc.

Husserl, Edmund. 1900-01. *Logische Untersuchungen*. Niemeyer, Tübingen. 6. Aufl.,

unveränd. Nachdr. d. 2., umgearb. Aufl.
Jackendoff, Ray. 1990. *Semantic Structures.* MIT Press, Cambridge Mass.
Kayser, Daniel and Hocine Abir. 1995. A non-montonic approach to lexical semantics. In Patrick Saint-Dizier and Evelyne Viegas, editors, *Computational Lexical Semantics.* Cambridge University Press, Cambridge, p. 303–318.
Kleiber, Georges. 1990. *La sémantique du prototype. Catégories et sens lexical.* PUF, Paris.
Langer, Stefan. 1995. Selektionsklassen und Hyponymie im Lexikon. Semantische Klassifizierung von Nomina für das elektronische Wörterbuch CISLEX.
Lehrer, Adrienne. 1990. Polysemy, Conventionality, and the Structure of the Lexicon. *Cognitive Linguistics*, 1:207–246.
Lenat, Doug and R.V. Guha. 1988. *The world according to CYC.* Technical report, Microelectronics and Computer Technology Corporation, Austin, Texas.
Levin, Beth. 1993. *English Verb Classes and Alternations. A Preliminary Investigation.* University of Chicago Press, Chicago.
Mahesh, Kavi et al. 1996. *An Assessment of Cyc for Natural Language Processing.* Memoranda in Computer and Cognitive Science302. Computing Research Laboratory, New Mexico.
Mahesh, Kavi and Sergei Nirenburg. 1995. A Situated Ontology for Practical NLP. In *In Proceedings of the Workshop on Basic Ontological Issues in Knowledge Sharing*, Montreal. International Joint Conference on Artificial Intelligence (IJCAI-95), Aug. 19-20, 1995.
Martin, Robert. 1972. Esquisse d'une analyse formelle de la polysémie. *Travaux de Linguistique et de Littérature*, 10:125–136.
Martin, Robert. 1979. La polysémie verbale. Esquisse d'une typologie formelle. *Travaux de Linguistique et de Littérature*, 17:251–261.
Martin, Robert. 1994. Dictionnaire informatisé et traitement automatique de la polysémie. In Evelyne Martin, editor, *Les textes et l'informatique.* Didier Erudition, Paris, p. 77–114.
Matthiessen, Christian. 1992. Lexicogrammatical Cartography: English Systems. Preliminary Draft Version. University of Sydney.
Mayo, Bruce. 1995. Describing Verbs of Motion in Prolog. In Urs Egli et al., editors, *Lexical Knowledge in the Organization of Language*, Current Issues in Linguistic Theory 114. Benjamins, Amsterdam & Philadelphia, p. 203–243.
Pause, Peter, Achim Botz, and Markus Egg. 1995. Partir c'est quitter une peu: A two-level approach to polysemy. In Urs Egli et al., editors, *Lexical Knowledge in the Organization of Language*, Current Issues in Linguistic Theory 114. Benjamins, Amsterdam & Philadelphia, p. 245–282.
Picoche, Jacqueline. 1994. A 'Continuous Definition' of Polysemous Items: Its Basis, Resources and Limits. In Catherine Fuchs, editor, *Continuity in Linguistic Semantics*, Lingvisticae investigationes Supplementa; 19. Benjamins, Amsterdam, p. 77–92.
Pustejovsky, James. 1995. *The Generative Lexicon. A Theory of Computational Lexical Semantics.* MIT Press, Cambridge, Mass.
Pustejovsky, James and Federica Busa. 1995. Unaccusativity and Event Composition. In Marco Bertinetto et al., editors, *Temporal Reference, Aspect and Actionality*, volume 1: Semantic and Syntactic Perspectives. Rosenberg & Seiler, Torino, p. 159–177.
Raskin, Victor and Sergei Nirenburg. 1995. *Lexical Semantics of Adjectives.* Memoranda in Computer and Cognitive Science288. Computing Research Laboratory, New Mexico.
Rastier, François. 1987. *Sémantique interprétative.* Presses universitaires de France, Paris.
Schmid, Helmut. 1996. *The YAP Manual.* Technical report, Institut für maschinelle Sprachverarbeitung, Universität Stuttgart.
Searle, John R. 1992. *The rediscovery of the mind.* MIT Press, Cambridge, Mass.
Steiner, Erich, Ursula Eckert, Birgit Roth, and Jutta Winter-Thielen. 1988. The devel-

opment of the EUROTRA-D system of Semantic Relations. In Erich Steiner et al., editors, *From syntax to semantics. Insights from machine translation.* Pinter, London, p. 40–104.

Touretzky, David. 1994. Continuity, Polysemy, and Representation: Understanding the Verb 'cut'. In Catherine Fuchs, editor, *Continuity in Linguistic Semantics*, Lingvisticae investigationes Supplementa; 19. Benjamins, Amsterdam, p. 231–239.

VELI. 1989. *Vocabolario elettronico della lingua italiana.* IBM Italia.

Victorri, Bernard and Catherine Fuchs. 1996. *La polysémie. Construction dynamique du sens.* Hermes, Paris.

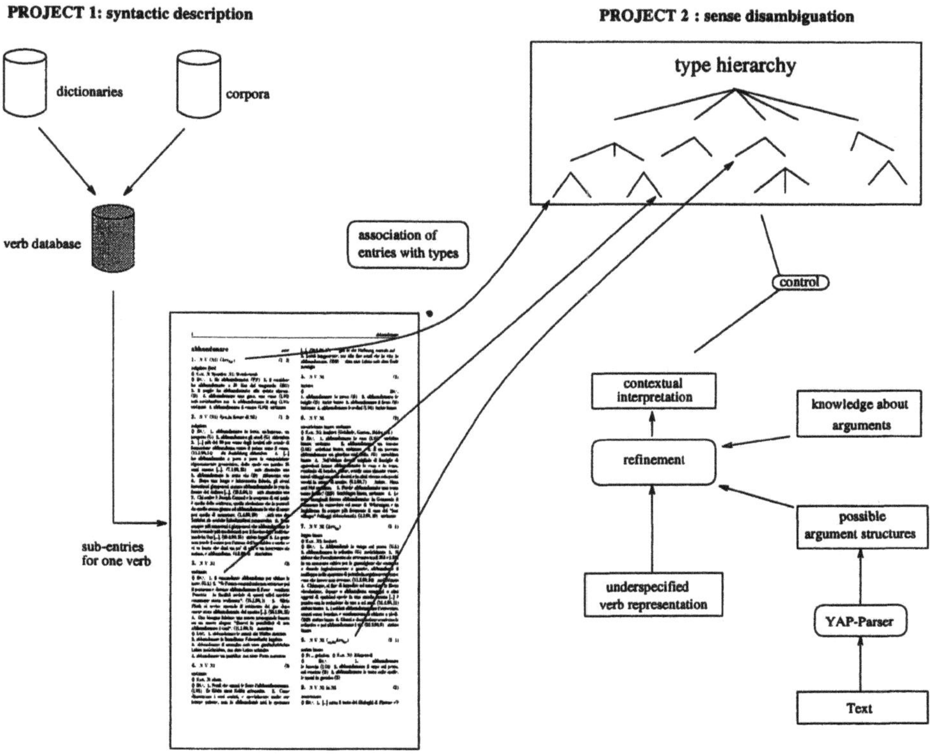

Figure 1. Italian verbs: general view of the project

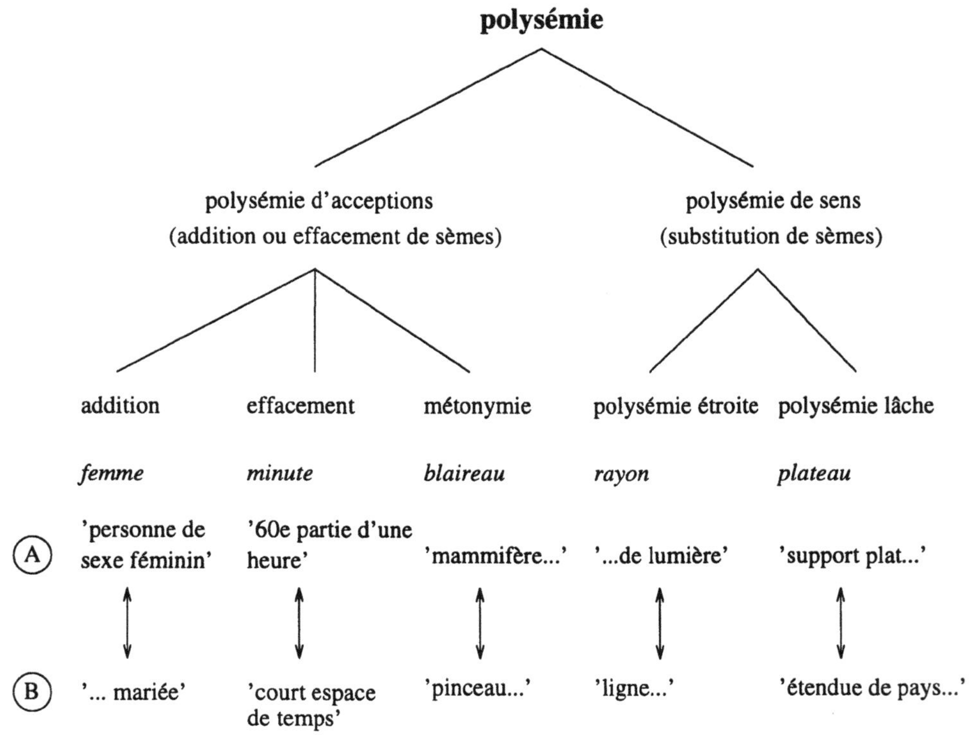

Figure 2. polysemy types (Martin 1972)

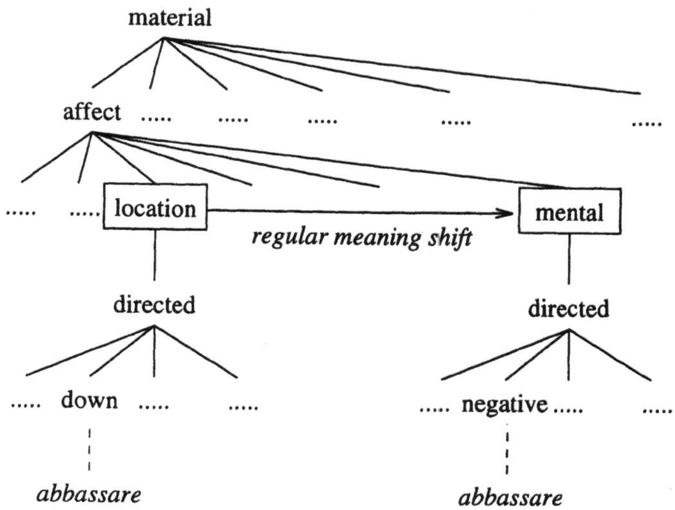

Figure 3. Semantic polysemy

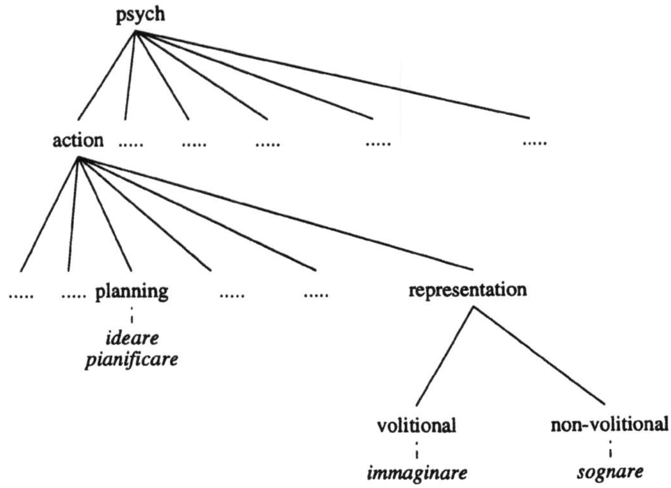

Figure 4. Psych verbs

	contexts of *abbandonare*	contexts in English
1.	~ *una gara, una corsa*	~ a competition, a race
2.	~ *lo yen in favore del dollaro*	~ the yen in favour of the dollar
3.	~ *per ultimo la nave*	be the last to ~ the ship
4.	~ *un progetto*	~ a project
5.	~ *la presa*	~ the grip
6.	~ *il giardino*	~ the garden
7.	~ *la vanga sul prato*	~ the spade on the lawn
8.	~ *le braccia*	~ one's arms ('bodyparts')
9.	~ *il messaggio nelle parole scritte*	~ the message into the written word
10.	~ *qn alla sua sorte*	~ somebody to his fate
11.	~ *una supplenza*	~ an employee
12.	~*si*	~ oneself
13.	~*si a terra*	~ oneself to the ground
14.	~*si a qn con piena fiducia*	~ oneself to somebody in absolute confidence
15.	~*si ai sogni*	~ oneself to one's dreams
16.	~*si a distorcere eventi*	~ oneself to corrupt the facts

Figure 5. the polysemy range of *abbandonare*

ALTERNATIONS AND VERB SEMANTIC CLASSES FOR FRENCH: ANALYSIS AND CLASS FORMATION

PATRICK SAINT-DIZIER
IRIT - CNRS
Université Paul Sabatier
118 route de Narbonne
31062 TOULOUSE Cedex FRANCE

Abstract. In this paper, we show how alternations (called here contexts) can be defined for French, what their semantic properties are and how verb semantic classes can be constructed from syntactic criteria following (Levin 93). We then analyze the global quality of the results in terms of overlap with classes formed from the same verb-senses using WordNet-like classification criteria. Finally, we propose a method which combines these two approaches to form verb semantic classes better suited for natural language processing.

1. Aims

Predicative forms are complex to describe (e.g. (Marantz 85), (Fellbaum 90)). It is indeed necessary to describe in detail their syntactic behavior; the different meanings they may convey by themselves and in composition with arguments, preferably at different levels of granularity (e.g. argument structure, thematic grids, conceptual representations); and the relations between syntactic forms and meaning(s) (Levin 93, 95), (Williams 94), (Baker 88), (Pinker 93). From a natural language point of view, it is also important to hierarchically organize these predicative forms so that syntactic and semantic descriptions are minimal and coherent.

In this document we briefly show how alternations (called here contexts) can be defined for French, what their semantic properties are and how verb semantic classes can be constructed from syntactic criteria following (Levin 93). We then analyze the global quality of the results in terms of

P. Saint-Dizier (ed.), Predicative Forms in Natural Language and in Lexical Knowledge Bases, 139–170.

overlap with classes formed from the same verb-senses using WordNet-like classification criteria. Classes formed on a strict syntactic basis do not exhibit a very high rate of semantic relatedness (about 50%). This is in particular due to the small number of syntactic alternations in French. We also show that about 16 contexts out of 47 really contribute to the formation of classes. The others are either too frequent, too rare or do not have any classification power. However, as shown also by Beth Levin, there is sufficient evidence to motivate classifications and the formation of classes. Classes obtained are based on specific criteria, which may not correspond to those elaborated directly from semantic criteria.

Finally, we show how a classification method better suited for natural language processing applications, which pairs syntactic and general-purpose semantic criteria, can be defined. This shows how syntax and WordNet-like criteria can converge to develop well-organized and well-founded verb semantic classes, which can then be used to form verb lexical semantics knowledge bases.

Our work is a much more modest entreprise than B. Levin's. The number of verbs studied is smaller and the depth of the analysis is more limited. This work has also a different perspective. It is basically oriented toward natural language processing and attempts to elaborate a more principled and methodological way of defining alternations and of producing verb semantic classes.

2. Verb semantic classes based on 'alternations'

2.1. THE CONTEXT SYSTEM

We have reformulated Beth Levin's notion of alternation into a more declarative one: the notion of *context* (Saint-Dizier and Marrafa 95), (Saint-Dizier 96). A context is a distribution frame where the category and some additional syntactic and semantic information are used to describe a precise form and position that the arguments of a verb may have in a sentence. Verb classes are then formed from verbs having similar sets of contexts.

The scope of a context is the proposition. Its description includes the description of the properties of the verb and of its arguments. These properties are expressed by means of constraints which, from a retrospective look at contexts (i.e. this is not theoretically-driven), can be classified into 3 groups:

- Local constraints on arguments or on the verb:
    - selectional restrictions,
    - thematic roles,
    - verb subcategorization frame,

  - arity of the verb,
- Adjunctions to the standard distribution form:
  - coordination,
  - some general modifiers (this is in general very delicate).
- Relations between arguments:
  - thematic role distributions,
  - modifier-modified relation between arguments (an argument can become a modifier of another argument),
  - semantic relations: container-contained, part-whole of various forms, an argument denotes a property of another argument

Very briefly, compared to the alternation system, our approach avoids having to define a basic form from which alternations are produced and to have to explain the relation between a basic and an alternated form. Moreover, it avoids having to account for changes in meaning provoked by alternations (e.g., by the adjunction of a preposition).

From a methodological point of view, contexts for French have been defined from a transposition of some English alternations (about 1/3 of our contexts), from French syntactic descriptions, among which (Gross 75), from corpora and from our own linguistic intuitions. These alternations have then been validated on corpora to ensure that we cover most of the syntactic behaviors of arguments with respect to predicates.

Defining contexts has led us to formulate the following principles:

- contexts must be of general purpose, which means that:
  - exceptional forms should be avoided (contexts must be productive),
  - only non-ambiguous and easy-to-use forms are acceptable,
  - theory-neutral descriptions must be used.
- contexts should minimally overlap,
- they must only describe lexical properties,
- contexts should contain as few semantic data as possible, otherwise the classification will also be based on these criteria,
- contexts must be defined at the most appropriate level of granularity by feedback and retro-evaluation on verbs; we must also consider generalizing two contexts into one, if their discriminatory power is low.

2.2. THE THEMATIC ROLE SYSTEM

We now present the hierarchy of thematic roles we use, since they are going to play a major role in the classification, following (Dowty 89, 91) and (Jackendoff 87). The classification we propose here emerged also from a project on intelligent indexing of documents where predicates and arguments were extracted from texts (Pugeault et al. 94, Pugeault 96). Arguments and some modifiers were marked with thematic roles identifying quite precisely the role they play w.r.t. the predicate. Since this classification looked satisfactory for a large number of cases and sufficiently fine-grained, we adopted it here.

Thematic roles are the following:

- agent (ag): a general notion of agentivity; the degree of agentivity may be more or less strong at this level.
    - effective agent (ae): the agent that carries out the action
    - initiative agent (ai): the agent that takes the initiative of an action
    - volitive agent (av): the agent that wants the action to be done.
- general theme (tg): themes may be any type of entity, including humans;
    - holistic theme (th): the theme not affected in its integrity.
    - incremental theme (ti)
        * incremental beneficiary theme (tib)
        * incremental victim theme (tiv), tib and tiv are used when there is a subtancial gain or loss for the theme
        * incremental theme of creation (tiec): the entity is created
        * incremental theme of destruction (tied): the entity is destroyed in some way.
    - causal theme (tc)
    - consequence theme (tcons)
    - theme experiencer (exp) someone who enjoys or suffers from something or from someone (somewhat close to tib, tiv for humans).
- localisation (loc) : may be subtyped spatial (spat), temporal (temp) or abstract (abs).
    - source (src)
    - position (pos) : a fixed location.
        * absolute position (pa)
        * relative position (pr)
    - goal (but)
    - destination (dest): the destination is effectively reached (as opposed to goal which is a direction, not necessarily reached),
    - possibly a notion of path can be included, which is different from just a source or a destination.

- means (moy)
  - instrumental means (mi)
  - non-instrumental means (or indirect means) (mni)
- manner (ma)
- identification (id)
- accompaniment (acp), indicates accompaniment, co-occurence and plural arguments
- amount or quantity (amount)

3. Description of the context system

3.1. CONTEXTS CORRESPONDING TO BASIC FORMS

These contexts describe forms which may be referred to as basic forms because they are related to the most 'direct' expression of the argument structure. They include the description of intransitive, transitive and ditransitive forms. Transitive verbs are verbs that select an object that is an NP or a PP (we won't call verbs selecting a PP intransitive verbs, this is a personal view on the semantics of 'intransitivity'). Arguments are either NPs, PPs or sentential forms. Some contexts include constraints on the thematic roles assigned to arguments. In general, these constraints are related to very specific notions, such as the notion of accompaniment. These constraints may also allow us to partition a large set of verbs into two subsets (such as verbs having a localization complements and those that do not).

Contexts are numbered from 1 to 200. Each ten describes contexts which are closely related. Contexts and their associated numbers are summarized in appendix 1 to this paper. In this section, the major contexts are described.

Intransitive contexts: We introduce here two types of intransitive constructions. First, a specific construction mainly dealing with 'weather' verbs (class number 57 in B. Levin) whose subject is the impersonal pronoun *il* (the dummy pronoun *it*) (no. 1). The second context describes standard intransitive forms (no. 2).

Transitive constructions Transitive constructions with NP or PP objects are treated separately because contexts describing non-basic forms associated with verbs selecting PP of NP objects are substantially different. Also, the relation between the head noun of a PP and the verb is mediated by the preposition whose semantics is crucial.

Context nb. 10 deals with the case where the subject is a causal theme and the NP object a general or a consequence theme (in verbs such as *entrainer, impliquer*), context 11 has a subject which is an experiencer and an NP object which is any type of theme. Context 12 is the general form for transitive constructions with an NP object. Context 13 considers agents which are either effective agents or themes and NP objects which are of the type accompaniment. Context 14 is similar to context 13 except that the object must be a PP. Context 15 is the general form for transitive verbs with a PP object while context 16 is a subtype of context 15 with a localization PP object. When several contexts are applicable, the most specific one is always preferred.

Ditransitive forms Here also a distinction is made on the basis of the nature of the object arguments: PPs or NPs. Context 20 is the general case, with an NP and a PP object. Context 21 adds the constraint that the second object must be assigned the thematic role accompaniment. Context 22 describes verbs having two PP objects. Context 24 is more specialized than context 20: it describes the case where one of the objects is of the type localization. Context 23 is the dative construction. It is quite different from the English one, since the dative PP cannot become a direct object in French (as it is the case with the dative shift).

3.2. CONTEXTS DESCRIBING NON-BASIC FORMS

In this section, we describe the main contexts (or alternations) proper to French. The description is informal, their TFS form is given in (Saint-Dizier 96b). We also give an idea of the number and type of verbs which accept each context. A short comparison is also made with the corresponding English alternation, whenever appropriate.

3.2.1. *Contexts with a middle reflexive 'se'*

Here are two contexts where a middle reflexive pronoun *se* is inserted before the verb. This construction is somewhat close to the alternation given in 1.1.1. in Beth Levin, but with no middle reflexive pronoun in English. The first context, number 50, can be used for transitive and ditransitive verbs, it has only as argument an NP subject, as in:
Le sport se pratique difficilement.
Les pizzas se livrent facilement.
while context 51 can only be used with verbs of arity 3, since the PP object is realized. This latter context has an additional constraint on the subject: it must show a certain degree of plurality for the construction to be acceptable. The resulting statement resembles a rule:
* *Une pizza se livre à Marie.*
Les pizzas se livrent à Marie.
L'argent de la recette se donne à Marie.

We consider that the introduction of the pronoun *se* is a lexical property of the verb. The construction corresponding to context 50 is more acceptable with an adverb modifying the proposition.

Both contexts 50 and 51 are very frequently accepted by verbs. 70% of the verbs of our sample accept context 50 and 32% accept context 51 (therefore 64% of the ditransitive verbs). The following verbs accept context 50: *abaisser, cuisiner, découvrir, emporter, examiner.* Verbs such as *accuser, acquitter, associer, aviser* do not accept it. For context 51, we have verbs such as : *amalgamer, accorder, dicter, flatter, fabriquer, payer, raconter,* whereas verbs like *concurrencer, aviser, parler, reprocher* do not accept it.

Partly because of the large number of verbs undergoing contexts 50 or 51, it is not possible to find common semantic features to these verbs. We noticed however that a number of communication and social interaction verbs do not undergo these alternations.

3.2.2. *Passive forms*

The two passive constructions, 60 and 61, respectively correspond to verbs of arity 2 and 3. The agentless passive, characterized by the deletion of the par-complement (by-complement) is not accounted for here since it is not a lexical property in French. We cover here fewer cases than covered under 5.1 in B. Levin. In particular, passives with a sentential complement are not taken into account here because they are rarely used in French. Similarly, we do not consider adjectival passives.

The agent in most verbs in French in the passive form is realized by a *par*-complement, a PP headed by the preposition *par*. However there are a few cases where the preposition is not *par*, but *de, en, à, au,* as in:
Jean est aimé de tous.
In our verb sample, a proportion of 37% of verbs accept context 60 (out of the 49% of transitive verbs, i.e. about 74%) while 44% of ditransitive verbs accept context 61 (i.e. 88% of ditransitive verbs). There are well-known sets of verbs which do not passivize, such as most of the verbs of movement (i.e. which have an NP or PP object of type localization). Also, most transitive verbs which have a PP object do not passivize.

3.2.3. *Inchoative forms*

Contexts 70 and 71 allow the expression of the result of an action. They are intransitive (context 70: 'la cloche sonne') or transitive (context 71: 'l'eau remplit la cruche') forms respectively corresponding to transitive and ditransitive verbs. The original subject is deleted and the object1 'becomes' the subject.

Cases where the object2 (originally a PP) becomes a subject as in:
Les bénéfices compensent les pertes. 'derived from':
Jean compense les pertes par les bénéfices.

are treated by context 72. This context focusses on the result. Compared to the closest alternation, the locatum alternation, given in Beth Levin under 3.5, we do not consider here locatum (which is not frequent in French) but just themes or instruments.

Context 70 is associated with about 3% of the verbs of our sample. In this context fall verbs such as *accellerer, augmenter, chavirer, cicatriser, claquer, diminuer, gonfler, jaunir, verdir, reculer*. In particular, this context includes verbs which have to do with coloring actions (*verdir, rougir, brunir*). Context 71 is not very frequent, about 1% in our sample. It is found in verbs such as *basculer, cadrer, dériver, rentrer* as in:
Le meuble bascule sur le sol.
The object in this context describes the resulting position or state of the subject. Context 72 is also not very frequently used (about 0.6%), it is associated with verbs such as *bourrer, coincer, remplir*.

3.2.4. *Inversion place-subject*

This context is applicable to very few verbs (about 0.5%) that express an idea of high concentration of entities in a place. It is the equivalent of the English *swarm* alternation. Verbs such as *pulluler, bourdonner, grouiller, fourmiller, résonner* accept this context. Context 80 is different from context 191 since in 191, the subject is a PP.

3.2.5. *Introduction of the semi-auxiliary faire*

We consider in this context (no. 90) the case of transitive verbs which can be preceeded by the semi-auxiliary *faire. Faire* introduces a stronger sense of causativity and/or agentivity, as in:
Edith cuit la viande.
Edith fais cuire la viande.
In this latter sentence, *faire* reinforces the agentivity of the subject with respect to the action described by the verb. Jean is indeed the initiative agent of the action (*cuire la viande*). But, in fact, it is the object that is most affected by the action. A number of verbs (5%) accept this context, among which cooking verbs and verbs such as *augmenter, varier*, and *diminuer*. From a syntactic point of view, we consider that the semi-auxiliary *faire* is treated as a regular auxiliary in the VP. This phenomenon has been studied in depth, we adopt here a relatively simple and superficial approach which seems to be sufficiently fine-grained from semantic classifications. In particular, as shall be seen below, this construction conveys a specific semantics which do not interact much with semantic features conveyed by other contexts.

Let us also consider another use of the semi-auxiliary *faire* where its presence provokes the introduction an additional argument, agent or cause,

subject of the verb. For example, the verb *atterir* is basically an intransitive verb as in:
L'avion atterit. An agent, e.g. the pilot, can be added:
Le pilote fait atterir l'avion. Similarly for:
Jean grossit.
Le fromage fait grossir Jean (The *fromage* (cheese) is the cause).

Verbs such as *agir, amerrir, capituler, croitre, danser, dormir, grossir, mourir, ressuciter* accept this context. These verbs are all initially intransitive verbs. In general the subject of the original form is not an agent but a theme or an experiencer. About 3% of the verbs in our sample accept this context, i.e. most intransitive verbs (90%), except verbs like *déceder* and verbs falling under the basic context 1.

3.2.6. *Contexts involving the nominalization of the verb*
There are in French a number of cases where the verb of the basic form can be nominalized. It is then introduced by a kind of 'support' (or light) verb such as *faire, pousser, porter, donner, prendre, provoquer, avoir, être*, etc. The introduction of this support verb does not change the meaning of the proposition. The argument structure of the original verb remains basically unchanged. In some cases an NP becomes a PP. A number of different prepositions may be used. There are cases that are relatively simple to identify whereas there are others that are subject to complex debates.

We present here the simplests and most frequent cases. These constructions have not been taken into account in English by Beth Levin; they may not be very frequent in English.

Here are the support verbs we have selected, with the associated context number and a few examples:

- 92: support verb: faire, example: *dessiner → faire un dessin* (instead of *faire*, we could also have expressions such as *réaliser, procéder à*).
- 93: support verb: pousser, example: *hurler → pousser un hurlement.*
- 94: support verb: porter, examples: *aimer → porter un amour; juger → porter un jugement.*
- 95: support verb: donner, examples: *assurer → donner l'assurance de; démontrer → donner la démonstration de.*
- 96: support verb: prendre, examples: *décider → prendre la décision de; défendre → prendre la défense de.*

Context 92 is frequently used (28% of our sample), and it is associated with a wide range of types of verbs describing actions such as *congeler, déconnecter, distribuer, imiter, guetter, narrer, photocopier, rencontrer, réserver.* To each of these verbs must obviously correspond a deverbal noun and the subject of the basic form must be an agent. Context 93 is associated with verbs of sound emission with an idea of pain, such as *crier, hurler* or animal sound emission *glapir, aboyer, meugler, hennir.* Context 94 is not frequently encountered (0.6%), it is associated with verbs describing a kind of positive or negative attachment of someone to someone else: *aimer, combattre, hair, soupconner, témoigner.* Context 95 is quite frequent (4.4%); it is often associated with verbs relating a communication act between people: *adhérer, alarmer, assurer, autoriser, commenter, conseiller, dégouter, démentir, féliciter, informer, inspirer, pardonner, renseigner, satisfaire.* Finally, context 96 is associated with few verbs (0.6%), it seems to be associated with some decision taking verbs: *décider, opter, défendre, risquer.*

3.2.7. *Contexts involving unexpressed objects*

These contexts describe constructions where an object argument has been deleted. Implicit objects are also at the origin of a large literature which won't be discussed here (see (Mittwoch 82), (Fellbaum, this volume), among many others). Let us present here simple and commonly admitted forms.

Context 100 (deletion of all the object arguments) is accepted by a large number of activity verbs, where the nature of the missing object can be inferred from the meaning of the verb (e.g. from *économiser* (to save), we can infer a generic object like money). In context 101, the deletion of a direct object entails that the indirect object becomes a direct object, and the preposition is dropped (je décharge le camion). The original indirect object must be an incremental theme or a localization, while the deleted object1 must neither be a localization nor a means.

Contexts 102 and 103 have a more general scope. Context 102 deals with the deletion of the second object (a PP in French: 'je conseille ce produit'). The deleted PP must not alter the meaning of the utterance. Context 103 describes the deletion of the first object, when it is a PP, for verbs of arity 3 (e.g. 'Anne collabore avec Marie'). The second object is also a PP.

Context 100 is accepted by a very large number of verbs, about 21%, (e.g. verbs such as *cuisiner, débarquer, débattre, débaucher, déborder, débuter, decentraliser, décevoir*). Some reflexive verbs such as *s'absenter* also accept this context. Verbs accepting this context usually have a precise and strong meaning by themselves. The object(s) plays a more modest role in the construction of the meaning of the proposition; we could say that it is partly incorporated into the verb (Baker 88). When objects are deleted, the proposition then focusses on the fact that the action is in progress.

About 2.4% of the verbs in our sample accept context 101. We have, for example, verbs such as: *atteler, bourrer, déleguer, extorquer, indemniser, redouter, rétribuer, semer, soulager.* These verbs often exhibit a part-of relation, or a possession or a spatial inclusion relation between the first object (included or possessed entity) and the second one (the possessor or the whole). A generalization is then possible by deletion of the first object; the preposition heading the second object disappears because its basic role was to make more explicit the relation holding between the two objects.

Context 102 is accepted by 37% of the verbs, since those verbs must be ditransitive. This means that, in fact, 74% of ditransitive verbs accept this context. Verbs such as *abonner, accroitre, additionner, aligner, amalgamer, ammasser, appuyer, articuler, associer, comparer, concentrer, concilier, délier, démettre, départager* do not accept this context. All these verbs include an idea of accompaniment between the two object arguments, or the idea that they must co-occur explicitly in the statement. Verbs can accept context 100 and not context 102. This is the case for a verb such as *plonger* (to plunge something into a liquid).

Context 103 is associated with 4.6% of the verbs of our sample. The types of verbs are very diverse, e.g.: transfer of possession, communication verbs and copy verbs.

3.2.8. *Contexts with a preposition change*

The two first contexts, 110 and 111, are quite similar to the with/against and to the spray/load alternations respectively. Either the location, the instrument or the theme may be objects of the verb. Context 110 introduces the preposition change sur/avec(e.g. 'Elisabeth frappe le baton sur la palissade / Elisabeth frappe la palissade avec le baton'), while context 111 introduces the preposition change sur-dans/de. In both contexts *sur* has a localization sense (e.g. 'je pulvérise la peinture sur le mur / je pulvérise le mur de peinture').

Context 112 introduces the preposition change de/à where the beneficiary and the theme are permuted (e.g. 'Jean profite d'un prêt / le prêt profite à Jean'). Context 113 describes the preposition change à-en where the two objects are permuted, without any change in meaning (e.g. 'Je fournis des livres à Jean / Je fournis Jean en livres').

The conative construction, realized by context 114, requires in French the introduction of the preposition *sur* (e.g. 'Je tire sur la chaise'). This preposition introduces here an abstract or a spatial localization. It applies to a smaller and different set of verbs than in English, probably because it is not exactly the same phenomenon which occurs, even if the syntactic form is identical to the English conative form. Its underlying semantics seems to be somewhat different (see (Pinker 93) for English).

Contexts 110 to 114 are not very frequently encountered in verbs, about 0.4% for each of them, except for context 114 which is accepted by 1.8% of verbs. Under context 110 fall verbs such as *aligner, frapper, frotter, taper*; associated with 111, we have verbs such as *charger, pulvériser, semer*; and with context 112, we have verbs like *apprendre* and *profiter*. Context 110 mainly concerns verbs which convey an idea of contact with motion. The verb *aligner* conveys an indirect notion of contact. Context 111 conveys the idea of spilling or spreading something on or into something. Context 112 is associated with verbs where the subject is not an agent but experiences something. Context 113 characterizes verbs which include an idea of delivering something to someone, such as *fournir, livrer, servir.* Context 114 includes verbs such as *bâtir, borner, butiner, cadencer, colorier, commencer, commenter, construire, débuter, découper.* The introduction of the preposition *sur* transforms the object1, which was initially a theme, into a position, or the starting point of an action.

3.2.9. *Reciprocal constructions*

The transitive reciprocal context (nb. 120) involves verbs with an object argument in the plural. The NPs in this object can be connected by a preposition like *avec* or by a conjunction like *et.* The preposition *avec* supports here a notion of accompaniment which makes the *avec/et* constructions meaningful (e.g. 'additionner A avec B / A et B'). This context is similar to alternation 2.5.1.

This construction concerns quite a large number of verbs, 2.7% verbs of our sample. It is associated with verbs such as *additionner, agglomerer, ajouter, aligner, alterner, amalgamer, articuler, associer, combiner, comparer, concilier, confondre, confronter, conjuguer, connecter, coordonner, croiser, déconnecter, départager, mélanger, méler, multiplier, opposer, séparer, serrer, soustraire.* Verbs in this set convey a notion of combining, putting together or mixing. Notice that verbs such as *harmoniser, diluer,* and *incorporer* do not accept context 120. This seems to be due to the fact that the two objects are not at the same 'hierarchical' level: something (object1) is affected by the fact that it is harmonized with something else (the object2).

The reciprocal subject, context 121 (similar to alternation 2.5.4), describes the case where an object argument introduced by the preposition *avec* is conjoined to the subject argument (e.g. 'Bruno argumente avec Edith / Bruno et Edith argumentent'). The verbs involved are of arity 2 and convey an implicit idea of reciprocity: the action performed by the subject is also realized, possibly to a lesser extent, by the object (which receives the thematic role accompaniment). In the case where the two arguments are coordinated, the subject is in the plural and both NPs are marked with the thematic role effective agent. Context 122 is similar except that the arity of the verb is 3. In contexts 121 and 122, the obj1 and the subject must be of comparable status. In context 122, the obj2 must be realized (e.g. 'Jean et Anne négocient un contrat').

Both contexts 121 and 122 are associated with a small number of verbs

(0.5%), which are semantically related. Associated with context 121, we have verbs such as *coexister, cohabiter, coincider, correspondre, rompre,* while the following verbs are associated with context 122: *collaborer, coopérer, débattre, déliberer, discuter.* These two sets include verbs which convey an idea of reciprocity or symmetry between the subject and an object (usually of type human or, metaphorical organization). The action described is expressed conjointly by the subject and the object.

The reflexive reciproqual, context 123 (alternation 1.2.4), coordinates the object with the subject argument and introduces a reflexive pronoun (if it is not already present), which indicates that the two protagonists are mutually necessary for the action described by the verb to take place (this reflexive pronoun could be paraphrased in English by 'each other'). These protagonists, as also shown in B. Levin, must have a comparable status (e.g. 'Jean et Julien se croisent').

Context 123 is associated with a quite large set of verbs (6%) such as *accoster, apprécier, associer, concurrencer, contacter, correspondre, croiser, épouser, fréquenter, rejoindre, rencontrer, ressembler, revoir, saluer, succéder.* The implicit semantics behind this context is that the subject and the object1 somewhat behave in a similar manner, either cooperatively or concurently. In some case (e.g. *accoster, apprécier*), the subject may however play a more prominent role than the object. Usually, the subject and the object1 are of type human, or a derived type.

3.2.10. *Body-part contexts*

Context 140 introduces the part-of relation restricted to inclusion and spatial position with respect to the whole, which are probably the most common types of part-of relations (e.g. 'Edith touche le dos du cheval / Edith touche le cheval sur le dos'). The part, initially the first object, becomes a second object, a PP of type localization. Functional relations are not included since they are not real parts. Context 140 is associated with verbs of arity 2 or 3, it is approximately similar to alternation 2.12.

Context 141 has some similarities with context 140: it is primarily based on a property-of relation (or a closely related relation) holding between the two objects. It describes the case where the 'part' or the property is deleted. The statement is then more general, (e.g. 'Je surveille les allées venues de Jean' implies 'Je surveille Jean'). This context has some similarities with alternation 2.3.3, but seems to be more general.

Context 140 is associated with a quite large number of verbs (about 5.4%), among which we have, for example, *aborder, amarrer, caler, caractériser, caricaturer, coincer, collaborer, consolider, déboucher, débuter, dérégler, salir, souiller, soutenir, tacher.* The action described by the verb is realized on a part of an object, but affects the whole. Context 141 is more general, it is accepted by about 20% of the verbs of our sample. Associated with this context are verbs such as: *abimer, aborder, allumer, alourdir, asperger, brancher, caler, connecter, coopérer, débarquer, découper, presser, soigner, soutenir, tacher.*

3.2.11. *Expression of the property of an object*

The following two contexts illustrate two ways of expressing a property of an object or of a person. Context 150 describes the preposition change de/en (e.g. j'admire le courage de Colin / en Colin'), it corresponds to the alternations 2.13.1, 2.13.2 and 2.13.3. Alternation 2.13.4 is treated separately by context 151 (e.g. 'Les grimaces de Jean terrifient Sophie'). In this latter context, the means or manner argument is an obligatory argument (it is part of the argument structure). This obligatory argument 'becomes' the head of the subject, while the first element of the argument structure is a noun complement expressing the means or manner. These two NPs are connected by a preposition that indicates a possession relation. The means must be closely related to the subject: it can be a quality or a property, or there must exist a direct possession relation, etc.

Context 150 is associated with relatively few verbs (2.9%), these verbs express admiration, approval or fear: *admirer, adorer, aimer, apprécier, approuver, corriger, craindre, détester, éviter, excuser, louer, mépriser, soupconner.* Context 151 is accepted by a larger set of verbs (9.6%), where the means or the manner substantially contributes to the sucess the action denoted by the verb. Verbs such as *acceuillir, aider, amender, amuser, contribuer, convaincre, dédommager, endormir, écraser, épuiser, louer, persuader, récompenser, ruiner, sauver, soulager, surprendre* accept this context. They convey an idea of improvement, help, causing damage to or reward.

3.2.12. *Contexts involving a means or an instrument*

In contexts 160 and 162, the means or the instrument becomes the subject of the verb. This means or instrument in context 160 is part of the argument structure of the verb and it must play an important role in the action described by the verb (e.g. 'Cette poutre consolide la charpente'). The verb is of arity 3. It is similar to alternations 3.3, 3.4, 3.5 and 3.8.

In context 162, this means or instrument is a modifier, and the verb may be of arity 2 or 3. The modifier must play an important role in the action decribed by the verb (e.g. 'Une lettre avertit les participants du danger'). It may also have a certain degree of causality. Similarly to alternations 3.3 and 3.4, which are included here into a single context, instruments are indirect, and means may be abstract means (or causes). The means or instrument must be a central participant in the action described by the verb. These two alternations are here integrated into context 162 because it is quite difficult to draw a precise line between them.

Context 161 is similar to the alternation 3.6. It must be associated with a di-transitive construction. The localization becomes the subject of the proposition (e.g. 'Cette pièce acceuille 5 participants' from 'Jean acceuille 5 personnes dans cette pièce'). In the basic form, the subject has a low degree of agentivity.

Context 163 is similar to the container-subject alternation 3.7. The subject is identified as a localization which may be spatial, temporal or abstract. The constraint of containment indicates in a general (but possibly fuzzy) way the relation between the subject and the object. The subject is then the container and the object is the contained (e.g. 'Edith inclut les nouveaux résulats dans cet article' / 'Cet article inclut les nouveaux résultats'). This context is more specific than context 161, where the subject is simply a localization. Finally, reflexive forms are not allowed in context 163.

Context 160 is accepted by about 12.6% of the verbs of our sample, this is about 2/3 of the verbs with an object2, which is a means or an instrument. These verbs describe actions where the instrument or the means play a central role, as in *accroitre, aider, alléger, attraper, coder, corriger, incliner, invoquer, jaunir, laver, peser, régler, salir, scier, soigner, tacher.* The subject may be just an initiative agent, leaving an important role to the means or instrument.

Context 161 is accepted by quite few verbs (about 1.4%) such as *acceuillir, attirer, capter, capturer.* It captures the idea that the localization experiences metaphorically the action or state described by the verb. Context 162 is accepted by a large number of verbs (about 32%), in which the instrument plays a major role. Verbs such as *abaisser, abandonner, aborder, abriter, abroger, accélérer, accentuer, acquitter, administrer.* Context 163 is quite unusual, it is accepted in our sample by only 1.2% of the verbs. These verbs are *conforter, integrer, inclure, amalgamer, omettre, enchâsser.* These verbs convey an idea of inclusion.

3.2.13. *Reflexive forms*

Context 170 describes a reflexive form where the original PP object is a reflexive pronoun. Context 171 describes the same case where it is the direct object which is reflexivized. These two contexts are relevant to our system since not all verbs accept reflexive constructions. Context 172 is quite close to alternation 7.6. However the action can be done intentionally or not by the subject (we believe that the notion of intentionality is too subtle and too hard to be efficiently considered in our framework). This context describes reflexive forms with no realized objects (e.g. 'l'entreprise se diversifie'). This context expresses the fact that the action realized on the object (related to the subject in some manner) also concerns directly the subject. The introduction of the reflexive pronoun *se* returns the object of the action on the subject.

Context 170 is accepted by a relatively large set of verbs in our sample (about 4.1%), for example, we have verbs such as *abimer, accorder, acheter, assigner, attribuer, avouer, laver, octroyer, offrir, piquer, refuser, regarder.* Context 171 is more widely accepted (21%), and includes verbs such as *abonner, concentrer, condamner, ennuyer, ravitailler.* Verbs that require the object to be different from the subject cannot accept this context (e.g. *accoster, inventer, perdre, rentrer*). Context 172 is also relatively frequently associated with verbs in our sample (about 7.5%). We have verbs such as *adapter, amarrer, analyser, arrêter, enregistrer, épargner, relire.*

3.2.14. *Absence of a quasi-incorporated argument*

The following context (no. 180) describes the case where an element usually incorporated into the semantics of the verb (and therefore not syntactically realized) becomes realized as a modifier. This 'des-incorporation' is necessary when the element is no longer standard but is modified by e.g. an adjective or a proposition. Verbs such as *beurrer* or *sourire* fall in this class. Sentences such as:

Florence beurre son pain de beurre.

are rejected because *beurre* is incorporated into *beurrer. Beurre* is made explicit, similarly to English, when it is modified, as in:

Florence beurre son pain de beurre salé.

Context 180 describes both transitive and ditransitive cases. Context 180 is similar to alternations 7.1 and 7.2. It appears in about 12% of the verbs of our sample. For example, it is associated with verbs such as *accueillir, blesser, courir, depayser, indemniser, réfléchir.*

3.2.15. *Change of perspective*

The following contexts describe a change of perspective. In context 190, a reflexive pronoun is introduced and the object becomes a PP of localization (e.g. 'Elle tient l'arbre / elle se tient à l'arbre'). Notice that the change in perspective is realized by the introduction of a preposition such as *a, contre, sur, de, avec.* In context 191, the subject and obj1 arguments are permuted, and the focus is then on the localization rather than on the agent (e.g. 'Des plantes poussent dans la serre' / 'Dans la serre poussent des plantes'). Notice that context 191 differs from context 80 since its subject is a PP. The arity of the verbs involved in context 190 may be 2 or 3, whereas it is restricted to 2 for context 191. These two contexts are also quite different from the conative form.

Context 190 occurs in about 5.6% of the verbs, as in *activer, agripper, alarmer, arrêter, cogner, coller, ennuyer, répandre, retrouver, tenir, venger*. Context 191 is associated with about 3.8% of the verbs as in *aboutir, camper, chuter, courir, grimper, plonger, résonner*. The semantic role of this context is to focus on the location rather than on the theme.

3.2.16. *There-insertion*

The following context is similar to the alternation 6.8, the There-insertion alternation. The subject moves to the first object position, whereas the initial object becomes the second object. This context is proper to verbs of arity 2. The term 'there' is in French the impersonal pronoun *il* (e.g. 'Un bateau apparait à l'horizon / Il apparait un bateau à l'horizon').

This context is quite unsual (about 0.7%), it is associated with verbs such as *apparaitre, découler, demeurer, disparaitre*, which convey the idea or appearing or of staying.

4. Semantic relations between contexts

In (Pinker 93) and (Goldberg 94), a description is given of the underlying semantics of syntactic constructions. It is shown that there are strong semantic constraints associated with most syntactic structures. This aspect is somewhat addressed in the above context description and analysis, it is also addressed below in Fig. 7. In (Goldberg 94), the semantic relations between alternations are studied and a number of links are defined.

In this section, we briefly show what are the semantic variations induced by the use of a context instead of the basic form. More precisely, we consider the basic forms 23 and 24 (see section 3.1), and show for a number of frequently used contexts what the changes in meaning are. These changes are characterized by a lexical semantic template (LST), as used in (Levin and Rappaport 95). As will be noticed, these changes are much less important for French than for English (see (Pinker 93)). In our study, contexts other than 23 and 24 cannot be considered because their implicit semantics is too vague or too general.

4.1. MEANING VARIATIONS FROM THE DATIVE FORM

Let us consider context 23, the dative form in French. We can represent it by the following LST:

A CAUSE B TO MOVE AT C(+hum) (or TO BE AT C(+hum)).

We then observe the following behaviors:

- The use of contexts 50 or 51 (middle reflexives) deletes the reference to the agent (A) and adds (1) an idea of a general rule and (2) a manner, introduced by the adverb, which is often required for the construction to be acceptable. The LST is: B TO MOVE AT C(+hum), +general rule ∧ +manner.

- The use of contexts 60 or 61 (passives) may also entail the absence of the agent: B TO MOVE AT C(+hum) (BY VIRTUE OF A). Focus of the sentence, from a more pragmatic perspective, is on B.
- Context 100 (deletion of objects) introduces an idea of iterativity, high frequency or habit, focussing on the activity of the agent A: A CAUSE undefined TO MOVE AT undefined(+hum).
- Context 102 (deletion of the obj2) also introduces a notion of habit, but directed only toward the target C: A CAUSE undefined TO MOVE AT C(+hum).
- Context 162 substitutes a modifier (a means) D to the agent A, where that modifier plays a prominent role in the action: D CAUSE(+indirect) B TO MOVE AT C(+hum).
- The reflexive constructions (no. 171) identifies C to A: A CAUSE B TO MOVE AT A.

4.2. MEANING VARIATIONS FROM THE DITRANSITIVE OF LOCALIZATION

Let us now consider context 24. It can be represented by the following LST: A CAUSE B TO MOVE AT C(+loc) (or TO BE AT C(+loc)). We then observe the following behaviors, which share some similarities with those observed above for context 23:

- Contexts 50 and 51 delete the reference to the agent and also introduce an idea of habit and a manner: B TO MOVE AT C(+loc), +general rule ∧ +manner.
- The passive construction involves only a few verbs where the agent can be omitted and where there is an idea of movement (not a fixed localization): B TO MOVE AT C(+loc).
- Context 71 (inchoative ditransitive) omits the agent to express the resulting state: B BE AT C(+loc).
- Context 102 (deletion of obj2) reinforces the role played by the agent: A CAUSE B TO MOVE AT undefined(+loc) or: A ACT ON B FOR B TO MOVE AT undefined(+loc).
- Context 103 (obj1 deleted) is somewhat like a reflexive form: A CAUSE probably(A) TO MOVE AT C(+loc). This is more evident with contexts 170 and 171.
- Context 161 focusses on the resulting state, given as a general rule: INCH(B TO BE AT C(+loc)) (INCH is the inchoative function of the LCS).

5. Construction of verb classes

The results presented below have been obtained on the basis of a sample of 1673 French verb-senses. These verb-senses are among the most frequently used ones. With each verb-sense, we have associated the list of contexts it accepts. The general form (in Prolog) is:

```
tverbe([verb],arity,[Thematic grid],[preposition],[basic context],
  [non basic contexts],[WordNet classes]).

tverbe([administrer],2,[[ae],[tib]],[],[12],[50,60,100,141,162,92],
  [[14,1,1],[14,2,3]]).
tverbe([admirer],3,[[ae],[tib],[tc,src]],[pour],[20],
  [50,51,61,102,150,171,180],[[10,1],[10,3,6]]).
tverbe([adopter],2,[[ae],[tg]],[],[12],[50,60],
  [[14,1,1],[14,2,1],[14,2,5]]).
tverbe([adorer],2,[[exp,src],[tib]],[],[11],[50,60,150,180],
  [[10,1],[10,3,7]]).
tverbe([adresser],3,[[ae,src],[tg],[dest]],[a],[23],[50,51,61,170],
  [[13,2,1],[14,1,5]]).
tverbe([aerer],2,[[ae],[pos]],[],[15],[50,60,100,162],[[2,6]]).
tverbe([affaiblir],3,[[ae],[tiv],[mni]],[avec,par],[20],
  [50,51,61,102,141,162,17,103],[[2,7],[14,1,4]]).
```

Verb classes are constructed on the basis of equivalence classes of verbs accepting exactly the same set of contexts (basic and non-basic ones). We obtain a total of 953 classes. We get a large number of classes with just one element. This is not surprising, however, since contexts can be combined in a large number of ways (but much less than $(n-1)!$ where n is the number of contexts, because of constraints holding on contexts such as arity of the verb). Results are summarized in the table below.

Fig. 1 - Size of verb classes

no. of elements/class	no. of classes	rate w.r.t. total no. of classes (%)
1	741	77.7
2	123	12.9
3	31	3.3
4	13	1.4
5	11	1.2
≥ 6	34	3.5

The number of classes is quite large compared to Beth Levin's results, however, our classes have been constructed on a strict equivalence class basis, without any exceptions, and all the contexts associated with verbs have been taken into account, independently of their relative importance. Our first goal is to evaluate the results we can obtain using a strict classification method, i.e. all the verbs in an equivalence class behave exactly identically w.r.t. contexts, they undergo exactly the same contexts.

It is clear that these results won't be as good as B. Levin's ones. In her approach, she has noticed sufficient evidence of semantic similarity or correlations being mirrored in syntax to allow for generalizations and the formation of verb sematic classes, despite numerous exceptions. Roughly, the classification method used by Beth Levin, which, in fact, can only be realized by hand, is to select a (small) set S of alternations and to form the class of verbs which all undergo this set of alternations. Given this set S and the identification of some of the semantic features shared by its elements, she then adds verbs not in this set but sharing also these semantic features, provided that they undergo a subset of the alternations selected in S. This approach is of much interest in the sense that it is much less systematic than ours, and attempts at constructing classes sharing meaning components, whatever they are. We are doing here something totally different, in fact. We want to have a classification method which is has systematic as possible and we want to evaluate the quality of the results. We then want to characterize in a rigorous way the additional elements which must be taken into account to form verb semantic classes useful for NLP.

Let us now come back to our classification method.

We have an average of 2 verbs per class. A similar result was also obtain by (Gross 75), on a different basis and with more criteria, where he gets 2000 classes from 3000 verbs, with an average of 1.5 verb per class. Notice however that about 56% of the verbs belong to classes with at least two elements. Similarly, 32.6% of verbs appear in classes which have at least 5 elements.

5.1. ANALYSIS OF THE GLOBAL QUALITY OF VERB CLASSES

We have carried out a comparative study with verb classes formed from WordNet (Fellbaum 90) principles. We have classified exactly the same set of verbs, with the same senses, according to WordNet-like criteria. These WordNet classes are formed from common-sense, ontological categories and properties. It is therefore of much interest to compare the classes formed from syntactic criteria (called VS classes) to the WN classes.

If we consider WordNet classes as a reference in terms of semantic classification, we can then make a global evaluation of the degree of semantic relatedness of VS classes. It is clear that there are other ways of evaluating the quality of our VS classes, but this approach seems to be one of the best and simplest.

In our WordNet classification, we consider 198 hierarchically organized classification criteria, instances of the *is-a* (or troponymy) relation, where the depth of the decomposition is 3 (Saint-Dizier 96b). We therefore get 198 verb classes (called WN classes) for levels 1 to 3. For example, a three level decomposition is: *movement* verbs (level 1), *directed motion*, *local motion*, etc. (level 2) and *explicit upward motion*, *explicit downward motion*, etc. (level 3).

Before analyzing the semantic relatedness of VS classes, let us recall the main characteristics of the WN classes. Let us first introduce the WN level 1 classes (these classes are those defined in (Fellbaum 90), with a few terminological changes and the adjunction of classes 16 and 17):

Fig. 2 - WN main semantic classes

WN class no.	name of class	(1)	(2)
1	Body care, life and death	56	2.72
2	Verbs of change	262	12.76
3	Verbs of communication	156	7.60
4	Competition verbs	62	3.02
5	Verbs of consumption	80	3.89
6	Contact verbs	133	6.48
7	Cognition verbs	153	7.45
8	Verbs of creation and destruction	132	6.43
9	Verbs of movement	252	12.28
10	Psychological verbs	118	5.75
11	State verbs, stative processes	29	1.41
12	Verbs of perception	33	1.61
13	Possession verbs	174	8.48
14	Verbs of social interaction	338	16.47
15	Verbs related to weather expressions	5	0.24
16	Verbs related to aspect and action	58	2.82
17	Verbs expressing causality	13	0.63

(1) Number of verbs concerned in our sample, (2) rate w.r.t. the total number of verbs.

Here is the average size of WN classes, for each of the three levels of decomposition:

Fig. 3 - Global analysis of the WordNet-like verb classification

WN level	no. of verb occ.	(1)	polysemy rate	(2)	(3)	(4)
1	2052	1673	1.23	17	-	120.7
2	3112	1078	2.88	75	4.4	41.5
3	1941	1076	1.80	106	4.8	18.31

Where: (1): number of verb senses considered, (2): total number of daughters from all mother nodes, (3): average number of daughters per mother, (4): average number of verbs per class in the network.

Let us now analyze the semantic relatedness of verbs in VS classes. We have, for each VS class, searched the class in WN with which it has the largest intersection. Smaller intersections have not been taken into account. For example, if a VS class with 7 elements has 2 elements in common with the WN class nb. 3 and 4 elements with the WN class nb. 12, then this latter intersection will be the only one considered and the overlapp rate is 4/7. This is summarized in Fig. 4, taking into account the size of VS classes:

Fig. 4 - Global quality of VS classes w.r.t WN classes

size of VS class	larger overlap rate (%) with a WN class
2	41
3	47
4	53.3
5	56.4
6	46.4
7 - 9	51.7
$\geq$ 10	43.9

If we take into account the number of elements in each class size, then the total overlap rate is 46.36%, which is not very high. From the above results, we can notice that **classes with 4 to 9 elements are those which have the highest semantic relatedness rate**. These are also, in average, the classes which are associated with the largest set of contexts. The correlation is of course not surprising. Classes with more than 10 elements are often formed from a very small number of contexts. They often contain several subsets of semantically related verbs, but this is much more difficult to evaluate. Finally, considering the size of our verb sample and the fact that it covers the most frequent verbs of French, a verb class with good semantic correlations among verbs should have a size between 4 and 10 elements.

We now show in the table below that the greater the number of contexts associated with a VS class is, the more homogeneous the class is:

Fig. 5 - Impact of the no. of contexts on semantic relatedness of VS classes

no. of contexts	semantic relatedness rate (%) w.r.t. WN classes
1 to 3	49
4	53
5	75
6	70
7 or 8	89

These results show that a minimum of 5 contexts is required to form classes with an average semantic relatedness of 75%. Taking into account that there are contexts that appear in almost every class (see next section), such as contexts 50, 51, 60, 61, etc., the definition of these classes is in fact based on about 2-3 less frequently used contexts, which is not much. A similar type of result, on different grounds, is also observed in (Gross 75). We can then consider that these less frequently used contexts are more 'selective'.

From these figures, we can conclude that contexts seem to have a relatively good classification power. However, we have fewer contexts in French than there are alternations in English (47 in French, about 80 in English, but some of them include semantic factors that we want to avoid). Therefore, there is a quite large number of verbs that accept only 1 or 2 significant contexts. It is then clear that such a small number of contexts does not allow *a priori* the construction of verbs semantic classes on a strong and large syntactic basis.

From these observations we can either develop a method close to B. Levin's, by allowing a number of exceptions (but we won't probably get results as good as those obtained for English) or develop a new method to construct effective verb semantic classes.

6. Analysis of contexts in verbs

Let us now analyze the distribution of contexts in verbs and the distribution of contexts with respect to WN classes in order to better identify the possible implicit semantics conveyed by each context. This study complements the descriptions given in section 4.

6.1. DISTRIBUTION OF CONTEXTS IN VERBS

There are 9 basic contexts that occur more than 1% in our sample. For non-basic contexts, the table below summarizes the proportion of contexts frequently used:

Fig. 6 - Synthesis of contexts occurences

occurence rate	number of contexts involved
≥ 10%	12
≥ 5%	22
≥ 60%	6
between 60% and 5%	16

These results show that there are about **16 non-basic contexts which are really useful to form verb semantic classes**. However, less frequently used contexts should not be neglected: they may give rise to a small number of very specific classes (this is the case e.g. for contexts 92 to 96). The non-acceptance of a frequently used context by a set of verbs may also be of interest.

6.2. DISTRIBUTION OF CONTEXTS WITH RESPECT TO WN CLASSES

It is clearly of much interest to study the distribution of contexts with respect to WN main classes (see fig. 2). It turns out that 22 non-basic contexts are very frequently associated with verbs belonging to one or two main WN classes. The table below shows a large sample of those contexts that neither appear too frequently nor too rarely (except for a few which we have estimated to be really very relevant):

Fig. 7 - WN main semantic classes

Context	(1)	WN class involved	(2) in %	(3) in %
70	51	16	15.7	45
		2	50	32
71	16	9	50	63
72	10	6	60	80
80	6	11	33	86
90	90	2	43.3	36
		16	7.8	29
91	51	1	29.4	80
93	8	3	100	100
94	9	10	33.3	63
95	76	3	25	30
		7	22.4	30
101	41	6	22	30
		13	26.8	30
103	78	10	16.7	30
		3	20.5	28
113	3	13	100	100
114	31	8	25.8	44
115	3	13	100	100
120	46	6	39	37
		16	8.7	19
		7	17.4	14

(Column 1) total number of verbs to which this context is assigned; (Column 2) rate: number of verbs concerned in this WN class by this context / total number of verbs concerned by this context; (Column 3) relative rate: same as (2) but with the taking into account of the size of the WN class (given in the fig. 2 above). This permits us to take into account the relative size of the class.

Notice that, in the last column, we have ranked from first to last the WN classes involved for each of the contexts presented. These results, as shall be seen in the next section, will prove to be extremely useful to form better semantic classes.

A similar, symmetric, array can be defined where entries are WN classes and where for each class we specify the contexts that are most frequently accepted by its members. This can be summarized by ranking, on the basis of the results in fig. 7, the WN categories with respect to their correlation with syntactic behaviors (i.e. with contexts):

Fig. 8 - Correlation WN classes / syntax	
Rank	WN class numbers
1	3, 13, 14
4	6, 10
6	2, 9
8	11
9	15
10	1, 7
12	8, 17
no influence	4, 5, 12, 15

The last line gives WN semantic classes that do not have any significant correlation with contexts. It should also be noticed that the correlation is quite strong down to rank 8 (context 11); below this rank it becomes either more specific to a group of verbs or to a very small set of contexts. Thus, **about half of the WN main criteria have a clear influence on syntax**. In particular, WN classes 3, 13, 14, 6 and 10 are strongly correlated with contexts.

7. An efficient method for building verb semantic classes: a closer cooperation between VS and WN

WordNet and syntactic behaviors are two approaches to construct verb semantic classes. They correspond to two very different strategies and perspectives, and the results they produce are quite different. However, these two approaches are not incompatible. The structuring principles that underlie VS and WN classes can cooperate in a very productive way to produce in a systematic way verb classes with a strong semantic basis. It is important to bear in mind that our main aim, beyond constructing verb semantic classes, is to use the power of such a classification system to define efficient, homogeneous and operational semantic representations of verbs.

If we want to associate partial semantic representations to verb semantic classes such as, for example, lexical conceptual structures or (LCS, the approach adopted by Bonnie Dorr in recent works) (Jackendoff 87, 90), we can then take advantage of the semantics associated with contexts. Thus, context 110 conveys the idea of contact with motion, which can be represented in a LCS. Some of these criteria are often totally orthogonal to the semantic criteria of WN, and would, in fact, complement them quite nicely.

Let us now briefly present the classification method we have defined. Our strategy for building verb semantic classes is organized as follows:

1. first select all the verbs in a given top-level WN class (see fig. 2) to forms a *family* of verbs,
2. consider those contexts which are frequently encountered (or very rarely for dealing with their absence) in that family (see fig. 7), and form

classes which are semantically homogeneous (this is evaluated by hand), from a single context or a set of contexts,
3. next, consider thematic grids and use partly instantiated grids to form additional classes (these classes are in general totally different in nature from classes built in 2.); they are useful to make classifications more systematic and to fill in 'holes' in the classifications,
4. add, if appropriate, classes based on WordNet properties, which would characterize domains (e.g. justice, education) or properties proper to the predicate.

As can be seen, our approach to build verb semantic classes involves a close cooperation between paradigms which are a priori orthogonal. Each paradigm introduces a certain view (or facet) on semantic classifications.

Let us now illustrate this strategy. Let us select the family of verbs of transfer of possession. The following context combinations produces semantically related verb classes:

[20, 101] (context numbers) : *notion of compensation*,

[23, 113] : *notion of service*,

[115] : *nominate verbs*, etc.

If we now consider verbs of communication, we get, for example:

[24] : *writing something to someone*,

[12, 95] : *commenting on something*,

[23, 95] : *advising someone*, etc.

Context combinations produce 11 different semantically related verb classes for this family, which is a very good rate. Contexts used in this manner provide relatively comprehensive and clear classes, but the classification remains partial. It will be made more comprehensive by using thematic grids.

In our approach, we have associated verbs with thematic grids (see section 2.2 above), where each argument is tagged by a thematic cluster, which reflects its major thematic dimensions. For example, verbs like *déposseder, dépouiller, dévaliser* (to rob, to strip) will be associated with the following thematic grid:

[[effective-agent, destination], [source, incremental-victim-theme], [general-theme]]

Classes can then be formed from verbs whose associated thematic grid is subsumed by a partial grid. For example, we can form verb classes along the following semantic dimensions for transfer of possession verbs:

transfer from agent to an object, grid: [[...*src*...], ..., [...*dest*...], ...]

transfer to agent from an object, grid: [[...*dest*...], ..., [...*src*...], ...]

subject is a victim, grid: [[...*tiv*...], ...]

transfer is made to a fixed location, grid: [[...*src*...], ..., [...*pa*...], ...], etc.

It is clear that this classification is superficial, but it generates significant fragments of semantic representations in LCS.

Finally, we can take into account some WordNet criteria for classifications which cannot be made explicit by other means (otherwise preferred because they have a more formal basis, in spite of the difficulty of some judgements), such as: for domains: business transactions, transfer of knowledge, etc.; for semantic properties of the predicate-arguments structure: balance between two transfers, notions of punishment, of making a present to someone, etc.

If we apply these procedures to the family of verbs of transfer of possession, we get about 25 verb semantic classes (see appendix) which, of course, may largely overlap. These classes constitute 'facets' of the meaning of the verbs, meanings which can, at least partly, be integrated into a semantic representation such as the LCS (Saint-Dizier 97).

8. Conclusion

In this paper, we presented the equivalent of alternations for French and investigated their implicit semantics. We then showed how verb semantic classes can be constructed following a rather quite simple method. We believe this approach is appropriate to our general aim, namely the construction of organized sets of verbs, well-suited for the definition of homogeneous and coherent semantic representations.

We have analyzed the global quality of the results in terms of overlap with classes formed from the same verb-senses using WordNet-like classification criteria. Classes formed on a strict syntactic basis do not exhibit a very high rate of semantic relatedness (about 50%). This is in particular due to the small number of syntactic alternations in French. We have also shown that about 16 contexts out of 47 really contribute to the formation of classes. The others are either too frequent, too rare or do not have any classification power.

It is also interesting to note, from appendix 1, the differences in terms of alternations between French and English. Preliminary studies show that alternations in other Romance languages are also quite different, and no generalization can really be drawn over Romance languages. In fact, French may be considered as a relatively particular Romance languages, Italian, Spanish and Portuguese, for the most well-known, seem to have more in common.

Our future investigations with focus on semantic representations, we will concentrate on thematic grids, the LCS and aspects of the Generative Lexicon (Pustejovsky 95) for adding flexibility in the use of verb-argument structures (Saint-Dizier 98).

9. Appendix 1: List of contexts

We give here the list of non-basic contexts. Whenever appropriate, the (approximately) equivalent alternation in Beth Levin's book classification is given.

Summary of contexts		
Number	Name	in (Levin 93)
50	middle reflexive intransitive	none
51	middle reflexive transitive	none
60	passive, 2 args	5.1, 5.2
61	passive, 3 args	5.1, 5.2
70	intrans. inchoative form	1.1.2.1, 1.1.2.3
71	trans. inchoative form	1.1.2.1, 1.1.2.3
72	inchoative of localization	3.5
80	place subject inversion	2.3.4
90	introduction of 'faire'	none
91	introduction of a support verb	none
92	support verb faire	none
93	support verb pousser	none
94	support verb porter	none
95	support verb donner	none
96	support verb prendre	none
100	unspecified objects	1.2.1
101	ditrans. with unspecified object	none
102	ditrans. unspecified PP	none
103	ditrans. object1 deleted	none
110	preposition change sur/avec	2.8
111	preposition change sur/de	2.3.1
112	preposition change de/a	3.10
113	preposition change a/en	none
114	conative	1.3
115	resultative S-bar	2.14
120	transitive reciprocal avec/et	2.5.1
121	reciprocal subject	2.5.4
122	ditransitive reciprocal	2.5.4
123	reflexive reciprocal	1.2.4
130	spatial reorganisation dans/en	2.4.1
140	contact	2.12

Summary of contexts (continued)		
Number	Name	in (Levin 93)
141	focus on NP complement	none
150	expression of property of object	2.13.1/2/3
151	possessor subject ascencion	2.13.4
160	means-instrument transitive	3.3/4/5/8
161	location subject	3.6
162	means-instrument ditransitive	3.3, 3.4, 3.8
163	container subject	3.7
170	reflexive ditrans. of iobj	none
171	reflexive ditransitive on obj.	none
172	reflexive transitive	none
180	cognate construction transitive	7.1, 7.2
190	change of perspective of subject	none
191	location perspective	6.2
192	change of perspective object	none
193	change of perspective object 'pour'	none
200	there-insertion	6.8

10. Appendix 2: Sample of transfer of possession verb classes

This class contains, in our sample, 178 verb senses. Let us first consider class formation by contexts (there are few context combinations for this family of verbs that produce semantically homogeneous classes):

Contexts 20 and 101: notion of retribution-help: dedommager, indemniser, ponctionner, remunerer, retribuer.

Contexts 23 and 113: notion of service: fournir, livrer, servir.

Let us now consider a sample of classes built from thematic grids:

transfer from agent to object: financer, communiquer, donner, fournir, offrir, prodiguer, alimenter, approvisionner, combler, dedommager, ravitailler, remunerer, indemniser, recompenser, retribuer, abandonner, accorder, adresser, allouer, amener, apporter, apprendre, attribuer, ceder, confier, distribuer, enseigner, envoyer, expedier, laisser, leguer, livrer, louer, octroyer, payer, poster, preter, procurer, redistribuer, redonner, rembourser, remettre, rendre, renouveler, restituer, retourner, retroceder, revendre, reverser, servir, sous, traiter, transferer, transmettre, vendre

transfer from object to agent: accepter, attendre, cambrioler, deposseder, depouiller, devaliser, heriter, interroger, pirater, recevoir, acheter, acquerir, emprunter, endosser, importer, obtenir, racheter, recuperer, reprendre, soutirer, arracher, confisquer, derober, enlever, extorquer, faucher, prelever, prendre, voler

notion of duplication from object1 to object2: copier, decalquer, dupliquer, enregistrer, photocopier, recopier

agent source alone, no target specified: debourser, depenser, egarer, perdre

object source alone: debarasser, escroquer, photographier, polycopier, reproduire, reprographier, ruiner, deduire, imputer, oter, ponctionner, retirer, soustraire, demunir, spolier

without any explicit source or destination: amasser, capitaliser, capturer, cloturer, devoir, disposer, economiser, epargner, frauder, fructifier, compenser, contrebalancer, masquer, meriter, jouir, manquer, posseder, marchander, negocier, partager, rajouter, repartir, garantir, indiquer, reserver, priver, restreindre, miser, remiser, retrouver, sauvegarder, titulariser, echanger, monnayer, troquer, collecter, commercialiser, comptabiliser, ecouler, escamoter, evaluer, gaspiller, inventorier, mensualiser, pourvoir, radiographier, renoncer, solder, subvenir, trouver, revaloriser, valoriser

object is a victim: cambrioler, capturer, deposseder, depouiller, devaliser, escroquer, priver, restreindre, ruiner, demunir, spolier, arracher, confisquer, derober, enlever, extorquer, faucher, prelever, prendre, voler

one object is a fixed location: amasser, capitaliser, cloturer, egarer, perdre, miser, remiser, retrouver, sauvegarder.

Finally let us present a few classes based on WordNet-like properties:

Money involved (as an incorporated argument): abonner acheter acquerir commercialiser comptabilise crediter debourser dedommager devoir economiser emprunter encaisser endosser enrichir epargner evaluer exonerer facturer financer fructifier imputer indemniser louer marchander miser monnayer negocier payer ponctionner racheter rembourser remunerer retribuer revendre reverser sous-louer subvenir valoriser vendre

Illegal transfer, robbery: arracher cambrioler capturer demunir deposseder depouiller derober devaliser enlever escamoter escroquer extorquer faucher frauder masquer pirater prendre priver rafler s'emparer spolier voler

Exchange of objects: compenser echanger troquer

Duplication of an object or of a document: copier decalquer dupliquer enregistrer photocopier photographier pirater polycopier radiographier recopier reproduire reprographier.

Acknowledgements I thank Christiane Fellbaum for her comments on this work and for kindly proofreading the text. I also thank 2 anonymous reviewers whose remarks helped improving this text.

References

Baker, M.C., *Incorporation: A Theory of Grammatical Function Changing*, Chicago University Press, 1988.

Dorr, B. J., Garman, J., and Weinberg, A., From Syntactic Encodings to Thematic Roles: Building Lexical Entries for Interlingual MT, *Machine Translation*, 9:3-4, pp.71-100, 1995.

Dowty, D., On the Semantic Content of the Notion of Thematic Role, in G. Cherchia, B. Partee, R. Turner (eds), *Properties, Types and meaning*, Kluwer, 1989.

Dowty, D., Thematic Proto-roles and Argument Selection, Language, vol. 67-3, 1991.

Dubois, D., Saint-Dizier, P., Construction et représentation de classes sémantiques de verbes: une coopération entre syntaxe et cognition, in proc. RFIA96, Rennes, France, January 1996.

Fellbaum, C., *English Verbs as a Semantic Net*, Journal of Lexicography, 1990. (see also paper in this volume)

Goldberg, A., *Constructions: A Construction Grammar Approach to Argument Structure*, University of Chicago Press, 1994.

Gross, M., *Méthodes en syntaxe*, Masson, Paris, 1975.

Jackendoff, R., The Status of Thematic Relations in Linguistic Theory, Linguistic Inquiry, vol. 18, 1987.

Jackendoff, R., *Semantic Structures*, MIT Press, 1990.

Levin, B., *English verb Classes and Alternations: A Preliminary Investigation*, Chicago Univ. Press, 1993.

Levin, B., Rappaport, M., *Unaccusativity, at the Syntax-Lexical Semantics Interface*, Linguistic Inquiry monograph no. 26, MIT Press, 1995.

Marantz, A. P., *On the Nature of Grammatical Relations*, MIT Press, 1985.

Mittwoch, A., *On the Difference between Eating and Eating Something: Activities versus Acomplishements*, LInguistic Inquiry 13, 113-122, 1982.

Pinker, S., *Learnability and Cognition*, MIT Press, 1993.

Pugeault, F., Saint-Dizier, P., Monteil, M.G., *Knowledge Extraction from Texts: a method for extracting predicate-argument structures from texts*, in proc. Coling 94, Kyoto, 1994.

Pugeault, F., Extraction dans des textes de connaissances structurées: une méthode fondée sur la sémantique lexicale linguistique, PhD dissertation, Université Paul Sabatier, IRIT, Toulouse, 1996.

Pustejovsky, J., *The Generative Lexicon*, MIT Press, 1995.

Reuland, E., Abraham, W., (eds) *Knowledge and Language, Vol II*, Kluwer Academic, 1993.

Saint-Dizier, P., Verb Semantic Classes in French, IRIT research report, December 1996b, also in proc. Coling96, pp 1127-1130, Copenhagen.

Saint-Dizier, P., Marrafa, P., A Context system for describing the syntax of French verbs, in *ACH-ALLC series*, vol. 6, Oxford University Press, 1995.

Saint-Dizier, P., Semantic Verb Classes and lexical Conceptual Structures for Enhancing the Conceptual Modelling and the Access to Databases, *Journal of Data and Knowledge Engineering*, North Holland, vol. 21, pp 197-211, 1997.

Saint-Dizier, P., Sense Variation and Lexical Semantics Generative Operations, in proc. NeMLaP'98, Sydney, January 1998.

Williams, E., *Thematic Structure in Syntax*, Linguistic Inquiry monograph no. 23, MIT Press, 1994.

SEMANTICS IN ACTION

EVELYNE VIEGAS, KAVI MAHESH, SERGEI NIRENBURG AND STEPHEN BEALE
Computing Research Laboratory
New Mexico State University
Las Cruces, NM 88003, USA

Abstract. The paper presents a concise description of a comprehensive approach to computational lexical semantics and focuses on the treatment of events. We reason about the semantic information that should be encoded in a lexicon entry to support the twin tasks of constructing Text Meaning Representations (TMRs) for input texts and generating texts off TMRs. As static knowledge sources cannot be expected to cover all textual inputs, we describe and illustrate how lexical entries can be changed dynamically to fit the textual context at processing time. On the very important issue of knowledge acquisition, our experience shows that determining the meaning of lexical items is not a trivial task for a team of human acquirers (who are, we believe, absolutely indispensable for the more complex decisions in lexical knowledge acquisition). We illustrate how one can overcome the subjectivity of acquirers partly through advanced methodology and partly by having the lexical-semantic model account for some of the combinatory and (semi-)productive principles of natural language.

1. Introduction

The ultimate goal of studying semantics is to investigate the mechanisms of obtaining a meaning representation given a text and vice versa. A complete semantic theory should account for both knowledge and processing. In this paper, we concentrate on "knowledge in action:" we discuss the type of information that should be encoded in the knowledge sources, mechanisms of enhancing the coverage of knowledge sources, and the interplay of the knowledge and the processors aimed at treating the complex phe-

P. Saint-Dizier (ed.), Predicative Forms in Natural Language and in Lexical Knowledge Bases, 171–203.

nomena found in natural languages. Our attitude to the static knowledge bases (the lexicon, the ontology, the grammar, the special "microtheory" rules) is almost strictly functional: the knowledge which is acquired is, from the outset, intended for use in extracting and representing meaning given a text or realizing a text given its meaning. Our approach to computational semantics has been developed most recently in the framework of the Mikrokosmos project.

Mikrokosmos is a knowledge-based machine translation (KBMT) system under development at New Mexico State University jointly with the US Department of Defense (e.g., Beale *et al.*, 1995; Onyshkevych and Nirenburg, 1994). The Mikrokosmos team[1] has been working on a wide variety of topics, such as the relation between underlying meaning and surface realizations, multilinguality aspects, syntactic and semantic dependency knowledge, acquisition of (lexical) semantic knowledge, and application to (multilingual) generation and machine translation. As its experimentation domain, Mikrokosmos has focused so far on translating Spanish and Chinese news articles to English. By the end of 1995, a core lexicon of approximately 7,000 Spanish word senses supported by an ontology of about 5,000 concepts was in place. After application of lexical rules,[2] the expanded Spanish lexicon contained about 40,000 word senses.

The Mikrokosmos analyzer has produced high quality semantic analyses of article-length Spanish texts in the domain of company mergers and acquisitions, see Section 2.1 for the word sense disambiguation results we obtained so far. We are now in the process of generating English texts off these TMRs, as reported in (Beale *et al.*, 1997). A set of graphical tools has been developed to support knowledge acquisition, system development, testing, and demonstration of each phase of language processing and machine translation. In addition, automated techniques and detailed methodologies have also been developed for testing lexicons and ontologies as well as assessing the performance of language processing components.

As this book is devoted to predicates, a terminological note is in order at this point. From a logical or syntactic perspective (the ones habitually assumed in Natural Language Processing (NLP)),[3] a predicate can be said

[1]See http://crl.nmsu.edu/Research/Projects/mikro for further information.

[2]To sketch this operation briefly, applying morpho-semantic lexical rules to the entry for the Spanish verb *comprar*, our acquisition system produced automatically 26 new entries (*comprador-N1*, *comprable-Adj*, etc). This includes creating a new syntax-semantics mapping with correct subcategorizations and also the right semantics; for instance, the lexical entry for *comprable* will have the subcategorization of an adjective and the semantics adds the attribute "feasibility-attribute" to the basic meaning of *comprar*, as reported in (Viegas *et al.*, 1996).

[3]The oldest notion of predicate from antiquity lies on a binary semantic opposition between what is called "theme/rheme" or "subject/predicate" and is still active in the-

to be a named n-ary relation between arguments, which concerns itself more with capturing the semantics of events or states than with the semantics of objects; or rather, since the distinction between natural language and the language of representation has not yet been appreciated enough by lexical semanticists, to verbs, not nouns. From this formal viewpoint, this notion seems to be merely equivalent to verbs or relational nouns or prepositions or relational adjectives viewed together with their syntactic dependency (subcategorization) frames.

For us, semantic *events* are defined by ontological (that is, non-syntactic) features (and their values) associated with them. Moreover, the relational ones among these features are not equivalent to subcategorization arguments (which constitute the arguments of a syntactic predicate), as they may have no surface realization at all.

The paper presents a concise description of a comprehensive approach to computational lexical semantics and focuses on treatment of events. We reason about the semantic information that should be encoded in a lexicon entry to support the twin tasks of constructing Text Meaning Representations (TMRs) for input texts and generating texts off of TMRs. A single frame-based representational formalism (e.g., Mahesh, 1996, Nirenburg and Defrise, 1991; earlier proposals included, seminally, Minsky, 1968; Luger and Stubblefield, 1992; Fillmore, 1985, 1993) is used in Mikrokosmos to encode information in lexicon entries and to build TMRs; in fact, a lexicon entry constitutes an unsaturated piece of TMR. As static knowledge sources cannot be expected to cover all textual inputs, we describe and illustrate how lexical entries can be changed dynamically to fit the textual context at processing time.

Our experience shows that determining the meaning of lexical items is not a trivial task in lexicon acquisition.[4] However, as humans still play a crucial role in lexicon acquisition (cf. Gross, 1984; Lenat *et al.*, 1986; Nirenburg and Raskin, 1987; Normier and Nossin, 1990; McNaught, 1990), this problem must be addressed centrally. We show that it is possible to reconcile the subjectivity of acquirers with the recognition of the "core" meaning of a word, if we have developed a theory of computational (lexical) semantics which can account for some of the combinatory and (semi-)productive phenomena in natural languages.

oretical linguistics and philosophy (Grimes, 1968; Searle, 1969).

[4]One could object that we do not need to determine the meaning of a lexical item but just the relations among lexical items, a common solution in lexical semantics (e.g., Cruse, 1986). However, for Natural Language Processing (NLP) applications, knowing the relations among lexical items is not enough, especially in multilingual applications. Consequently, many researchers working in computational lexical semantics have added a conceptual layer to their lexicons (e.g. Sanfilippo, 1992).

The Mikrokosmos semantic theory addresses the following types of ambiguity - homography, homomorphy and polysemy. [5] Briefly, homography refers to words of perceivably different meaning which have the same spelling, such as [bank-N1, financial-institution] and [bank-N2, river-bank]. We treat such cases at the lexical level in Mikrokosmos, using different entries ([bank-N1] and [bank-N2]). Homomorphy refers to words which have the same form and different parts of speech, such as [fast-Adj] and [fast-Adv], we treat such cases at the syntactic level in Mikrokosmos, with two lexicon entries [fast-Adj1] and [fast-Adv1]. Finally, polysemy refers to words which have the same form and different meaning which is usually not perceived as such by humans, such as [record-N1, music - artifact]. We treat such cases in Mikrokomos at the semantic (or ontological) level, and we thus have one lexicon entry for [record-N1]. It is important to remember that polysemy in this sense (often called regular polysemy) is a lexicographic or computational concern and as such it is a virtual (or non-existent) ambiguity for people.

In section 2, we give an example of how we use lexical-semantic knowledge to produce the TMR of a text. In section 3, we address representational issues with respect to the lexicon, specifically, the (lexical) semantic information that should be included in the lexical items signifying events, in order to support multilingual NLP applications. In section 4, we introduce the methodology of acquisition within a multilingual environment, paying particular attention to the trade-offs between language-dependent and language-independent information, and variations among the acquirers. In particular, we show that performing acquisition in a "situated environment" (that is, geared at particular tasks) and allowing the static knowledge sources to be changed dynamically leads to improved processing of meaning in context. Finally, in section 5 we compare our approach to other major trends in computational lexical semantics; in particular, we address the issue of sense enumeration.

2. Semantics in Action

We illustrate our approach to processing text meaning by discussing our approach to one of the most difficult tasks in NLP, namely, word sense disambiguation. We begin by presenting the results from the Mikrokosmos semantic analyzer, and then explain how they were obtained.[6]

[5]See Weinreich (1964) on contrastive and complementary ambiguities; Cruse (1995) on facets; Pustejovsky (1995) on logical polysemies.

[6]See also (Mahesh *et al.*, 1997b) for a description of the word sense disambiguation task in Mikrokosmos.

2.1. RESULTS

Experiment 1: Mikrokosmos semantic analyzer applied on 4 Spanish texts from real-world texts (news articles on company mergers and acquisitions from the EFE newswire), in which there were no unknown words to Mikrokosmos.

Text	1	2	3	4	**Mean**
1 - words	347	385	370	353	**364**
2 - words per sentence	16.5	24.0	26.4	20.8	**21.4**
3 - open-class words	183	167	177	177	**176**
4 - ambiguous open-class words	57	42	57	35	**48**
5 - ambiguous words resolved by syntax	21	19	20	12	**18**
6 - ambiguous words correctly resolved	89%	98%	79%	97%	**91%**
7 - total words correctly resolved	51	41	45	34	**43**
8 - % of total words correctly resolved	97%	99%	93%	99%	**97%**

TABLE 1. Mikrokosmos Results in Disambiguating Open Class Words in Spanish Texts.

Table 1 shows sample disambiguation results from the Mikrokosmos analysis of four texts.[7] The average text was 17 sentences long; average sentence length was over 21. For evaluation purposes, correct senses for all the open class words were tagged by a native speaker. Mikrokosmos selects the right sense of open-class words about 97% of the time. Syntactic analysis contributed to about 38% of word sense disambiguation (that is, these were the cases when different senses of a lexeme had different subcategorizations). The performance on Text 1 and Text 3 was inferior to that on the other two texts. Texts 1 and 3 had longer sentences (see line 2 in Table 1), and many

[7]These texts were selected at random from about 400 texts which were used in building the core Spanish lexicon in Mikrokosmos. This, however, does not at all mean that the data were in any sense cooked. Unlike the case with the training procedures for statistical methods, in Mikrokosmos the acquisition of the lexical and ontological knowledge has been performed by humans, and the corpus served only as a source of lexeme strings, not individual senses. For each word string from the training corpus the acquirers acquired ***a full set*** of meanings and recorded them either in the lexicon or in the special rules for treating regular polysemy, realized, however, imperfectly, through weights on ontological arcs. In short, our acquisition or "training" procedure was in no way directed at processing only the given corpus. It resulted in static knowledge sources which contained many more meanings than those present in the corpus.

more ambiguous words (see line 4 in Table 1) and constructions that make disambiguation hard (e.g., ambiguous words embedded in appositions).

Note that most researchers using corpus-oriented methods for word sense disambiguation also get impressive results. However, they are typically concerned with disambiguating homographs only (e.g., getting about 92% of accuracy on an average 3-way sense distinction, counting all open class words in (Yarowsky, 1992)) and mainly homomorphs (getting about 92% counting all open class words in (Wilks and Stevenson, 1997)). This is why lines 6 in Table 1 and Table 2 are absent from the results given by these researchers. All this means that Mikrokosmos addresses a more comprehensive task. However, to be fully precise, the numbers shown here should be refined by indicating the percentage average per n-way sense distinction. We have not done such an experiment on a large scale (this was easier for Yarowsky and others who consider only a relatively small set of ambiguous words; we dealt with running text, and in our sample 48% of open-class words were ambiguous). Note that the Mikrokosmos lexicon has only about 1.2 senses per lexeme. This number is misleadingly low because it covers only homography and homomorphy, while polysemy is dealt with dynamically, based on ontological processing. As a result, the lexicon entries are kept underspecified, leaving the task of coming up with the correct sense in context to the semantic analyzer.

Experiment 2: Mikrokosmos semantic analyzer applied on a Spanish text not used in knowledge acquisition.

Table 2 shows the Mikrokosmos Spanish analyzer results on a previously unseen text. The results were essentially similar to those for the texts in Table 1.

1 - number of words	390
2 - number of words per sentence	26
3 - number of open-class words	104
4 - number of ambiguous open-class words	26
5 - number of words resolved by syntax	9
6 - % of ambiguous words correctly resolved	88.9%
7 - total of words correctly resolved	23
8 - % of all open-class words correctly resolved	97.1%

TABLE 2. Mikrokosmos Results on an Unseen Text.

The unseen text used in the experiment contained 19 words missing from the Mikrokosmos Spanish lexicon. In such cases, the Mikrokosmos analyzer generates dummy lexicon entries, arbitrarily marked as nouns and semantically mapped to ALL, the root concept in the ontology (this has the effect of having no semantic constraints in the definition). No changes were made in the lexicon, ontology, or the programs. There were many syntax-to-semantics binding (mapping) problems with this text. For instance, lexicon entries ruled out some of the subcategorizations found in the text. We did not fix any of them. We could get even better results if we assumed perfect syntactic output and fixed all the binding problems. Unknown words (which were mapped to ALL) were treated as unambiguous. Twelve of the 19 unknown words appeared to be proper names and only 3 unknown words were, in fact, ambiguous. Among the 85 words present in the lexicon, only 13 words had occurred in previously analyzed texts.

Overall, one can say that the Spanish core lexicon is of a good quality, as it provided enough constraints to help in the task of disambiguation even in the presence of unknown words and deficiencies in syntactic information. This experiment seems to validate our methodology, where the "training" of the static resources only concerns correcting errors in the lexicon or the ontology. This includes such revisions as correcting wrong syntactic class assignments or wrong mappings to an ontological concept or constraints on a concept, all of which are part of the lexicographic loop in the acquisition process. Revisions are not aimed at "hard-coding" lexico-semantic information in the entries so that they give the best results for a particular corpus.

We now illustrate our approach using the following input:[8]

Consider the Spanish sentence:

(1) El grupo Roche, a través de su compañía en España, adquirió Doctor Andreu.

(The Roche group, through its company in Spain, acquired Doctor Andreu.)

[8]For reasons of space, we work with a simplified version of an actual corpus sentence: *El grupo Roche, a travès de su compañía en España, adquirió el laboratorio farmacéutico Doctor Andreu, se informó hoy aquí.* (The Roche Group acquired the pharmaceutical laboratory Doctor Andreu through its company in Spain, it was announced here today.) In reality, the processing often involved situations which required additional kinds of processing, not a simple process of instantiation and combination, as described in the example. See (Beale *et al.*, 1995), for a more detailed description of the Mikrokosmos analyzer.

Many words in this sentence are ambiguous: *a través de* can have the semantic roles of Path or Instrument; *compañía* can be CORPORATION or SOCIAL-EVENT; *en* is multiply ambiguous; and *adquirir* can mean either ACQUIRE or LEARN. In this illustration, we focus on disambiguating *adquirir* although our system resolves all of the above ambiguities. The desired meaning of the sentence is the following (extremely simplified) TMR:

```
ACQUIRE-1
        Agent: ORGANIZATION-1
        Theme: ORGANIZATION-2
        Instrument: ORGANIZATION-3
ORGANIZATION-1
        Object-Name: Grupo Roche
        Agent-Of: ACQUIRE-1
ORGANIZATION-2
        Object-Name: Doctor Andreu
        Theme-Of: ACQUIRE-1
ORGANIZATION-3
        Location: NATION-1
        Instrument-Of: ACQUIRE-1
NATION-1
        Object-Name: Espaa
        Location-Of: ORGANIZATION-3
```

Note that all of the labels used in the above TMR are well-defined concepts in the Mikrokosmos ontology (see the next section for more detail). This TMR, in the absence of microtheories such as focus or theme/rheme oppositions, can be generated in English as any of the following:

The Roche group, through its company in Spain, acquired Doctor Andreu.
The Roche group acquired Doctor Andreu, through its company in Spain.
The acquisition of Doctor Andreu by the Roche Group was made through its company in Spain.

2.2. GENERATING CONSTRAINTS

The first step for the Mikrokosmos analyzer is to collect all the possible lexicon entries for each of the words. For instance, for *adquirir*, the two lexicon entries *adquirir-V1* and *adquirir-V2* are retrieved, which correspond to the conceptual meanings ACQUIRE and LEARN. For each word sense, the syntactic mapping, derived via a variable binding process, must be examined to see if it fits the current sentence. For *adquirir*, both word senses have identical syntactic mapping and can be correctly applied to this sentence. After variable binding, the semantic analyzer examines the semantics of each word sense in order to construct a list of constraints that must be satisfied for that word sense.

2.3. APPLYING CONSTRAINTS

Mikrokosmos employs an ontological graph search mechanism, Onto-Search, (Onyshkevych, 1997) to check constraints. Onto-Search, determines relevant paths between any two concepts and returns a score based on their degree of closeness. For example, the command **check-onto-con**(ACQUIRE EVENT)[9] returns a score of 1.0 (out of 1.0) since ACQUIRE is a type of EVENT. However, **check-onto-con**(ORGANIZATION HUMAN) returns a score of 0.9 along with the path (ORGANIZATION HAS-MEMBER HUMAN). This indicates that ORGANIZATION can stand in the place of HUMAN because it has HUMAN members. This and other types of metonymy are frequent in natural language and are detected automatically by the Mikrokosmos analyzer.

2.4. CONTEXTUAL DISAMBIGUATION

Each combination of word senses activates the applicable constraints, which are combined into a total score for that combination. The combination with the best total score is chosen as the basic Semantic Dependency Analysis, the core TMR to which other microtheories (such as aspect or coreference) can be applied. In the example sentence, the following choices were made:

1. *a-través-de* is INSTRUMENT, since its LOCATION meaning would require *adquirir* to be a PHYSICAL-OBJECT.
2. *en* is LOCATION, since its TEMPORAL meaning would require *españa* to be a TEMPORAL-OBJECT.
3. *adquirir* maps into ACQUIRE, since its LEARN sense would require *Dr-Andrew* to be INFORMATION.
4. *Dr-Andrew* is an ORGANIZATION, since its HUMAN meaning cannot be the theme of an ACQUIRE concept.
5. Mikrokosmos currently has trouble choosing between the CORPORATION and SOCIAL-EVENT meaning of *compañía*, the object of the *á-traves-de* PP-adjunct. Both can have locations in Spain, and both can be INSTRUMENTS of EVENTs. At this point, Mikrokosmos needs to add information into the ontology that ORGANIZATIONs can typically fill the INSTRUMENT value of ACQUIRE acts, but SOCIAL-EVENTs cannot. Another alternative could consist in tuning our lexicons towards the domains using statistical techniques.

To summarize, disambiguation decisions for a word can rarely be done independently of the decisions on other words. The Mikrokosmos analyzer, therefore, operates as follows:

[9]Which asks "Is ACQUIRE an EVENT?"

– Derive selectional constraints from the lexicon and the ontology for each pair of syntactically dependent words, in both directions.
– Check each constraint by finding the "distance" between the pair of concepts in the ontology.
– Combine the results in an efficient constraint satisfaction algorithm (Beale, 1997; Beale *et al.*, 1996) to select the best combination of senses for all the words in a sentence.

The above description gives a general view of our approach to the application of selectional constraints during processing, and some insight into our approach to computational semantics. We illustrate the analysis process and the static knowledge sources in the figure below:

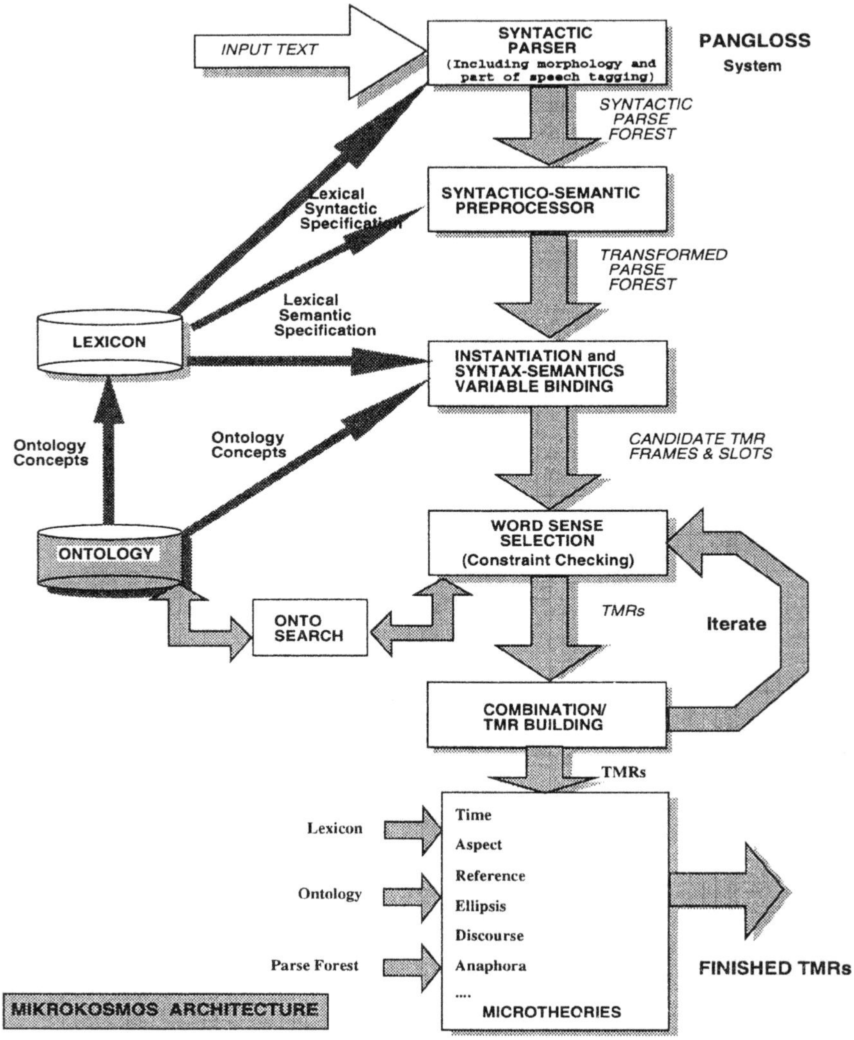

Figure 1. The Mikrokosmos Analysis Architecture.

Input texts are passed through a multi-stage morpho-syntactic analyzer for Spanish called Panglyzer (Farwell *et al.*, 1994). The resulting syntactic trees are transformed to an LFG-like syntactic structure (Bresnan, 1982). Lexical entries from the Mikrokosmos lexicon are instantiated for each of the root forms in the syntactic structure. Syntactic variables in the instantiations of the various lexicon entries are bound to one another using the syntactic patterns in the lexical entries to establish syntactic dependencies and map them to semantic dependencies. In addition, ontological concepts referred to the semantic zones of the lexical entries are instantiated.

In the next step, selectional constraints are retrieved from the ontology and added to those encoded in the lexicon. Individual selectional constraints are checked by an ontological search program called Onto-Search (Onyshkevych, 1997). The resulting preference values for each constraint are combined in an efficient control and search algorithm called Hunter-Gatherer that combines constraint satisfaction (Tsang, 1993), branch and bound, and solution synthesis techniques (Freuder, 1978) to pick the best combination of word senses of the entire sentence in near linear time, as described in (Beale, 1997), (Beale *et al.*, 1996). Chosen word senses are assembled into TMR frames using lexical semantic representations from the lexicon. Finally, a variety of microtheories are applied to further analyze elements of text meaning such as time, aspect, propositions, sets, coreference, and so on, to produce the final TMR.[10]

A perceived limitation of our approach is the cost of comprehensive evaluation on a large corpus of inputs, due to high start-up costs of knowledge acquisition and to the inevitable complexity of any realistic application-oriented evaluation scheme. However, our experience shows that it is possible to reduce the acquisition costs. In Mikrokosmos, we acquired an ontology of about 5,000 concepts (Mahesh, 1996; Mahesh and Nirenburg, 1995) and a Spanish core lexicon of about 7,000 word senses (see Viegas and Raskin, 1998) in about four person-years of effort. The core lexicon has been automatically extended by about 32,000 new entries, using morpho-semantic lexical rules (derivational morphology), as reported in (Viegas *et al.*, 1996).

[10] A few microtheories are already in place and several are currently being developed by the Mikrokosmos team.

3. Representational Issues

3.1. THE LEXICON

The lexicon is linked with the ontology and the onomasticon (a special-purpose lexicon of proper names such as toponyms, names of corporations or products). It serves as a connection between lexical units in texts and the TMR.[11] Each lexical entry contains a semantic specification which uses terms from the ontology, in addition to other, non-ontological, features, e.g., to reflect speaker attitudes or modalities. The advantages of connecting the lexicon to an ontology are twofold: 1) it allows one to capture the semantic properties of natural language units in a "language-neutral" way; and 2) it is cost-effective for multilingual NLP applications, as only the "language-dependent" properties have to be acquired when adding new natural languages to the system. See (Nirenburg *et al.*, 1995) for further discussion.

It is important to note that we adopt a transcategorial approach, according to which semantic or ontological categories are largely independent of syntactic categories. Even such a basic approximation as the connection between verbs and EVENTS and nouns and OBJECTS is not a real rule. This approach helps us to automatically establish the semantics of the words derived through the use of derivational-morphological lexical rules (LRs).

In what follows we describe some of the more important zones in the lexical entry, specifically those that play a part in semantic analysis.

3.1.1. *SYN Zone*

The information contained in the SYN zone essentially amounts to an underspecified piece of a syntactic parse of a sentence using the lexeme. For instance, in our example sentence, while processing the *adquirir* lexicon entry shown in Figure 1, *Grupo Roche* will be bound to [1] as the SUBJ, while *Dr. Andreu* will be bound to [2] as the OBJ.

3.1.2. *SEM Zone*

The SEM zone contains a possibly constrained mapping to an ontological concept, called the lexical item's Lexical Semantic Representation (LSR). Each SEM zone constitutes an "unsaturated" TMR fragment which includes as much meaning as can be extracted from the word alone. The interaction of the contents of the SEM zones from all the words in the sentence, (along with information added by other microtheories), results in the

[11] For a recent review on computational lexical semantics, the reader can consult (Nirenburg and Raskin, 1996), where the authors describe the different approaches to lexical semantics; (Wilks *et al.*, 1995) where the authors discuss computational dictionaries.

final TMR. The formalism for the lexical semantic specification in this zone has been discussed in detail in e.g., (Onyshkevych and Nirenburg, 1994).

3.1.3. *SYN-SEM Zone*

The SYN-SEM zone provides the syntax-semantic linking. For instance, in Figure 2, we know that the SUBJ [1] is linked to the semantics AGENT [11] of type HUMAN.[12]

We illustrate the SYN, SEM and SYN-SEM notions through the relevant fragments of the Spanish lexicon entry for *adquirir*, shown in Figure 2.[13]

```
adquirir-V1
  syn:      [root: [0]
             subj: [1][cat: NP]
             obj:  [2][cat: NP]]
  sem:      [acquire
             agent:  [11] human
             theme:  [21] object]
  syn-sem:  [subj: [1][sem: [11]]
             obj:  [2][sem: [21]]]
adquirir-V2
  syn:      [root: [0]
             subj: [1][cat: NP]
             obj:  [2][cat: NP]]
  sem:      [learn
             agent:  [11] human
             theme:  [21] information]
  syn-sem:  [subj: [1][sem: [11]]
             obj:  [2][sem: [21]]]
```

Figure 2. Sense (Partial) Entries for the Spanish lexical item *adquirir*.

Figure 3 shows the SEM zone of adquirir-V1. It contains a link to the ACQUIRE concept (figure 3). The latter concept contains the AGENT and THEME properties (slots) that will be filled with the TMR fragments pro-

[12]Previously, we only had 2 zones, the SYN zone and the SEM zone, and the linking was added inside these zones, such that we specified for each SYN feature its correspondent SEM feature inside the SYN zone. This new representation is equivalent to the former we used, it facilitates acquisition by modularizing the process of acquisition per zone.

[13]We use the typed feature structures (tfs) as described in (Pollard and Sag, 1987).

duced by deriving the semantics of *Grupo Roche* (1) and *Dr. Andreu* (2), respectively. Other words in the sentence can add information to the nascent TMR by filling or further constraining property slots or even by adding new frames (e.g., for speaker attitudes). One of the meanings of *a través de*, treated as a phrasal entry, will add an INSTRUMENT slot to the concept instance. The adquirir-V2 SEM zone links to the LEARN concept in figure 2 with a theme of type INFORMATION. The information shown in the SYN zone here is partial. In fact, adquirir-V1 has optional pp-adjuncts, lexicalizing the semantics of such properties of ACQUIRE as SOURCE (in which case it may expect in the input a prepositional phrase introduced by *from*).

3.1.4. *Other Lexicon Zones*

The SYN zone, SEM zone and SYN-SEM zone are the main zones used to produce the TMRs. The lexicon entry includes many other zones, some of which participate in the production of the TMR, some which provide information used mostly in generation, and finally some which serve knowledge management purposes. Briefly, a lexeme is minimally described via zones corresponding to various levels of lexical information, relevant to phonology, orthography, morphology, syntax, semantics, syntax-semantic linking, stylistics, and paradigmatic and syntagmatic information.[14]

```
Concept Name:
    ACQUIRE
------------------------------------------------------------
DEFINITION
   VALUE
      "the transfer of possession event where the agent
 transfers an object to its possession."
IS-A
   VALUE
      TRANSFER-POSSESSION
SUBCLASSES
   VALUE
      INHERIT
SOURCE
   SEM
      HUMAN PLACE
THEME
   SEM
      OBJECT (NOT HUMAN)
AGENT
   SEM
      ANIMAL
   DEFAULT
      HUMAN
DESTINATION
   DEFAULT
      HUMAN
   SEM
      ANIMAL PLACE
------------------------------------------------------------
Inherited Slots
------------------------------------------------------------
BENEFICIARY
   SEM
      HUMAN
------------------------------------------------------------
```

Figure 3. Conceptual frame for ACQUIRE.

```
Concept Name:
    LEARN
------------------------------------------------------------
DEFINITION
   VALUE
      "to take information into your brain"
IS-A
   VALUE
      ACTIVE-COGNITIVE-EVENT
EFFECT
   SEM
      UNDERSTAND
THEME
   SEM
      INFORMATION
PURPOSE-OF
   SEM
      ACADEMIC-EVENT READ
CAUSED-BY
   SEM
      TEACH
AGENT
   SEM
      HUMAN
------------------------------------------------------------
```

Figure 4. Conceptual frame for LEARN.

[14] See (Viegas and Raskin, 1998), (Onyshkevych and Nirenburg, 1994), (Meyer *et al.*, 1990) for reasons for inclusion of these zones, and on how to acquire them.

3.2. THE ONTOLOGY

An ontology is a large collection of information about EVENTs, OBJECTs and PROPERTYs in a domain (Mahesh, 1996; Mahesh and Nirenburg, 1995). The Mikrokosmos ontology consists of concepts (named sets of property-value pairs) organized heterarchically on subsumption links. Figure 2 and figure 3 illustrate the content of the event-type concepts ACQUIRE and LEARN respectively.

4. Acquisition and Multilinguality

In this section, after briefly discussing the methodology for acquisition, we turn to the decisions which have to be made by the acquirers of ontologies and lexicons to build lexical entries which will contribute best to 1) capturing the meanings of words as well as 2) constituting the building blocks of the meaning of a text. In particular, we will address the issues of a) "language-neutrality" versus "language-dependency" within a multilingual environment, and b) semantic ambiguity. We argue that these are mainly lexicographic concerns, and not computational-semantic concerns: the static knowledge sources should be allowed to be changed dynamically at processing time, to a) fit the linguistic context in which they appear and b) reconcile different acquirers' points of views on the data.

4.1. ONTOLOGY ACQUISITION

The ontology is being acquired incrementally, relying on continuous interactions with lexicon and semantic analyzer teams. A series of negotiations between lexicographers, ontologists, and developers of Onto-Search (see Section 2.3.) leads to the best choice of meaning representation in each case. It also ensures that every entry in each knowledge base is consistent, compatible with its counterparts, and has a purpose towards the ultimate objective of producing quality TMRs. We refer to this method as *situated development* of ontologies and lexical resources. This method is ideal for a multilingual situation such as in Mikrokosmos where it ensures that representational needs of more than one language are taken into account.

Ontology acquisition is a very expensive empirical task. Situated development is a good way to constrain the process and make it attainable. For example, the acquirer must focus on concepts in the domain of the input texts and thereby increase the ratio of the number of concepts that are actually used in processing a set of texts to the total number of concepts present in the ontology. The best example of a large ontological database acquired with enormous efforts but entirely out of any situation is CyC (Lenat and Guha, 1990). While the utility of CyC in a particular situation such as large scale NLP is yet to be demonstrated (Mahesh *et al.*, 1996), it is also true that most projects cannot afford to spend as many resources as it has taken to develop CyC and must strive to constrain acquisition significantly or share existing ontologies.

Methodology and Guidelines The basic methodology for concept acquisition employed in the Mikrokosmos project involves a fine-grained cycle of requests for concepts from the lexicon acquisition team and the resulting responses which may involve pointing out an existing concept, adding a new concept, enhancing the internal structure of one or more concepts, or suggesting a different lexical mapping for the word in question. If it is determined that a word sense requires a new concept in the ontology, the "algorithm" applied for adding the new concept hinges on viewing the ontology as a discrimination tree. The acquirer discriminates from the top down until at some point there is no child that subsumes the meaning in question. A new concept is added as a child at that point. In the Mikrokosmos project, sets of guidelines have emerged for making various kinds of decisions in ontology acquisition. These guidelines, some of which are shown in Figure 5, collectively define the methodology for ontology building (Mahesh, 1996).

4.2. LEXICON ACQUISITION

Acquiring a large-scale computational semantic lexicon is a very expensive enterprise, this is why it is advantageous to build lexicons which are reusable for other domains or applications. We need lexicons that are multi-purpose, supporting the three following paradigms:

- **multi-lingual**: French, English, Japanese, Russian, Spanish, etc..., *(encoding of characters)*
- **multi-media**: containing linguistic information for natural language processing, phonological information, essentially for speech recognition and production, and graphics, motion for visual processing, ... *(structure of the lexicons)*
- **multi-process**: applicable for analysis, generation (both mono- and multilingual), MT, summarization, information extraction, or speech processing,... *(reversibility of the lexicons)*

In other words, we develop one lexicon per language, and this lexicon can be used for different applications by automatically reindexing the lexicon as we did for instance in (Viegas and Beale, 1996) to create an English generation lexicon from an existing Spanish analysis lexicon. Having as a goal a multi-purpose lexicon saves a lot of time in acquisition.

1. Do not add instances as concepts in the ontology. Rules of thumb for distinguishing an instance from a concept are:
   - **Instance-Rule1:** See if the entity can have its own instance. Instances do not have their own instances; concepts do.
   - **Instance-Rule2:** See if the thing has a fixed position in time and/or space in the world. If yes, it is an instance. If not, it is a concept. For example, SUNDAY is a concept, not an instance, because it is not a fixed position in time ("last Sunday," "the first Sunday of the month," etc.).
2. Do not decompose concepts further into other concepts merely because you can. It is important to focus on building those parts that are needed immediately for the Mikrokosmos task. For example, though EVENTs like BUY or MARKETING can be decomposed to a great extent, unless there is an indication that detailed decompositions are needed for the task, do not decompose such EVENTs.
3. Do not add a concept if there is already one "close" to it or slightly more general than the one being considered. Consider the expressiveness of the representation provided by gradations (i.e., attribute values) before adding separate concepts. For example, we do not need separate concepts for *suggest*, *urge*, and *order*. They are all gradations of the same concept, a DIRECTIVE-ACT, with various degrees of force which can be captured in an appropriate attribute.
4. Do not add specialized EVENTs with particular arguments as new concepts. For example, we do not need separate concepts for "walk to airport terminal" and "walk to parking lot."
5. Certain elements of text meaning such as aspect, temporal relations, attitudes, and so on, that are instance-specific belong only in the TMRs. For example, BREAKFAST is probably a concept in the ontology (and a subclass of MEAL, say) but a meal that happened at 3 O'clock on a particular day is not a separate concept in the ontology.
6. If any part of a meaning representation is specific to a particular language that part does not belong in the ontology.
7. Mikrokosmos representations have a very expressive **set** and **subset** notation. Hence, there is no need to create ontological concepts for collections of different types of things in the world.

Fig 5. Guidelines for Deciding What Concepts to Add.

The process of acquisition itself has been reported in many documents by the members of our team. People interested in the detailed process of acquisition can consult (Viegas and Raskin, 1998), (Viegas and Nirenburg, 1996). Briefly, lexicon acquirers (whether lexicographers or terminologists) have access to various on-line resources, such as **corpus search**, **look-**

up dictionary, **ontology browser** tools. The acquisition is done semi-automatically; in other words we advocate human intervention to develop the lexicons.

Apart from the tools to help acquiring the data, we have also developed programs to check the semi-automatically acquired data. Using this approach, we have acquired about one-fifth of our lexicon, and have developed a morpho-semantic acquisition program, which has allowed us to expand our core Spanish lexicon (containing 7,000 word senses) to 40,000 word senses, entirely automatically, through the application of lexical rules using a morpho-semantic generator (Viegas *et al.*, 1996).

The acquisition of large-scale computational semantic lexicons is a very time consuming task, and the trade-offs between interlingua and lexical knowledge (or "language-neutral" versus "language-dependent" knowledge) on one hand and, between different senses within a lexicon, is not always easy to determine as we will see in next two sections.

4.3. ONTOLOGY-LEXICON TRADE OFFS

Word meanings in Mikrokosmos are represented in the lexicon, with meanings partly anchored in the ontology. The ontology has been built for NLP purposes and as such its acquisition involved continual trade-offs between the ontology and the lexicon.[15] (Mahesh and Nirenburg, 1996) argued for an intermediary position between a highly minimalist and a highly excessive number of primitives. The question of primitives is often left unspecified in NLP systems. Semantic lexicons are sometimes built by introducing a number of primitives as needed for representing word meanings. Neither the set of primitives, nor the taxonomic or other relationships between the primitives is specified in such systems. The best known example of an extremely minimalist position can be seen in Schank's (1973) Conceptual Dependency theory (CDs), an ontology of 11 events.[16] Such a small number of primitives is not practical for building large scale systems that attempt to capture the full richness of meaning that is necessary for MT or other NLP tasks in a domain. When we attempt to decompose complex events such as "a takeover bid for a company" in CDs, the resulting meaning representations will be lengthy, convoluted, and hard to acquire on a large scale. Moreover, they are unsuitable for MT since it is very hard to generate the equivalent word(s) in a target language from such highly decomposed meaning representations. The other extreme position, a popular one in NLP, makes

[15]Note that it could be possible to develop the ontology to support inferencing if needed.

[16]Other well-known minimalist approaches include Jackendoff's (1990) lexical-conceptual structures (LCS). See also (Dorr 1993), (Wilks, 1992) or (Onyshkevych and Nirenburg, 1994) for criticisms of the minimalist approach to meaning representation.

almost every word sense in a natural language a primitive by itself. Examples of such "ontologies" are WordNet (Miller and Fellbaum, 1991) and Sensus/Pangloss (Knight and Luk, 1994). In this approach, the large set of primitives is necessarily tied to a particular language (English in the above systems), which may be desirable for MT if the target language is always the same (say English).

We take an intermediate approach and propose a set of primitives that is much bigger than CDs or LCSs but significantly smaller than the typical size (of the order of 50,000) of a "word sense taxonomy" such as WordNet (Miller, 1990). Our experience in Mikrokosmos and its predecessor projects shows that fewer than 10,000 primitives are sufficient for building practical MT systems in a nontrivial domain, such as company mergers and acquisitions.[17] For successful, multilingual MT, such a system must be provided with a rich compositional structure in its meaning representations. The 6000-8000 primitives must also be organized in a highly interconnected ontological network. Using such a scheme, we have built a Spanish core lexicon with over 7000 word senses that use less than 2500 primitives in their meaning representations.[18]

A smaller ontology is not only cheaper to acquire, but we can also ensure better quality of concepts and inter-conceptual relations when the size is small. However, a smaller number of concepts necessitates a greater degree of decomposition in meanings in representing word senses in the lexicon. This not only explodes the cost of training and lexical acquisition, it also creates problems in analysis and generation. In Mikrokosmos, we have strived to achieve an intermediate grain size of meaning representation in both the lexicon and the ontology. Many word senses have direct mappings to concepts in the ontology; many others must be decomposed and mapped indirectly through composition and modification of ontological concepts. In a multilingual situation, the set of primitives must not be anchored in any natural language. A more compositional meaning representation with a smaller number of primitives is much more practical for constructing large-scale semantic lexicons for multiple languages.

For example, the English word *acquire* can be considered to be just the beginning phase of the concept for the word *own* so that the two words can be mapped to the same event in the ontology. However, it is often desirable to add separate concepts for the two words even from a strictly ontological perspective. For example, in the domain of company mergers and acqui-

[17]It must be noted that we do not propose to encode only those meanings of words that are in the chosen domain. We in fact encode all meanings of words in a corpus using much less than 10,000 primitives. This number excludes any onomasticon entries needed.

[18]Construction of lexicons for Japanese, Chinese and English using the same set of primitives has also begun recently.

sitions, there may be a need to further classify the concept ACQUIRE into types of acquisitions. This cannot be done if the word *acquire* was mapped to OWN in the ontology with no separate concept for acquisitions. Similarly, certain attributes or relations in the ontology may be applicable only to events involving transfers of possession and hence cannot be applied to the word *own*. There is no algorithm to determine when to decompose a word meaning and when to make it a new concept in the ontology. We have developed a set of guidelines and a training methodology that results in acceptable quality and uniformity in lexical and ontological representations.

Heuristics: In principle, the separation between ontology and lexicon is as follows: "language-neutral" meanings are stored in the former; language-specific information in the latter.

In a multilingual situation, as we saw earlier, it is not always easy to determine the boundary between ontology and lexicon. As a result, ontology and lexicon acquisition involves a process of daily negotiations between the two teams of acquirers.

For instance, if we consider the English verbs *cook* and *bake*, which translate "approximately" into the French equivalents *cuire [+/- sur le feu]* (cook on the fire) and *cuire [+/- au four]* (cook in the oven) respectively, then there are three different ways of doing the mappings, depending on the information available in the ontology:

- One-to-one Mapping between Ontology and Lexicon
- Unspecified Lexicon Entries
- Lexicon Ontology Balance

One-to-one Mapping between Ontology and Lexicon For instance, one could consider that we have two concepts BAKE and COOK, subtypes of PREPARE-FOOD as in Table 3 to which correspond the English verbs *bake* and *cook* and the French expressions *cuire [+/- sur le feu]* and *cuire [+/- au four]* respectively. This solution seems more artificial for French than for English, because the verb *cuire* by itself is not ambiguous in French, it is just unspecified with respect to English.[19]

[19]The issue of actually translating *bake* and *cook* into French involves more than representational issues as discussed in (Viegas, 1997) and (Beale and Viegas, 1996).

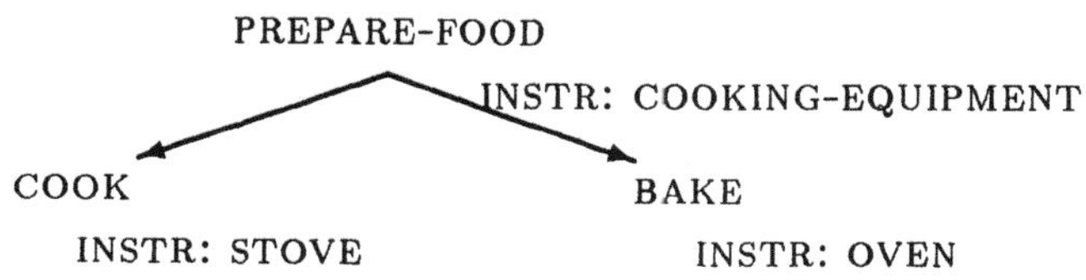

Concepts	COOK	BAKE
English words	*cook*	*bake*
French words	*cuire [+/- sur le feu]*	*cuire [+/- au four]*

TABLE 3. Lexicon Ontology Trade-offs in a Multilingual Environment: one-to-one mapping

This solution is the easiest, so it seems, for lexicographers who do not have to worry about changing the ontological constraints inside the lexicon. Here, the distinctions will be made in the ontology, with COOK and BAKE having STOVE and OVEN as their respective instruments. There are two problems with this solution though, lexical and ontological. First, we do not necessarily want a one-to-one mapping between concepts and lexemes, or in other words, we do not consider that every lexeme in every language constitutes a primitive in the ontology, as we discussed earlier. Second, such a representation introduces an unnecessary computational ambiguity in the French lexicon, as *cuire* by itself is not ambiguous, it is just unspecified with respect to English for the type of cooking involved such as (*cook, bake*).

Unspecified Lexicon Entries Another solution would be to map all the entries to just one concept, labeled PREPARE-FOOD as in Table 4 with different constraints on the INSTRUMENT value of the concept, as shown below:

PREPARE-FOOD
INSTR: COOKING-EQUIPMENT

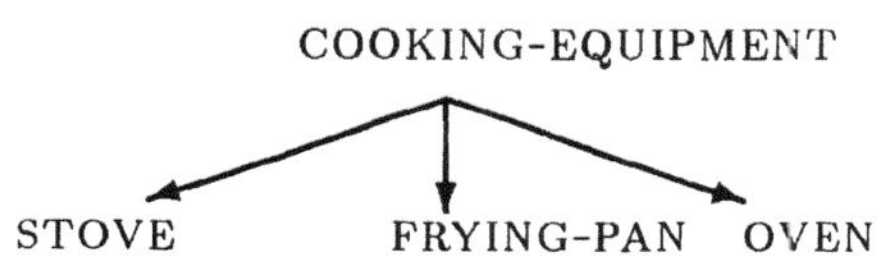

Concepts	PREPARE-FOOD	PREPARE-FOOD
+ instrument		OVEN
English words	*cook*	*bake*
French words	*cuire [+/- sur le feu]*	*cuire [+/- au four]*

TABLE 4. Lexicon Ontology Trade-offs in a Multilingual Environment: Unspecified Lexicon Entries

Note that in this case we do not constrain the instrument of the English *cook* to be of type STOVE, as one can also *cook* with a barbecue for instance. Also note that STOVE and OVEN are subtypes of COOKING-EQUIPMENT.

Although it is possible to have unspecified entries such as in Table 4, it may be desirable to have BAKE as a separate event in the ontology. Without such an event, it is not possible in the ontology to add constraints on instruments of baking, for instance, due to the limited expressiveness of ontological representation in Mikrokosmos, a highly desirable feature from the point of view of ontology acquisition (Mahesh, 1996). As noted earlier, the BAKE node in the ontology is necessary also if there is a need to further classify baking events.

Lexicon-Ontology Balance

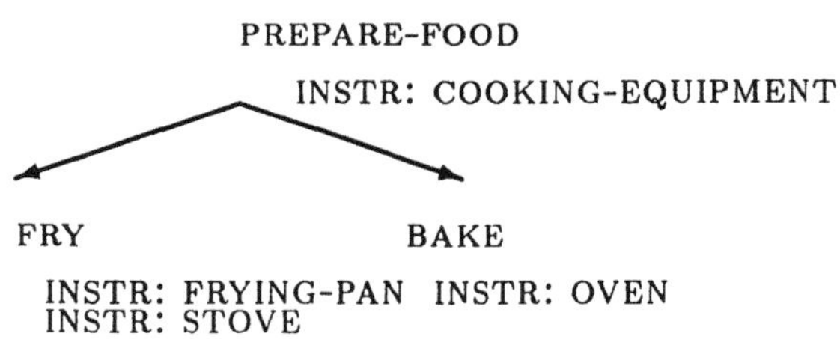

Table 5 presents the best solution from the lexicon-ontology trade-off perspective as it limits unnecessary decomposition in the lexicon, as would have been the case otherwise in Table 4. It also allows the treatment of language gaps, by lessening the gap between the two types of mismatch and divergence for *cuire* [+/- au four], (Viegas, 1997), which was not the case in Table 3. Finally, it acknowledges the fact that the lexical items *fry*, *bake* are more specific than the lexical item *cook*, and that *cuire* by itself is only unspecified with respect to other languages.

Concepts	PREPARE-FOOD	BAKE
+ instrument	COOKING-EQUIPMENT	
English words	*cook*	*bake*
French words	*cuire*	ϕ

TABLE 5. Lexicon Ontology Trade-offs in a Multilingual Environment: Lexicon Ontology Balance

4.4. RECONCILING ACQUIRERS' VIEWPOINTS

We have discussed so far the lexicon-ontology trade-offs, we now turn to the choices a lexicon acquirer has to face when creating a new lexicon entry. In the case of the lexical items discussed here, *cook, bake*, and the French *cuire*, and assuming the ontological knowledge illustrated below,[20]

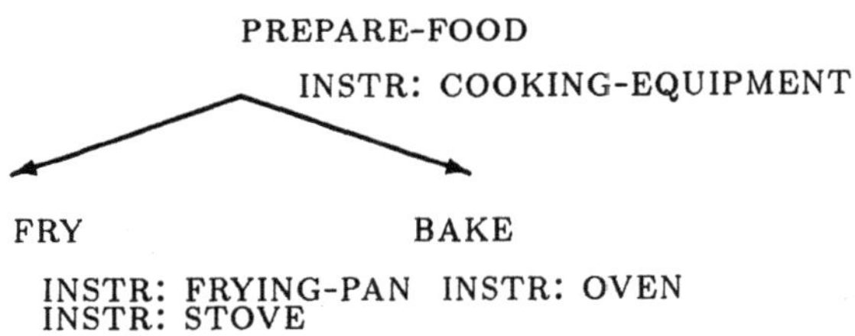

the lexicon acquirer is faced with the two distinct choices:

- the type of ontological mapping, which can be either direct or constrained
- the number of entries to create per lexical item

Our experience shows that different acquirers, who have been trained, will arrive at the same number of entries and/or mapping. What might vary is how vague or unspecified an entry can be. This is due mostly because

[20]In reality and as pointed out earlier, there is a continuous interaction between the developers of ontologies and the developers of lexicons; for expository purposes, we will assume however the ontological information as shown in the last conceptual diagram.

of the presence of a rich well structured and organized ontology,[21] which supports the lexicon.

In the following paragraphs, we discuss the two types of mappings between ontology and lexicon and then we address the issue of semantic ambiguity within a lexicon.

Direct Mapping In a direct mapping (Table 6), the lexicon acquirer, just maps the lexical item to a concept in the ontology, checking that all the features and constraints on these features fit the lexical item.

In our case, this is the solution adopted for English (as described formerly in Table 4), where we have the following mappings:

Concepts	PREPARE-FOOD	BAKE
English words	*cook*	*bake*

TABLE 6. Direct Mapping

Constrained Mapping In a constrained mapping (Table 7), the lexicon acquirer maps the lexical item to a concept where the values of the features attached to the concept have to be constrained.

For instance, in our example, we could constrain the INSTRUMENT of PREPARE-FOOD to be STOVE in the lexicon entry for COOK if *cook* in English was always used with such a restriction, which is actually not the case as you can also cook on a barbecue or grill.

Concepts	PREPARE-FOOD
Instruments	STOVE
English words	*cook*

TABLE 7. Constrained Mapping

[21]On theoretical discussions about ontologies, see (Nirenburg *et al.*, 1995).

Number of Entries Another concern in lexical acquisition is the number of entries to create. If we look at data from a corpus, without relying on some underlying semantics, then we could come up with as many as the entries in Table 8.

Words	Concepts	Examples
cook-V1	PREPARE-FOOD	*Cook the meat rare*
cook-V2	PREPARE-FOOD INSTRUMENT: OVEN	*Cook the pasta au gratin for 35 min. in the oven*
bake-V1	BAKE	*Bake the pasta au gratin 35 min.*
cuire-V1	PREPARE-FOOD	*Cuis les pates al'dente* (cook the pasta al'dente)
cuire-V2	BAKE	*Cuis le gâteau 45 min.* (bake the cake 45 min.)
cuire-V3	PREPARE-FOOD INSTRUMENT: OVEN	*Cuis les pâtes à four moyen.* (bake the cake at medium heat)

TABLE 8. Entries for lexical items.

Table 9 presents our solution where we wrote one sense per mapping and other meanings (such as the BAKE meaning for *cook in the oven*) are created at processing time, in the context of other constraints. Moreover, we use mechanisms such as generalization and specialization, as described in (Mahesh *et al.*, 1997a) to dynamically go from PREPARE-FOOD (along with contextual constraints on the values of the features attached to the concept) to BAKE and vice versa, to account for all the examples given in Table 8 for *cuire*.

cook-V1	PREPARE-FOOD
bake-V1	BAKE
cuire-V1	PREPARE-FOOD

TABLE 9. Trade-off Output

Another issue we address in Mikrokosmos is to take into account the lexicon acquirers' points of view.[22] If we look at the following data and their subcategorizations:

[22]This work is in progress and has not yet been implemented.

(2) I fixed the meal - [NP1, NP2]
(3) I fixed a sandwich for you - [NP1, NP2, PP1]
(4) I fixed you a sandwich - [NP1, NP3, NP2]

where *fix* means PREPARE-FOOD, then our system must allow the analysis and generation of any of sentences (2), (3) and (4), whether we have a mapping of *fix* onto the concept A or on the frames B or C:[23]

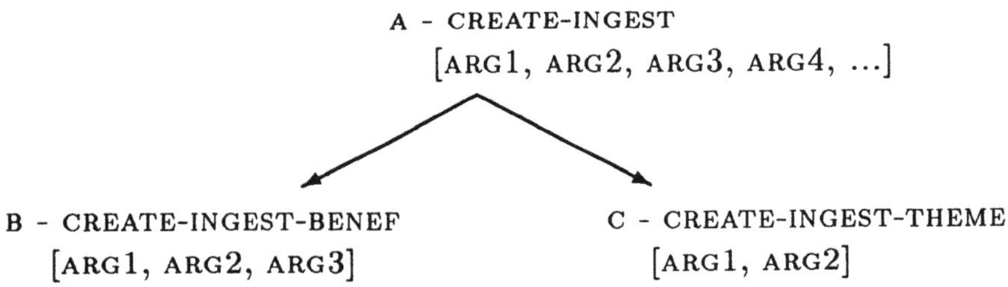

Looking at example (2), and in absence of examples (3) and (4) in the corpus, *fix* could easily be mapped into A or C, whereas with examples (3) and (4), and in absence of example (2) in the corpus, it could be easily mapped into A or B by the acquirers. We claim that this is of no importance as far as we have the mechanisms to go from one to the other. The diagram above is a computational linguist construct and has no "reality" per se; A, which is a concept in the ontology, also belongs to a truly multilingual hierarchy of semantic classes (Cahill and Gazdar, 1996) subsumed by the ontology.[24] B and C are constructs which provide for every semantic class the different semantic patterns that a particular semantic class accepts, such as the pattern CREATE-INGEST-BENEF requires 3 semantic arguments (AGENT, BENEFICIARY and THEME), whereas CREATE-INGEST-THEME only requires 2 semantic arguments. This diagram can be further specified for a particular natural language, where the required arguments are mapped to syntactic arguments and where lexical rules for a particular language provide the link between the different semantic patterns for a semantic class. The diagram below is for English where subcategorizations in between square brackets are associated to the lexical items mapped to A, B and C.

[23]We give the general diagram for CREATE-INGEST events, as PREPARE-FOOD is a subtype and will inherit all the properties of the semantic class CREATE-INGEST.

[24]The difference between the ontology and the multilingual hierarchy of semantic classes, is that there are less nodes in the latter; FRY, BAKE, GRILL are equivalent to the semantic class PREPARE-FOOD, and as such do not appear in the multilingual hierarchy.

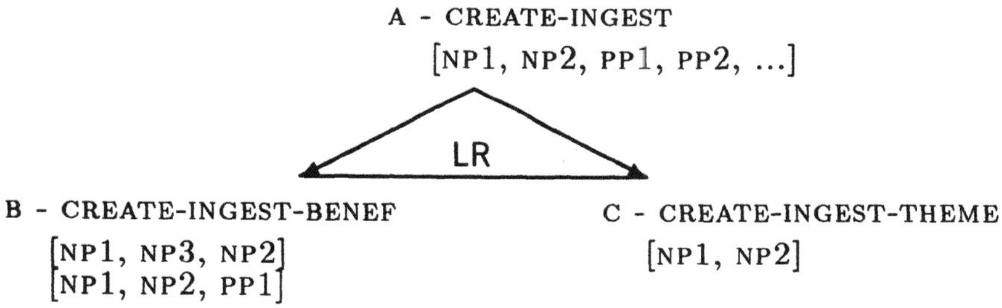

The corpus can indeed influence the way a lexicon acquirer will do the mapping, so if a lexicon acquirer creates an unspecified entry (mapping *fix* on (A), as opposed to (B) or (C)), dynamic mechanisms such as specialization or generalization (Kameyama *et al.*, 1991) would enable the system to get to (B) and (C) from (A) and vice versa (to (A) from (B) or (C)). Moreover, if a lexicon acquirer decides to map to (B) instead of (C) or vice versa, then a lexical rule (LR) between (B) and (C) will enable the system to go from (B) to (C) and vice versa.

In other words, although there are 3 potentially different ways of writing the lexicon entry for *fix* for example sentences (2), (3) and (4), these different ways of encoding *fix* should remain a virtual difference at processing time. In other words, the system must encode mechanisms and rules to reconcile the different points of view of different acquirers, so that the system can treat sentences (2), (3) and (4) from any of the three potential lexicon entries.

In this section, we have outlined the needs for allowing lexicon entries to be dynamically changed to fit different linguistic contexts and different acquirers' analysis of the data. In next section, we compare our approach to the main trends in computational (lexical) semantics.

5. Comparison with Other Major Approaches

The comparison below is partial and incomplete: partial, because it is difficult to compare and do justice to all approaches as they differ in their goals; incomplete, because we highlight from other theories only some of the points addressed in this paper.

The main difference between Mikrokosmos and other approaches is that in Mikrokosmos the lexicon is only one part of the program of studying semantics. Other approaches to computational semantics concentrate on issues connected with economy of lexicon acquisition based on exploiting lexical regularities, such as, for instance, derivational morphology (e.g., Viegas *et al.*, 1996) or standard metonymies (such as, e.g., *creator* for *creation*, as in *Rachmaninov played Chopin*), (cf. (Ostler and Atkins, 1992), (Briscoe and Copestake, 1991), (Copestake and Briscoe, 1992), (Briscoe *et al.*, 1993), (Pustejovsky, 1995)). The main research goal of workers in this approach is to eliminate the need for specifying some word senses altogether and instead have rules for deriving the necessary word senses automatically from the main (and, often, only) sense, which most of the time consists of an underspecified meaning.[25] From the point of view of a processing application, the question is rather when to apply the sense expansion rules. For instance, if the derivational morphology-oriented rules are applied at load time, then the resulting system lexicon includes word senses. If it is done "when needed," the only real difference is in the trade-off between space (to store the expanded system lexicon) against time (to apply lexical rules at need time). An important point to be made is that lexical rules constitute a powerful conceptual tool to generate the semantics of a word on the fly as illustrated in (Viegas *et al.*, 1996). Our approach is different in that we discuss the nature and content of the semantic knowledge sources only in the context of the discussion of the mechanisms of extraction, representation and use of meaning.

We claim that a sense enumeration approach is neither too restrictive nor totally inadequate or inflexible to handle sense disambiguation even for new uses of words in novel contexts. It is indeed true that it is impossible to secure the exhaustive list of meanings for every single word, or complex expression (for instance what about the treatment of metonymies or the whole range of metaphors), as it is true that there is no theory of context available to help constructing meanings on the fly. Our position is to take advantage of most of the information listed in lexicons and look at processing methods to dynamically find new meanings which were not listed in a lexicon entry.

There is not yet a single semantics-based operational system which can avoid some form of enumeration. Whether *dejar* in Spanish has 7 senses, as in (Nirenburg *et al.*, 1995), or the 54 meanings listed in the Collins Spanish-English dictionary; whether *break* has 7 meanings in (Palmer and Wu, 1995) or the 18 in LDOCE, or a single underspecified meaning, as claimed in (Pustejovsky, 1995), should be determined by the existing organization of an ontology or conceptual domain or lexical conceptual paradigms. Meaning underspecification as an alternative to a sense enumeration approach is a very appealing idea, although until we are able to operationally define *contextual meaning* it will remain but an idea. Saturating some underspecified lexical representation by context is still today a very difficult task.

[25]See (Van Deemter and Peters, 1996) for computational treatments on underspecification.

The advantage of underspecified lexicon entries, is that it obviously limits sense enumeration plus avoids spurious ambiguity. The problems with such an approach are worth mentioning. It becomes very difficult to expand a lexicon through lexical rules (which constitute a powerful conceptual tool for rapid acquisition); it requires the availability of a theory of context for computational semantics (not yet available); finally, the underspecified entries often seem very far away from the linguistic data (thus complexifying the task of acquisition). Lexical semantic underspecification is still today difficult to consider for NLP applications, not because it is the wrong way to go, on the contrary, but because as of today, we still lack the mechanisms to produce the specified meaning of a word in context. In that sense, we believe that underspecification of lexical items should not constitute a goal in itself for NLP applications.

Finally, another main difference in our approach to computational lexical semantics is that we try to reconcile computational linguistic theoretical concerns (in terms of understanding the combinatory and productive principles involved in the analysis and generation of natural languages) with NLP concerns (developing efficient non-toy working systems). From this perspective, one should mention that a computational (lexical)-semantic approach to NLP, although feasible (as demonstrated in Section 2), would be considered an overkill, due to high costs in acquisition, for applications such as Information Retrieval, where a statistics based approach would be more cost-effective. The future of computational (lexical)-semantics for NLP applications will rely on the ability to create knowledge at a much lower cost.

6. Conclusion

In this paper, we presented our approach to computational (lexical) semantics, focusing on the dynamics of semantics. From our viewpoint, we must discuss the nature and content of the semantic knowledge sources only in the context of the discussion of the mechanisms of extraction, representation and use of meaning.

We discussed in this context the lexicon ontology trade-offs, or in other words, the language related versus language neutral knowledge. In the case of EVENTs realized as verbs, we advocated a truly semantic approach, with an event hierarchy which can help predict the semantic behavior of a verb, rather than a syntactically-driven classification of verbs, as is usually done.

We stressed that the debate on sense enumeration should remain a legitimate lexicographers' concern, but is not the central concern of theories and applications of semantic processing.

Finally, we want to emphasize that in terms of knowledge acquisition, once we have developed the core lexicon for a natural language, more attention and work should be devoted to the entirely automatic generation of (lexical) semantic data at run time, to overcome some virtual incompleteness due to lexicon acquirer's points of views.

Further research in computational lexical semantics includes investigating how to bypass the closed world assumption under which knowledge-base systems work, i.e. under the simplistic assumption that the knowledge sources are complete.

Acknowledgments:

We would like to thank the Mikrokosmos team for many discussions we had on the research presented here. We are also grateful to the anonymous reviewers for their useful comments. Research reported in this paper was supported in part by Contract MDA904-92-C-5189 from the U.S. Department of Defense.

References

Beale, S., Nirenburg, S. and K. Mahesh (1995). Semantic Analysis in the Mikrokosmos Machine Translation Project. In Proceedings of the *Second Symposium on Natural Language Processing*, Bangkok, Thailand, 297-307.

Beale, S. (1997). *HUNTER-GATHERER: Applying Constraint Satisfaction, Branch-and-Bound and Solution Synthesis to Computational Semantics.* Ph.D. Dissertation, School of Computer Science, Carnegie Mellon University.

Beale, S., Nirenburg, S. and K. Mahesh (1996). HUNTER-GATHERER: Three Search Techniques Integrated for Natural Language Semantics. In the *Thirteenth National Conference on Artificial Intelligence*, Portland, Oregon, 1056-1061.

Beale, S. and E. Viegas (1996). Intelligent Planning meets Intelligent Planners. In Proceedings of the Workshop on *Gaps and Bridges: New Directions in Planning and Natural Language Generation, at ECAI'96*, Budapest, 59-64.

Beale, S., Viegas, E. and S. Nirenburg (1997). Breaking Down Barriers: The Mikrokosmos Generator. In Proceedings of *Natural Language Processing Pacific Rim Symposium*, Phuket, Thailand, 141-146.

Bresnan, J. (Ed.) (1982). *The Mental Representation of Grammatical Relations*. Cambridge MA: MIT Press.

Briscoe, T. and A. Copestake (1991). Sense extensions as lexical rules. In *Proceedings of the IJCAI Workshop on Computational Approaches to Non-Literal Language.* Sydney, Australia, 12-20.

Briscoe, T., de Paiva, V., and A. Copestake (eds.) (1993). *Inheritance, Defaults, and the Lexicon.* Cambridge: Cambridge University Press.

Briscoe, T., Copestake, A. and A. Lascarides (1995). Blocking. In P. Saint-Dizier and E. Viegas (Eds.), *Computational Lexical Semantics.* Cambridge University Press, 273-302.

Cahill, L. and G. Gazdar (1996). Multilingual Lexicons for Related Lexicons. In Proceedings of *AISB'96 Workshop on Multilinguality in the Lexicon*, Brighton, UK, 69-75.

Copestake, A. and T. Briscoe (1992). Lexical operations in a unification-based framework In J. Pustejovsky and S. Bergler (eds), *Lexical Semantics and Knowledge Representation.* Berlin: Springer, 101-119.

Cruse, D.A. (1986). *Lexical Semantics.* Cambridge Texbooks in Linguistics.

Cruse, D.A. (1995) Polysemy and Related Phenomena. In St-Dizier P. and E., Viegas (eds), *Computational Lexical Semantics.* Cambridge University Press.

Dorr, B.J. (1993). *Machine Translation: A View from the Lexicon,* MIT Press, Cambridge, MA.

Dorr, B., and J. Klavans (eds.) (1994/1995). Special Issue: Building Lexicons for Machine Translation I. *Machine Translation 9: 3-4.*

Dorr, B., and J. Klavans (eds.) (1995). Special Issue: Building Lexicons for Machine Translation II. *Machine Translation 10: 1-2.*

Farwell, D., Helmreich, S., Jin, W., Casper, M., Hargrave, J., Molina-Salgado, H. and F. Weng (1994). PANGLYZER: Spanish Language Analysis System. In Proceedings of *the 1st Conference of the AMTA,* 56-64.

Fillmore, C.J. (1985). Frames and the Semantics of Understanding. *Quaderni de Semantica,* VI.2.

Fillmore, C.J. (1993). Frame Semantics and Perception Verbs, contribution to the Daghstuhl Seminar on Universals in the Lexicon, in: Hans Kamp, James Pustejovsky (eds.), *Universals in the Lexicon: At the Intersection of Lexical Semantic Theories,* 1993, ms, Daghstuhl.

Freuder, E.C. (1978). Synthesizing Constraint Expressions. in *Communications ACM* 21(11): 958-966.

Grimes, J.E. (1968). *The Thread of Discourse.* ERIC Document ED-019669.

Gross, M. (1984). Lexicon-Grammar and the Syntactic Analysis of French. In Proceedings of *the 10th International Conference on Computational Linguistics,* Stanford, Ca.

Jackendoff, R. (1990). *Semantic Structures.* Cambridge, MA: MIT Press.

Kameyama M., Ochitani, R. and S. Peters (1991). Resolving Translation Mismatches With Information Flow. In Proceedings of the *29th Annual Meeting for the Association for Computational Linguists,* 193-200.

Knight, K. and S.K. Luk (1994). Building a Large-Scale Knowledge Base for Machine Translation. In Proceedings of the *Twelfth National Conference on Artificial Intelligence.*

Lenat, D., M. Prakash and M. Shepherd (1986). CYC: Using Common Sense Knowledge to Overcome Brittleness and Knowledge Acquisition Bottlenecks. *AI Magazine, VI.*

Lenat, D. B. and R.V. Guha (1990). *Building Large Knowledge-Based Systems.* Reading, MA: Addison-Wesley.

Levin, B. (1993). *English Verb Classes and Alternations.* The University of Chicago Press.

Luger, G.F. and W.A. Stubblefield (1992). Rule based expert systems. In *Computer Engineering Handbook,* C.H. Chen, ed. New York, NY: McGraw-Hill.

Mahesh, K. (1996). *Ontology Development for Machine Translation: Ideology and Methodology.* Memoranda in Computer and Cognitive Science, MCCS-96-292. Las Cruces, N.M.: New Mexico State University.

Mahesh, K. and S. Nirenburg (1995). A situated ontology for practical NLP. In the Proceedings of *IJCAI'95* Workshop on Basic Ontological Issues in Knowledge Sharing. Montréal.

Mahesh, K., Nirenburg, S., Cowie, J. and D. Farwell (1996). *An Assessment of CyC for Natural Language.* Memoranda in Computer and Cognitive Science, MCCS-96-302. Las Cruces, N.M.: New Mexico State University.

Mahesh, K., Nirenburg, S. and S. Beale (1997a). If You Have It, Flaunt It: Using Full Ontological Knowledge for Word Sense Disambiguation. In Proceedings of *the 7th International Conference on Theoretical and Methodological Issues,* 1-9.

Mahesh, K., Nirenburg, S., Beale, S., Viegas, E., Raskin, V. and B. Onyshkevych (1997b). Word sense disambiguation: why statistics when we have these numbers? In Proceedings of *the 7th International Conference on Theoretical and Methodological Issues in Machine Translation,* Santa Fe, 151-159.

McNaught, J. (1990). Reusability of Lexical and Terminological Resources; Steps towards Independence.

Meyer, I., Onyshkevych, B. and L. Carlson (1990). *Lexicographic principles and design for*

knowledge- based machine translation. Technical Report CMU-CMT-90-118, Center for Machine Translation, Carnegie Melon University, Pittsburgh, PA.
Miller, G. A. (1990). nouns in WordNet: A Lexical Inheritance System. International Journal of Lexicography 3.4, 245-264.
Miller, G.A. and C. Fellbaum (1991). Semantic Networks of English. *Cognition 41*, 197-229.
Minsky, M. (1968). *Semantic Information Processing.* MIT Press, Cambridge, MA.
Nirenburg, S., Raskin, V. and B. Onyshkevych (1995). Apologiae Ontologia. In Proceedings of the *International Conference on Theoretical and Methodological Issues*, Leuven, Belgium.
Nirenburg, S. and C. Defrise (1991). Application-oriented computational semantics. In Rosner and Johnson (eds.) *Computational linguistics and formal semantics*, 223-256.
Nirenburg, S., Mahesh, K. and S. Beale (1996). Measuring semantic coverage. In Proceedings of *the 16th International Conference on Computational Linguistics*, Copenhagen, 83-88.
Nirenburg, S. and V. Raskin (1987). The Subworld Concept Lexicon and the Lexicon Management System. In *Computational Linguistics*, Volume 13, Numbers 3-4, 276-289.
Nirenburg, S., Beale, S., Helmreich, S., Mahesh, K., Viegas, E. and R. Zajac (1996). Two principles and six techniques for rapid MT development. In Proceedings of *the Second Conference of the Association for Machine Translation in the Americas*, Montréal, Canada, 96-105.
Nirenburg, S. and V. Raskin (1996). *Ten Choices for Lexical Semantics. Memoranda in Computer and Cognitive Science*, MCCS-96-304. Las Cruces, N.M.: New Mexico State University.
Normier, B. and M. Nossin (1990). GENELEX Project: EUREKA for Linguistic Engineering. In *Proceedings of the international Workshop on Electronic Dictionaries*, OISO, Kanagawa, Japan.
Onyshkevych, B. and S. Nirenburg (1994). *The lexicon in the scheme of KBMT things.* Memoranda in Computer and Cognitive Science, MCCS-94-277. Las Cruces, N.M.: New Mexico State University. Reprinted as: A lexicon for knowledge-based machine translation, in: Dorr and Klavans 1995 (eds), 5-57.
Onyshkevych, B. (1997). *An Ontological-Semantic Framework for Text Analysis.* Ph.D. Dissertation, School of Computer Science, Carnegie Mellon University.
Ostler, N. and B.T.S. Atkins (1992). Predictable meaning shift: Some linguistic properties of lexical implication rules. In J. Pustejovsky and S. Bergler (eds), *Lexical Semantics and Knowledge Representation.* Berlin: Springer, 87-100.
Palmer, M. and Z. Wu (1995). Verb Semantics for English-Chinese Translation. *Machine Translation*, Volume 10, Nos 1-2, 59-92.
Pollard, C. and I. Sag (1987). *An Information-based Approach to Syntax and Semantics: Volume 1 Fundamentals.* CSLI Lecture Notes 13, Stanford CA.
Pustejovsky, J. and P. Anick (1988). On the Semantic Interpretation of Nominals. In *Coling 1988*, vol.2: 518-523.
Pustejovsky, J. (1995). *The Generative Lexicon.* Cambridge, MA: MIT Press.
Sanfilippo, A. (Ed.) (1992). *The (Other) Cambridge Acquilex Papers.* Technical Report No. 253. Computer Laboratory, University of Cambridge, New Museums Site, Pembrocke Street, Cambridge.
Schank, R. (1973). Identification of conceptualizations underlying natural language. In *Computer Models of Thought and Language*, Schank, R. and Colby, K., editors, San Francisco: W. H. Freeman Co.
Searle, J.R. (1969). *Speech Acts. An Essay in the Philosophy of Language..* Cambridge University Press.
Tsang, E.P.K. (1993). *Foundations of Constraint Satisfaction.* London: Academic Press.
Van Deemter, K. and S. Peters (1996). *Semantic Ambiguity and Underspecification.* CSLI Publications.

Viegas, E. and S. Nirenburg (1996). The Ecology of Lexical Acquisition: Computational Lexicon Making Process. In Proceedings of *Euralex96*, Göteberg University, Sweden, 489-496.

Viegas, E. and S. Beale (1996). Multilinguality and Reversibility in Computational Semantic Lexicons. In Proceedings of the 8th *International Natural Language Generation Workshop*, Sussex, England, 49-52.

Viegas, E., Onyshkevych, B., Raskin, V., and S. Nirenburg (1996). From *Submit* to *Submitted* via *Submission*: on Lexical Rules in Large-scale Lexicon Acquisition. In Proceedings of *the 34th Association for Computational Linguistics*, University of California, Santa-Cruz, California, 32-39.

Viegas, E. and V. Raskin (1998). *Computational Semantic Lexicon Acquisition: Methodology and Guidelines.* Memoranda in Computer and Cognitive Science. MCCS-98-315. New Mexico State University: Computing Research Laboratory.

Viegas, E. (1997). Mismatches and Divergences: the Continuum Perspective. In Proceedings of *the 7th International Conference on Theoretical and Methodological Issues*, Santa Fe, 216-224.

Yarowsky, D. (1992). Word sense disambiguation using statistical models of Roget's categories trained on large corpora. In *Proceedings of the 14th International Conference on Computational Linguistics*, Nantes, France, 454-460.

Weinreich, U. (1964). Webster's Third: A Critique of its Semantics. In *International Journal of American Linguistics* 30: 405-409.

Wilks, Y. (1992). Review of Ray Jackendoff's *Semantic Structures, Computational Linguistics*, vol. 18:1, 95-97.

Wilks, Y., Slator, B. and L. Guthrie (1995). *Electric Words: Dictionaries, Computers and Meanings.* Cambridge, MA: MIT Press.

Wilks, Y. and M. Stevenson (1997). Sense Tagging: Semantic Tagging with a Lexicon. In Proceedings of the SIGLEX workshop *Tagging Text with Lexical Semantics: Why, What, and How?*, Washington DC.

CORPUS-BASED ARGUMENT IDENTIFICATION USING A STATISTICALLY ENRICHED VALENCY MRD

DIMITRIOS KOKKINAKIS
Språkdata/Dept. of Swedish Language
Göteborg University
SE 405 20
Sweden

Abstract. This chapter describes a system for automatically acquiring subcategorization information for Swedish verbs from corpora by using available lexical resources in order to reduce the effects of the knowledge acquisition bottleneck. I emphasize the need of such knowledge and focus on the automatic validation and qualitative and quantitative completion of the content of these resources, as far as that is possible. The aim of this work is to maintain a complete knowledge-base, which can be easily updated and extended with minimal human supervision. This knowledge is of vital importance in a broad range of applications, such as lexically specified grammars, e.g. HPSG, parsing and language learning.

1. Introduction

Nearly all contemporary syntactic theories and grammars incorporate large quantities of information in the lexicon, reflecting a shift of focus in the area of computational linguistics dealing with grammar creation to a strongly lexicalized view of it. Parsing and generation systems make extensive use of valency or subcategorization information in order to constrain the number of analyses or generations that are produced. Dictionaries for native speakers contain a whole range of grammatical information, helpful not only to native but to non-native users of the language as well. Therefore, this type of information, regardless of how it is labelled, is the key component required by all the above mentioned applications. One way or another, applications such as these, have to deal with acquiring this information and organiz-

P. Saint-Dizier (ed.), Predicative Forms in Natural Language and in Lexical Knowledge Bases, 205–228.

ing it for the satisfaction of individual needs, e.g. lexicography, parsing, generation and language learning.

In this chapter, a description of an implemented system for the automatic acquisition of arguments for some common verbs from labelled texts will be presented. The acquisition is carried out with the help of a valency database containing subcategorization information. The content of the database is verb lemmas with associated argument frames, plus cues that lead a matching algorithm towards the identification of the verbal arguments. The labels on the arguments should be seen as a way to define surface oriented, explicit relations between sentential components. With the help of a matching algorithm, and a simple regular expression grammar, the valency information found in the database can be semi-automatically validated with respect to a corpus and can be verified, rejected or updated with empirical results. The information in the database has been automatically acquired by using information taken from a verb valency machine readable dictionary (VMRD) and the Gothenburg Lexical Database (GLDB). In turn, these two sources of information have been manually created, and to a certain extent verified relative to authentic corpus texts. The system has been tested on material from the Stockholm-Umeå Corpus (SUC) and newspaper articles downloaded from the Internet.

The author's intention in building the system presented is a step towards the construction and rapid development of a lexically specified, wide coverage Swedish, comprising of, among other things, valency information that had to be verified and completed on authentic corpus material prior to the development of the grammatical model.

2. Background

2.1. THE NEED AND IMPORTANCE

An overall trend in NLP is to incorporate large quantities of information in the lexicon and at the same time gain access to large quantities of empirical data (corpora) in order to support or reject linguistic introspection. This is a task that started to become more and more demanding and challenging with the rise of complex lexically specified grammars, which require the lexicon to become much broader syntactically (cf Briscoe *et al.* (1994), Hajicova 1994a). Nearly all the modern syntactic theories and grammars, 'project' syntax and semantics from the lexicon (cf Pollard *et al.* 1994). Learner's dictionaries contain grammatical information, syntactic patterns and structural information for the major word classes. Dictionaries for native speakers contain a whole range of grammatical information, useful not only to native but also non-native speakers of the language as well. This is also a trend that has been strengthened by the work of Sinclair and his colleagues, Sinclair (1987). Text generation, text understanding and parsing systems require subcategorization information. This is an almost indispensable requirement for transfer-based Machine Translation systems. Rule-based parsers need a constraint mechanism for reducing the parsing results - this can mainly be accomplished by subcategorization information.

The usefulness and importance of a structured and tractable way to represent and store subcategorization information of verbs, as well as other word classes, has been the focus of a number of language engineering projects funded by the European Union. For instance, large scale multilingual projects, such as the LE2-4017 PAROLE (*Preparatory Action for Linguistic Resources Organisation for Language Engineering*); LE1-2111 SPARKLE (*Shallow PARsing and Knowledge extraction for Language Engineering*) and ECRAN (*Extraction of Content: Research At Near-market*), are aiming, among other things, at the standardization, automatic acquisition and development of subcategorization lexica for a large number of European languages, thus emphasizing the importance for this information in language R&D. More information about these projects can be found at the following internet address: `http://www.anite-systems.lu/langeng/en/le1`.

For the purpose of training stochastic parsers and providing a consistent database for use in linguistic research there is also a large scale work attempting to annotate the Penn Treebank with predicate-argument structure: Marcus *et al.* (1994). The desired level of annotation proposed would make explicit at least the logical subject and object of the verb.

2.2. THE ACQUISITION

The problem of acquiring this type of syntactic knowledge and providing it for systems and human readers motivated NLP researchers to try to extract the knowledge from the text themselves and thus create large valency lexicons either manually, semi-automatically or automatically. Likewise, researchers are developing methods for acquiring that knowledge from existing MRDs, such as the Oxford Advanced Learner's Dictionary (cf Hajicova *et al.* 1994b), or the Longman Dictionary of Contemporary English and the Longman Lexicon of Contemporary English (cf Sanfilippo 1996).

The reason for using corpora in valency research is evident: the methodological basis is that inherent criteria of the language are most capable of assuring a consistent description (cf Daugaard *et al.* 1992). Intuition or introspection are not enough for identifying all valencies of verbs, especially when new senses, syntactic structures and constraints steadily emerge in different sublanguage domains. For instance, Daugaard *et al.* use authentic texts in order to elucidate valency patterns for the purpose of creating a large dictionary for Danish. Their first attempts were carried out using a limited, closed class lexicon for parsing the corpora. They report higher successful analyses figures for mono- and divalent verbs, 60-80%, and lower correct analyses rates, 50%, for divalent verbs with a noun phrase as the second argument. This idea was abandoned due to the poor results, and hence, they used an MRD instead, reporting *'sufficiently promising'* results.

The reason for using MRDs in valency research is also evident. It is often the case that the required knowledge for NLP systems is present in such dictionaries, but in an unstructured way, unsuitable for immediate use in NLP applications (Ide *et al.* 1995).

Adriaens *et al.* (1992) discuss the conversion of a valency dictionary to a format suitable for a large-scale Machine Translation system, METAL. The acquisition of the lexicon in METAL, as in other large NLP applications, is a major bottleneck and Adriaens *et al.* convert existing MRDs to the frame representation in METAL, pointing out the importance of subcategorization/valency lexicons for the reading discriminations of sentences during analysis and for disambiguation purposes.

One of the best described and documented works on subcategorization acquisition is the work of Brent (1991, 1993, 1994), which deals with the acquisition of valency information from unlabelled texts. Brent uses simple heuristic cues, referred to as *observations*, like pronouns, proper names, conjunctions, punctuation markers, as well as tokens ending in "-ing" and not preceeded by a determiner or preposition, as indicators for identification of verbs.

Although Brent's method distinguishes only six different types of complement types for subcategorization frames, (infinitive and tensed clauses, NP, NP followed by NP, NP followed by an infinitive clause, and NP followed by a tensed clause), his results are very promising. It is unclear, though, how other possible types of complements could be distinguished, except prepositional ones, since prepositions can also be incorporated into the system as cues.

Manning (1993) extends Brent's ideas by acquiring more patterns from corpora and including a statistical component to filter the result from a finite state parser, which in turn runs on the output of a stochastic tagger. The parser scans a text until it reaches a verb and then parses complements following the verb until something is recognized as a terminator of a subcategorized argument. The parser does not distinguish between arguments and adjuncts. The statistical filtering attempts to determine whether or not one can be confident that a frame found by the parser is a correct frame or not. The filtering is based on binomial frequency data. This means that a verb is shown to have a frame `+F` by assuming that it has `-F` and then showing that if this was the case, the observed pattern of co-occurrence of the verb with cues for `F` would be extermely unlikely; this method is described in detail in Brent (1993). Manning finally distinguishes 19 classes of complement types. The presented results, compared with published dictionaries, give a rate of precision of 90% and a recall rate of 43%.

Briscoe *et al.* (1994) present a system for acquiring argument structure using six main components, i.e. a part-of-speech tagger, a lemmatizer, a tag sequence grammar, a chart parser, a pattern-set extractor and an evaluator. Their preliminary results, using 14 common verbs which are compared with manually created MRDs containing valency information, are that the recall rate of the manually created dictionaries is by no means perfect, while the precision is considerably better than their corpus system. They suggest that with some modifications of their system and more corpus data, an acceptable performance can be accomplished.

Aone *et al.* (1996), acquire predicate-argument information from multilingual texts, using a knowledge base with verb senses. Each verb, in their lexicon, encodes three things: the default predicate-argument type, i.e. default and prohibited thematic roles and number of permitted NP or S arguments; any word-specific idiosyncrasies, i.e. typical pre/postpositions; and finally semantic concepts or frames. All the information is acquired automatically from texts using statistics and heuristic algorithms. Their application areas are machine translation and data extraction. No performance rates are given.

Although this work is primarily considering automatic or semi-automatic methods to lexical acquisition, we should not ignore the fact that manual methods are still the most frequently used, (cf Järborg (1990) for Swedish texts).

3. Resource Reuse

In this article I present a system for the automatic acquisition of subcategorization information, based on a combination of available, and sometimes incomplete, lexical resources. The motivation for using such resources has been the need to drastically reduce the effects of the knowledge acquisition bottleneck. At the same time I try to build on previous research at the department on the topic of (manual) valency acquisition, this time by employing automatic means.

Although some researchers have emphasized the need of acquiring information from raw texts, without the *a priori* application of grammatical knowledge, we cannot ignore the fact that a whole range of lexical and textual resources as well as software are becoming more and more available to the linguistic community. These tools serve the purpose of making life easier for language researchers, although sometimes these resources might push the problem of knowledge acquisition and rapid development of NLP tools back a bit. This is due to inconsistencies, incompatibilities and difficulties in integrating different tools and consequently mapping the resources to the tools or to each other. Nevertheless, it is obvious that a detailed and accurate description of a complex linguistic problem requires all possible knowledge and expertise that is available.

4. System Components

The most important components of the presented system are: (i) valency information which has been semi-automatically acquired from the GLDB and an existing VMRD, incomplete in coverage; and (ii) a grammar, built on the simple, compact and powerful regular expression technique. The valency information has been structured as a database. A large subset of its content has been already tested and updated using large corpora, annotated with detailed morphosyntactic information, preprocessed by a rule-based tagger and a lemmatizer, for easy identification and database look-up of the verb lemmas.

A summary of the components of the overall system architecture is given below. The components are:

1. a rule-based part-of-speech tagger;
2. a (verb) lemmatizer;
3. a valency database containing:
   (a) verb lemmas with their lexemes, i.e. their different senses;
   (b) associated valency frames for every lexeme;
   (c) strong and weak cues;
4. a regular expression grammar.

4.1. TAGGING AND LEMMATIZATION

The rule-based tagger used has been implemented by Eric Brill (1995), and trained on the Swedish manually post-edited SUC corpus, Ejerhed *et al.* (1992), by Johansson-Kokkinakis *et al.* (1996). The morphosyntactic annotation scheme used in the tagset follows the EAGLES recommendations (Leech *et al.* 1996). The tagger achieves over 96% accuracy on the part-of-speech assignment in previously unseen texts, and over 93% correct assignment of part-of-speech tags for unknown words found in these texts. The tagger uses a relatively small lexicon of 50000 entries and returns only one tag per token.

The lemmatizer uses a full-form lexicon and is described in Kokkinakis *et al.* (1997). The lemmatizer and the tagger use the same tagset. This makes it possible to disambiguate the results obtained by the lemmatizer with the help of the result returned by the tagger.

Both tools benefit from a rather advanced tokenization process, and thus can recognize multiword expressions, such as continuous and discontinuous phrasal verbs, and lexicalized multiword adverbials, as one-token sequences. This is of vital importance for the good performance of any system processing Swedish, since phrasal verbs in Swedish are frequent. For instance *se ned på*, literally: 'see down att', i.e. 'despise', must be analyzed as one token; the same principle applies to discontinuous phrasal verbs, such as *bry sig inte ens om*, literally 'care not at all about', i.e. 'doesn't care', which consists of the phrasal verb *bry sig om* 'care/worry about' and the adverbs *inte ens* 'not at all'.

Lexicalized multiword adverbials are also very common and without the proper identification of them the error rates in all levels of the linguistic analyses increase dramatically. Consider for instance the adverb *i går*: 'yesterday' where its individual components are the preposition *i* literally 'in' and the homograph *går* which may be recognized either as the uninflected noun lemma *går* or as the present tense of the verb lemma *gå* literally 'to go'.

Apart from the tagger, simple statistics, collected from the manually inspected SUC corpus, are used for disambiguation. These are statistics for the most frequent homograph function words in Swedish, such as prepositions, articles, adverbs and pronouns.

The morphosyntactic tagset contains 13 part-of-speech categories; these are the conventional ones such as verbs, nouns, adverbs, adjectives etc. as well as punctuation as a category of its own.

As an example of that scheme consider the annotated phrase:
I_går/R greps/VMISP han/PPU efter/S att/CI ha/VMN0A
sålt/VMU0A amfetamin/NCNSNI från/S en/DIUS0 korvkiosk-
/NCUSNI -/F bland_annat/R till/S minderåriga/A0P0PN0 ./F
Literally: 'He was arrested yesterday, after he had sold amfetamin from a hot-dog stand - to among other minors.'

I do not intend to describe the annotation shown in the above examples in detail. A very thorough description is given in the EAGLES project information site: (`http://www.ilc.pi.cnr.it/EAGLES96/`). The important thing to notice here is that in position 1 in the tag `V` stands for verb, (`VA` auxiliary, `VM` main), `P` pronoun, `R` adverb, `D` article, `A` adjective, `N` noun, `S` preposition and `F` punctuation. `CI` is the infinitive marker and `0`, i.e. zero, denotes underspecified information. In the example sentences that follow, I will only use the first position of each tag for reasons of clarity.

4.2. VALENCY DATABASE

The information in the valency database consists of verb lemmas, dictionary forms, subdivided in the case of polysemous verbs to lexemes, i.e. different senses, denoted with a lexeme number. The lemma-lexeme model as applied in the database is described in Järborg (1988).

The database has been constructed by automatically merging a VMRD consisting of 1521 verb lemmas, 2661 lexemes and 1130 different types of valencies, described in Toporowska-Gronostaj (1991) and valency information taken from the GLDB, which covers 7236 verb lemmas, 9445 lexemes and 1839 different types of valencies. The final average complexity of the valency patterns for the verbs in the database is 2.3 valency frames per verb lemma.

Both resources have been semi-automatically converted to a uniform format. This format is now suitable for various programming tasks, such as parsing, as well as a bootstrapping mechanism for the acquisition of valency information for lexicographic purposes from new domains.

Furthermore, generation and language learning are two other possible extensions where the content of the database can be applied.

Table (1) illustrates some examples of the content of these two resources before their merging to the new uniform format.

Table 1: The original content of the valency resources

Res.	Verb	Sense	Valency
GLDB	bläddra	1/1	(i/igenom ngt)
	flyga	1/1	(ngn/ngt) (ADVL)
	flyga	1/2	på ngn, & av ngt, & upp/ut (ADVL), & (ur ngn/ngt)
	trycka	1/1	(in/på/upp/ut/ner) ngn/ngt (mot/på) ngt, & på ngt/att+SATS
	trycka	1/2	ngn
	trycka	1/3	ngt
	trycka	1/4	(ADVL)
	växa	1/1	(fram/in/till/upp)
	växa	1/2	ADVL
	växa ifrån	1/1	ngn/ngt
VMRD	bläddra	vb.1	A SS, PO:i/1#3 ;;;~ fram och tillbaka;;;
	flyga	vb.1	A SS
	flyga	vb.2	i ;;;~ av, ~ fram, ~ på (ngn);;;
	trycka	vb.1	A SS, OO, FO#3
	trycka	vb.2	A SS, OO
	trycka	vb.3	P SS, OO, LO/SO#4
	trycka	vb.4	A SS
	växa	vb.1	I SS ;;;~upp;;;
	växa	vb.2	A SS, LO#2
	N/A	N/A	N/A

The verbs shown in the examples are: *bläddra* (1 sense): 'to browse through/skim'; *flyga* (2 senses): 'to fly'; *trycka* (4 senses): 'to press/ squeeze'; *växa* (2 senses): 'to grow'; and *växa ifrån* (1 sense): 'grow out of', a verb+ particle construction.[1]

In the notation of the GLDB in table (1), 1/1, 1/2, 1/3 and 1/4 denote the lemma and lexeme (sense) number of the verb in question; ADVL stands for adverbial; att+SATS denote a clause headed by the subordinate conjunction att; ngt denotes animal, thing or collective as the object of the verb; ngn denotes human as the object of the verb and finally "&" refers to a shorthanded denotation of the verbal entry.

In the VMRD vb.1, vb.2, vb.3, vb.4 denote the combined lemma and lexeme number of the verb in question; SS denotes subject; PO prepositional object, (PO:i denotes the most frequent preposition used as the head of the prepositional object, in this case *i*); LO locational object; SO manner object; FO transition object; A denotes active form; P passive form; I infinitive form; i infinitive construction and "∼" refers to the verbal entry.

After the semi-automatic conversion of the above valency features, a total number of 19 different types of arguments could be distinguished within the two resources. Since there is no standard and acceptable categorization of verbal arguments, it is hard to claim that these agree in number and content with any known classification scheme. Nevertheless, these arguments were kept and might be refined in the future.

The valency frames for each verb vary between 1 and 23 for some common polysemous verbs with great variation in valency. Most of the discrepancies between the number of lexemes and the number of valency frames can be explained by reference to the fact that many lexemes also incorporate lexicalized sub-senses, which must often be assigned separate valency frames. Thus in the most valency-rich verbs, for instance, such as *ta* (16 senses): 'to take' and *hålla* (11 senses): 'to hold/keep', 23 different valencies could be distinguished, while for the verbs *dra* (9 senses): 'to pull/stretch' or 'draw' and *se* (14 senses): 'to see/look' or 'notice' there were 18 valencies distinguished.

The distinguished argument types with associated examples are listed in table (2) below. The argument type referred to as CCLA in table (2) captures a whole range of obligatory sentential parts, usually realized as adjectival or noun phrases, not appropriate for categorization with the rest of the argument types listed.

[1]In contrast, consider the WordNet (Miller *et al.* 1993) representation of the English counterpart of the verb *bläddra*: 'browse'. 'browse' is described having 4 senses and 7 sentence frames: i)(shop, browse) sbd -s & sbd -s PP, ii)(crop, browse, graze, range, pasture) sth -s, iii)(browse, search randomly) sbd -s & sbd -s sth, iv)(browse, graze) sbd -s & sbd -s PP, where 'sth' stands for something, 'sbd' for somebody and 'PP' for prepositional phrase.

Table 2: The arguments used in the database

Label	Description	Example
ARGO	Direct object	De åtalade riskerar *livstids fängelse.*
ARGP	Prepositional object	A&B avstod *från deltagande.*
ARGL	Locational object	Ensemblen ska även turnera *i Norrland.*
ARGR	Reflexive object	Den befattar *sig* inte alls med kriget,
ARGI	Indirect object	De ska visa *göteborgslirare* hur man ...
ARGV	Verbal object	... men hon hade lovat *att komma*
ARGK	Comparison object	vill synas *gentemot våra medlemmar.*
ARGF	Transition: FROM	... rekrytera allmänläkare *från Norrland*
ARGT	Transition: TO	Diagnosen kan dröja *ett par veckor.*
ARGM	Measure object	Man kan tjäna *flera tusen per år (minst).*
ARGS	Manner object	folk försörjer sig *på stölder*
SUBJ	Subject	*Jag* ska spara ihop pengarna.
SUBO	Impersonal Subject	*Det* regnade en skvätt också.
SUBS	Subject in plural	*42 procent* grälade aldrig med sina fäder.
PREDS	Subject Predicative	*Reglerna* är absurda, förklarade hon.
PREDO	Object Predicative	X betecknade nederlaget *som smärtsamt*
IFS	Interrogative Sentence	Sahlin spelade inte *när IFK förlorade.*
INFO	Infinitive Construction	De ansåg oss *vara en progressiv region.*
CCLA	Complement Clause	Övriga kan påstå *vad de vill*
MOD	Modifier	han flög *för 10 år sen, den 1a maj,*

Apart from the 19 types discussed previously, one more should be added, namely modifier, **MOD**, which is a label motivated by the need of having a category that captures a whole range of non-obligatory or "non-essential" sentential parts, usually realized as adverbial phrases. This category is equivalent to the adjunct category, and is used as a default category when no other argument labels can be assigned in a construction in a text under analysis.

4.3. STATISTICS AND 'CUES'

The strong cues listed in the database are function words, like adverbs, fixed expressions, syntactic markers and prepositions with which verbs tend to appear very often. The weak cues, used only when the strong cues have failed for each frame, consist of certain function words, short characteristic and frequent expressions, different types of adverbials marked in the database as **MOD** and syntactic markers, such as pronouns, often appearing in the immediate context of a particular verb. The decision on which words can appear in the 'strong' and 'weak' cue slots in the valency database is made partly by calculating the mutual information statistics, (MI), for the different verbs, from large corpora (Church *et al.* 1991). A sample is given in table (3).

Table 3: MI statistics for the verbs *bläddra* (82 occ.), *flyga* (247 occ.), *trycka* (48 occ.), *växa* (546 occ.) and *växa ifrån* (21 occ.) right context +1 word, in a 2,9 million word corpus.

(x)	(y)	f(y)	f(xy)	MI(x,y)
bläddra	igenom	422	7	9,18
bläddra	bland	733	6	8,16
bläddra	fram	3311	4	5,40
bläddra	i	49762	34	4,57
flyga	förbi	568	5	6,67
flyga	upp	6657	29	5,66
flyga	över	5986	17	5,04
flyga	bort	1824	5	4,99
trycka	mot	982	8	8,94
trycka	ner/ned	1092	7	8,59
trycka	upp	6657	6	5,76
trycka	på	31019	21	5,35
växa	upp	6657	105	6,37
växa	fram	3311	27	5,42
växa	ur	2522	10	4,38
växa	ut	7462	17	3,58
växa ifrån	PRONOUN	34080	8	-

The formula used for MI is defined as:

$$MI(x,y) \equiv log_2 \frac{P(xy)}{P(x)P(y)}$$

The probabilities `P(xy)`, `P(x)` and `P(y)` have been normalized by the corpora size `(N)`, i.e. `P(xy)/N`, `P(x)/N` and `P(y)/N`. Since small window sizes, (distance in words from the key word), ±1, ±2, usually produce a lot of noise, lemmatization of the first word form (x), as well as summarization of the inflected forms for (y) -when this is feasible- into the appropriate lemmata is performed. By collecting items[2] that belong together, a substantial amount of noise reduction has been achieved.

MI is a simple and powerful statistical technique that has been used for word sense disambiguation (Brown *et al.* 1991), as well as extended to full parsing (Magerman *et al.* 1990).

4.4. THE NEW VALENCY DATABASE

A small fragment of the new valency database can be viewed in table (4). Here **`Lemma`** is the verb lemma; **`LxNr`** is the lemma-lexeme number, the first digit being the lemma number and the second (or third) the lexeme or sense

[2]Some of the tokens shown in column (y) of table (3) can, in certain contexts, behave either as prepositions or as particles; one of the most difficult problems in parsing written Swedish. Heuristics used during tagging and lemmatization can, in many cases, determine the appropriate annotation.

Table 4: A small portion of the verb database

Lemma/LxNr	Frame/Observed Env.	Strong	Weak
bläddra/11	SUBJ, ARGP/(A)	i	fram och tillbaka, fram, igenom, bland
flyga/11	SUBJ/(A)	till	+MOD,över, med
flyga/12	SUBJ,ARGO/(i)	i	av, på, ur, upp, ut
	SUBJ/(i)	fram	bort, upp, ut, förbi, +MOD
trycka/11	SUBJ, ARGO, ARGF/(A)	på	in, upp, ut, i
	SUBJ, ARGP/(A)	mot	över, ner/ned
	SUBJ, ARGO/(A)	på	mot, ner/ned
	SUBJ, ARGF/(A)	att+CCLA	på
trycka/12	SUBJ, ARGO/(A)	+PRON	-
trycka/13	SUBJ, ARGO, ARGL/(P)	i	-
	SUBJ, ARGO, ARGS/(P)	i	på,om
trycka/14	SUBJ/(A)	-	+MOD
trycka/14	SUBJ, ARGO/(A)	till	+MOD
växa/11	SUBJ/(I)	upp	fram,in,till,i,ur,ut
växa/12	SUBJ,ARGL/(A)	i	+MOD
växa_ifrån/11	SUBJ, ARGO/(_)	+PRON	för

number; `Frame` denotes the grammatical valency of the verb; `Observed Env` denotes the syntactical environment on which the manual observations for the original creation of the resources have been based. These values are: active voice '`A`', passive voice '`P`', deponence form '`D`', infinitive form '`I`', infinitive construction '`i`' and imperative '`IMP`'; `Strong` stands for strong cue, `Weak` for a list of weak cues and (_) denotes unspecified information in the database. The labels within the frame column are described in table (2).

4.5. REGULAR EXPRESSION GRAMMAR

The actual content for the argument structure of the verbs has been acquired by using a compact but very powerful deterministic regular expression grammar. This grammar[3] is intended to to perform partial parsing (Abney 1994), and identification of the relevant phrases and clauses. There is a threshold between partial and full parsing, namely efficient analysis vs depth of analysis (Abney 1996). Partial parsing of phrases give more accurate and detailed results than full parsing and at the same time it takes far less time and resources to develop.

Five major classes of phrases (sometimes referred to as chunks) are recognized:

[3]Chart parsing could be used as an alternative option for the same goal, i.e. partial parsing, especially when dealing with *real-life* texts and incomplete grammar coverage, but such a developing approach would have exceeded the available time resources.

(i) The noun-phrase chunk, marked as **NPx**, consisting of nominal, adjectival, numerical and participle constructions, or combinations of these.

(ii) The prepositional-phrase chunk, headed always by a preposition, marked as **PPx**.

(iii) The infinitive construction chunk, headed often by the infinitive marker *att*, as well as other non-finite verbal constructions, marked as **INFx**.

(iv) The finite verb-phrase chunk, marked as **VPx**.

(v) The adverbial construction chunk, marked as **MODx**.

Conjunctions, punctuation markers and most of the modal and auxiliary verbs do not belong in any chunk.

The different structural environments modelled by the grammar rules, which can recognize, among other things, the five types of phrases described above, have been acquired by manually inspecting part-of-speech annotated sentences. These empirical observations have been completed with certain ideas borrowed from the description of a surface oriented grammar for Swedish, presented in Ejerhed (1993).

Consider for instance the following sentence:

Han hade bläddrat i några gamla böcker med vänstra handen.

Lit: 'He had paged through some old books with his left hand.'

which after the tokenization and tagging phase is annotated with the following part-of-speech categories:

Han/P hade/VA bläddrat/VM i/S några/D gamla/A böcker/N med/S vänstra/A handen/N ./F

Consider also the oversimplified regular expression rules:

NP0x → A* N+
NPx → P | D? NP0x
PPx → S (NP0x|NPx)
VPx → VM | VA+ VM

which can produce the following phrase annotation:

$_{\text{NPx}}$[Han/P] $_{\text{VPx}}$[hade/VA bläddrat/VM]
$_{\text{PPx}}$[i/S $_{\text{NPx}}$[några/D $_{\text{NP0x}}$[gamla/A böcker/N]]]
$_{\text{PPx}}$[med/S $_{\text{NP0x}}$[vänstra/A handen/N]]
./F

According to the regular expression mechanism ''|'' stands for alternative readings; ''?'' stands for 0 or 1 occurrence of the previous pattern; ''*'' stands for ≥ 0 occurrences of the previous pattern; and ''+'' stands for ≥ 1 occurrences of the previous pattern. Note also that it is the longest match found in the rules that will be returned by the grammar. In conflict situations where two or more rules can be applied, the one that matches the longest pattern will be favoured. If for instance there were three rules:

R1 → N, R2 → D and R3 → D N

and all three could be applied to a phrase such as *en/D pojke/N*: 'a boy', then R3 would be favoured and applied, since it matches the longest pattern.

Grammars denoted by regular expressions, such as the small fragment shown, are equivalent to and can be converted to finite state automata (Hopcroft *et al.* 1979).

5. The Method

The basic approach used for the identification of the arguments is rather simple. There are three main sources of knowledge used by the matching algorithm of the system, i.e. the built-in heuristics, mostly cues for word order alternations, the information induced by the grammar and the content of the database.

First the texts are tokenized and tagged by the rule-based tagger with detailed morphosyntactic information, as shown previously. Then the grammar is used for partially parsing the texts and grouping the annotated tokens into chunks.

Then, since all tokens are annotated, the verbs can easily be detected and lemmatized, and the database consultation performed. For each verb found in every sentence all the associated frames are returned. Then, depending on the strong and the weak cues in the database, and the heuristics of the algorithm, they are matched against the chunks found in the previous step, and at least one of these frames is suggested by the algorithm as the most plausible one. All these steps are carried out sequentially.

Thus, in the example sentence discussed previously, the strong cue of the verb *bläddra* signals that the first PPx phrase headed by the preposition *i* will be assigned the role of the prepositional object **ARGP** given in the frame slot of the database entry. The second PPx phrase will be regarded as an adjunct and will be annotated by the default frame **MOD**. **MOD** is used simply because there is no other available information in the database for the verbal entry **BLÄDDRA** which could signal the requirement for another annotation and thus another argument type. Furthermore, since the strong cue is matched, the content of the weak cue-list will be ignored. The whole input sentence will then be marked appropriately with the argument labels found in the frame slot of the examined verb[4].

[4]The similarity between the shown annotation and the **VALENCE PRINCIPLE** in HPSG, (cf Pollard *et al.* 1994, pp. 348), is evident. Automatic means can easily convert that notation to a 'HPSG-like' one, where the **ARGP** will be specified as **COMPS** of the **HEAD BLÄDDRA**

SUBJ[Han/P] [hade/VA bläddrat/VM]{bläddra}
ARGP[i/S [några/D [gamla/A böcker/N]]]
MOD[med/S [vänstra/A handen/N]] ./F

In other possible scenarios, where syntactical alternations of the form: *'Han hade bläddrat med vänstra handen i några gamla böcker.'*, or *'Med vänstra handen hade han bläddrat i några gamla böcker.'*, a correct identification of the arguments would again have been made possible due to the heads of the prepositional phrases in the sentences.

A discussion about the strong and weak cues was given in section 4.3. The content of the cues is statistically motivated, as well as based on the assumption that there are collocational contraints between the verbal predicate and its arguments.

The heuristic part of the algorithm consists of different ways to detect some of the phenomena responsible for word order alternations which could make the automatic detection of the arguments difficult to accomplish. Since the order of the arguments in the `Frame` column in table (4) presupposes *normal* word order, the program used for the identification of the verbal arguments starts by looking at the tokens preceding the detected verb. The reason for this heuristic is due to the fact that there is a whole range of common phenomena that through specific tokens, signal word order alternations in Swedish. These phenomena include cleft sentences beginning with the often occuring subject *det*: 'it', unbounded dependencies headed by the (wh-)adverbs, *när*: 'when', *hur*: 'how', *vart*: 'where' etc. which signal interrogative sentences. Subordinate conjunctions such as *sedan*: 'then, afterwards', *eftersom*: 'because' may also signal word order alternations, especially in subordinate clauses. Passive constructions can be easily detected using the rich morphological annotation returned by the tagger, which distinguishes active from passive[5] voice.

The last aspect mentioned, i.e. the usefulness of the rich annotation, is applicable also to pronouns, since these are tagged as subject or object by the tagger if certain constraints are fullfilled, and thus give reliable clues regarding the position of the subject and object of the verb.

Also clause boundaries like commas, as well as other punctuation markers and conjunctions, are taken under consideration, since they usually match the border between candidate arguments.

[5]Passive inflected forms in Swedish can be confused with the deponent or -s form, *färdas*: 'travel', *lyckas*: 'succeed', fortunately their number is manageable and limited to 23 entries in the database.

6. Experiments and Results

The system has been tested with sentences taken from the SUC corpus and contemporary Swedish newspaper articles. SUC also consists, to a great degree, of newspaper articles. I chose here to focus my experiment on five verbs, *bläddra, flyga, trycka, växa, växa_ifrån*, and thus examine and evaluate a large set of sentences chosen randomly from large text bodies. There were three requirements taken into consideration, prior to the extraction of the test sentences. The first was that the examined verbs should appear in any of their inflected forms, for instance *växa: växer, vuxit, växte, växt, växas, vuxits, växtes, växts*; literally 'grow, grew, grown' etc, with or without possible particles (non-lexicalized forms). Second, the total length of each sentence extracted should be less than 20 tokens long. Finally, no more than 2 main verbs should appear in every extracted sentence. No participle forms were used, since they have an adjectival behaviour in Swedish.

These requirements might seem restrictive, but it should be kept in mind that it is very difficult to detect frames not presented in the verb argument database, since a conservative approach to frame acquisition was followed. This means that the system, at least at this stage of its development, does not try to *guess* frames if these cannot be detected by the database content, the cues and the heuristics. In such cases a human posteditor must manually decide whether there has been a miss by the system or whether the coverage of the database is insufficient, and then take appropriate measures by evaluating the reason for the result. These measures often take the form of adding new frames to a certain verb in the database. This way a substantial improvement of the quality of the content in the database is accomplished.

Table (5) gives some example sentences for some not very straightforward syntactical constructions. For reasons of simplicity only the first position of the tags has been retained, as well as the last for some pronouns. Sentences preceded by ** gave erroneous results during the valency identification process.

In the first example, example (a), there is an emphatic contruction where the token *Henne*: 'Her', the object of the sentence, has been moved to the beginning. In this case, the morphosyntactic annotation returned by the tagger, i.e. PO pronoun in object position, makes the detection of the word order alternation by the system fairly easy to detect, and thus a correct argument annotation can be made.

Example (b) is an elliptical construction, in which the subject of the second clause, *bläddrat i arkiven*: 'has searched through the archives', is not explicitly given. Heuristically, since this is a sequence of two conjoined clauses with only one subject found, the subject of the second clause can

Table 5: Example Sentences

a) Henne/PO skulle/VA han/P förmodligen/R snart/R växa_ifrån/VM ./F
NPx[Henne/PO] skulle/VA NPx[han/P]
MODx[förmodligen/R snart/R] VPx[växa_ifrån/VM] ./F
ARGO[Henne/PO] skulle/VA
SUBJ[han/P] MOD[förmodligen/R snart/R] [växa_ifrån/VM] ./F
He should probably grow out of her soon.

b) Han/P har/VA rest/VM runt/R bland/S olika/A tidningar/N ,/F
och/C bläddrat/VM i/S arkiven/N ./F
NPx[Han/P] VPx[har/VA rest/VM] MODx[runt/R] PPx[bland/S
[olika/A tidningar/N]] ,/F och/C VPx[bläddrat/VM] PPx[i/S
[arkiven/N]] ./F
SUBJ[Han/P] [har/VA rest/VM] ARGL[runt/R] ARGP[bland/S
[olika/A tidningar/N]] ,/F och/C SUBJ[ø] [bläddrat/VM]
ARGP[i/S [arkiven/N]] ./F
He has passsed by many newspapers and searched through the archives.

c) Den/D svenske/A diplomaten/N Mikael/N Glas/N flög/VM i/S
måndags/N från/S Kenya till/S Egypten/N ./F
NPx[Den/D svenske/A diplomaten/N Mikael/N Glas/N] VPx[flög/VM]
MODx[i/S [måndags/N]] MODx[från/S [Kenya/N]]
MODx[till/S [Egypten/N]] ./F
SUBJ[Den/D svenske/A diplomaten/N Mikael/N Glas/N] [flög/VM]
MOD[i/S [måndags/N]] MOD[från/S [Kenya/N]]
MOD[till/S [Egypten/N]] ./F
The Swedish diplomat Mikael Glas flew on monday from Kenya to Egypt

d) Min/P mor/N hade/VA vuxit/VM med/S uppgiften/N
att/C hjälpa/VM oss/PO ./F
**[Min/P mor/N] [hade/VA vuxit/VM] [med/S [uppgiften/N]]
[att/C [hjälpa/VM] [oss/PO]] ./F
The task of helping us made my mother mature.

e) Genom/S kablar/N och/C paraboler/N har/VA det/P vuxit/VM fram/S
hybrider/N som/C ZTV/N ./F
**[Genom/S [kablar/N och/C paraboler/N]] [har/VA]
[det/P] [vuxit/VM] [fram/S [hybrider/N]] [som/C [ZTV/N]] ./F
Through cable TV and satellite dishes, there have arisen hybrids like ZTV.

be recovered and understood as the same subject of the first clause, i.e. the verb *rest*: 'passed (by)'.

Example (c) is a simple sentence with a long noun phrase functioning as subject and three adjunct phrases.

The second and the third of these three phrases can be also interpreted as locational arguments.

Example (d) is a problematic case, since the database content can not correctly identify the relevant argument structure, partly because the preposition *med*: 'with' is not included in the strong or weak cues of the verb *vuxit/växa*, and partly because there is no frame giving an adequate argument description. On the other hand, the partial parsing distinguishes the two phrases following the verb, namely the prepositional phrase *med uppgiften*: 'with the task (of)' and the infinitive clause *att hjälpa oss*: 'to help us'. In this example, the system failed to give a correct assignment of the argument structure, and it is up to a human posteditor to update the frames for *växa/11* with the possibility of a prepositional and infinite phrase construction, with a strong cue being the preposition *med.*

Example (e) is also a problematic case in two ways. First, the prepositional phrase *Genom kablar och paraboler*: 'Through cable TV and satellite dishes', which is an adjunct, introduces the sentence. Second, the token *fram*: 'ahead' has been wrongly assigned the tag S, i.e. preposition, by the tagger; it should be a particle. This example illustrates that mistakes can arise on different levels of the text processing, in this case during tagging.

Some cumulative results for the discussed verbs for the correctly assigned frames are given in the appendix.

The first column of the table given in the appendix, denotes the inflected forms found in the sentences, while their respective occurrences, total and correct, are presented in parenthesis. **Sense** denotes the semantic category of the verb, i.e. the lexeme number, in case of polysemic verbs. **Subcat. Pattern and Frame** gives the most commom subcategorization and frame pattern for a verb form. Patterns containing "ø" denote that some of the arguments were not explicitly present in the construction, i.e. elliptical clauses. For instance, *bläddra: ø att ARGP(PPx(i))*, denotes that the subject of the predicate *bläddra* was not explicitly present in the examined constructions.

In the 221 sentences extracted from the available texts, 125 were assigned the right frame, 65 were wrongly assigned a frame and in 31 cases no frame could be assigned since neither the strong nor the weak cues, nor the heuristics could help towards the identification of a frame. The frame assignment errors depend on tokenization, tagging and omissions in the database.

Given these results, the performance of the system can be measured by a precision rate of 65.8%, (relevant retrieved frames/all retrieved frames), and a recall rate of 56.6%, (relevant retrieved frames/totally desired frames).

7. Remarks and Discussion

The presented system belongs to the class of systems for the automatic acquisition of subcategorization frames from unrestricted texts, with manual evaluation and post-editing. Among other methods towards the same goal are, for instance, the unsupervised Brent (1993); automatic Brent (1991) and Manning (1993); and those that use a combination of manual acquisition with automatic refinement (Grishman *et al.* 1994). The content of the present study exploits certain similarities with the work presented by Basili *et al.* (1992), that is, apart from the database instance, a similar methodological approach has been used for the extraction of word associations with the help of so called *elementary syntactic links*, (esl), such as subject-verb from two subcorpora domains, legal and economics. Their esl concept appears to have many similarities with the notion of chunks. Poznański *et al.* (1996) present also a body of programs, following some of Basili *et al.*'s methodology, for acquiring, in this case, semantically tagged frames from bracketed texts. They also emphasize the idea that text processing techniques, such as shallow syntactic analysis, is an important preprocessor to detailed lexical acquisition; to this should be added the importance of the interaction with some kind of lexicon with lexical-semantic content. These attempts and the one presented here point to the fact that the application of similar techniques to other languages is likely to need modifications.

In this work I have tried to integrate available hand-built lexical resources, and automatically find support in textual data for the semi- automatic validation and expansion of the lexical knowledge content of them (Calzolari 1991). There has been a considerable amount of human interaction with the different parts of the system, especially for the evaluation of the results, since there is no built-in way to evaluate them automatically. Methods do exist, though, and they will be explored in future reasearch(Briscoe *et al.* 1994).

A note on the cues is worth mentioning here. These cues are not always correct, as Brent (1994) has shown, and they can mislead an algorithm. Brent gives the example *"John drove* **to fish** *markets all over town"* as an exception to the fact that *"to"* followed by a word that can be interpreted as an uninflected verb usually marks an infinitive. Similar problems are also discussed in Hindle *et al.* (1994). On the other hand, these cues, if properly expanded, can give valuable information regarding the semantic disambiguation of different verb senses, or different syntactic readings within the same sense. This can be accomplished by, for instance, looking at the nearby context of the predicate (cf Yarowsky 1992). This is a typical situation for the frames of the verb *trycka/13*: 'to press/squeeze' in table (4). Both of its frames appear with the strong cue *i* and the possibility of marking the second object with either ARGL or ARGS. The subtle difference between these two can be grasped by the fact that in the most cases ARGL is used only when in the nearby context of the verb there exists an indication of location (proper) noun; and ARGS otherwise. Thus, an appropriate list of cues, maybe in the synset fashion like in WordNet, could easily be used to as a clue for the proper assignment of one of the two frame markers.

Another problem faced when working with the system arose when trying to make a clear distinction between arguments and adjuncts. As mentioned earlier, it is very difficult to detect frames that are not known to the database. The choice I had was either to use the MOD or the CCLA categories as default frames and then manually inspect and evaluate these results for the cases not captured by the available lexical content. This is a problem that certainly needs further investigation. Similarly, if a frame identified by a native speaker in the lexical resources fails to appear in the corpus, this does not mean that the frame is invalid, the frame is preserved since one of the main roles of the automatic processing of corpora in such research is to support and extend the native speaker's intuitions in the development of the subcategorization frames, but it is also the only promising method for rapid development of such knowledge into new domains.

8. Future Perspectives

Work on the evaluation of the argument frames in the database, as well as the refinement of the system performance, will continue in the future. Due to time limits and the huge effort required to evaluate the result returned by the system, only a relatively small amount of verbs have been evaluated. As mentioned earlier, the main purpose of the system is to get the valency information found in the database verified with corpus material as thoroughly as possible, then use the content of the extended valency database in the construction of a lexically specified wide coverage grammar. As mentioned earlier, the generated specification so far can be automatically converted into the syntactic slot, SYN, of the typed feature representation formalisms, (tff), Pollard *et al.* (1994), and thus used as the appropriate format for parsing purposes of systems using this type of notation. Adding semantic information, the SEM slot on the tff, is the next step, already under development. For that purpose a set of 36 semantic arguments or roles described in Järborg (1997) are used as fillers of the SEM slots. The most common of these semantic arguments include the groups:

1. AGENT, CAUSE, EXPERIENCER, THEME (focused object);
2. ANTAGONIST, PATIENT, RESULT, OBJECT, ASSOCIATION, PROCESS, PHENOMENON etc.
3. RECIPIENT;
4. INSTRUMENT, MATERIAL, METHOD, SOURCE, PATH, GOAL etc.

To these roles the operations: negation NON-; completion CO-; oppposition COUNTER- should be added. Automatic methods with post-editation are capable of identifying these roles, which to a great degree correspond to the syntactic arguments presented in table (2). This correspondance can be exemplified by the fact that group (1), above, usually matches the subject of a sentence, group (2) the direct object, group (3) the indirect object and group (4) the prepositional object.

Of course, the content of the database becomes a valuable resource for other kinds of research activities related to NLP, such as a constraint mechanism for overgeneration during and after parsing, a source for the derivation of thematic knowledge (Zernik *et al.* 1990), the automatic clustering of verbs exhibiting similar subcategorization tendencies to semantic classes, (cf Dorr *et al.* 1996). Future modifications of the system will include statistical information associated with every structural environment (see Briscoe *et al.* 1994). Furthermore, data gained from the evaluation of the results returned are, and will be, continuously added into the system. In that way, the qualitative and quantitative content of the database will continue to grow.

Acknowledgment

Many thanks to Jerker Järborg, Sofie Johansson Kokkinakis, Celia Roniotes and the two anonymous reviewers for their comments on earlier versions of this paper.

References

Abney, S. (1994) Partial Parsing, *Tutorial ANLP-94*, Stuttgart

Abney, S. (1996) Part of Speech Tagging and Partial Parsing, In *Corpus-Based Methods in Language and Speech*, Young, S. and Bloothooft, G. (eds), Vol. 2 of the Text, Speech and Language Technology series, Kluwer Academic Publishers, Dodrecht

Adriaens, G., de Braekeleer, G. (1992) Converting Large On-Line Valency Dictionaries for NLP Applications: from PROTON Descriptions to METAL Frames, *Proc. COLING 92*, Nantes, France, Vol. 3, pp. 1182-1186

Aone, C., McKee, D. (1996) Acquiring Predicate-Argument Mapping Information from Multilingual Texts, In *Corpus Processing for Language Acquisition*, Boguraev B. and Pustejovsky J. (eds), pp. 191-202, Bradford MIT Press

Basili, R., Pazienza, M.T., Velardi, P. (1992) A Shallow Syntactic Analyser to Extract Word Associations from Corpora, *Journal of the Association for Literary and Linguistic Computing*, Vol. 7.2, pp. 113-123

Brent, M. (1991) Automatic Acquisition of Subcategorization Frames from Untagged text, *Proc. of the 29th ACL*, Univ. of California, Berkeley, pp. 209-214

Brent, M. (1993) From Grammar to Lexicon: Unsupervised Learning of Lexical Syntax, *Journal of Computational Linguistics*, Vol. 19.2, pp. 243-262

Brent, M. (1994) Surface Cues and Robust Inference as a Basis for the Early Acquisition of Subcategorization Frames, *Lingua 92*, pp. 433-470

Brill, E. (1995) Transformation-Based Error-Driven Learning and Natural Language Processing: A Case Study in Part of Speech Tagging, *Journal of Computational Linguistics*, Vol. 21.4, pp. 543-565

Briscoe, T., Carroll, J. (1994) Towards Automatic Extraction of Argument Structure from Corpora, *Rank Xerox Technical report MLTT-006*

Brown, P., Della Pietra S., Della Pietra, V., Mercer, R. (1991) Word-Sense Disambiguation Using Statistical Methods, *Proc. 29th ACL*, Univ. of California, Berkeley, pp. 264-270

Calzolari, N. (1991) Lexical Databases and Textual Corpora: Perspectives of Integration for a Lexical Knowledge-Base, *In Lexical Acquisition: Exploiting On-Line Resources to Build a Lexicon*, Zernik U. (ed), chap. 8, pp. 191-208, Lawrence Erlbaum Assoc. Publ.

Church K.W., Gale W., Hanks P., Hindle D. (1991) Parsing, Word Associations, and Typical Predicate-Argument Relations, In *Current Issues in Parsing Technology*, Tomita M. (ed), chap. 7, pp.103-112

Daugaard, J. Kirchmeier-Andersen, S. Schosler, L. (1992) Parsing Large Scale Corpora for Valency Information, Holmboe, H. (ed.) *SPS Nr 6* Aarhus Univ., pp. 181-194

Dorr, B.J., Jones, D. (1996) Role of Word Sense Disambiguation in Lexical Acquisition: Predicting Semantics from Syntactic Cues, *Proc. COLING 96*, Copenhagen, Denmark, Vol. I, pp. 322-327

Ejerhed, E., Källgren, G., Wennstedt, G., Åström, M. (1992) The Linguistic Annotation of the Stockholm-Umeå Corpus Project, *Technical Report No 33*, University of Umeå

Ejerhed, E. (1993) En ytstrukturgrammatik för svenska, *Publication 27*, Department of General Linguistics, University of Umeå

Grishman R., Macleod, C., Meyers, A. (1994) Comlex Syntax: Building a Computational Lexicon, *Proc. COLING 94*, Kyoto, Japan, Vol. I, pp. 268-272

Hajicova, E. (1994a) Grammatical Data in the Lexicon, In *Computational Approaches to the Lexicon*, Atkins, B.T.S. and Zampolli, A. (eds)

Hajicova, E., Rosen, A. (1994b) Machine Readable Dictionary as a source of Grammatical Information, In *Current Issues in Computational Linguistics, In Honour of Don Walker*, Zampolli, A., Calzolari, N., and Palmer, M. (eds), Kluwer Academic Press, Boston

Hindle, D., Rooth, M., (1994) Structural Ambiguity and Lexical, *Journal of Computational Linguistics*, Vol. 19.1, pp. 103-120

Hopcroft, J.E., Ullman, J.D., (1979) *Introduction to Automata Theory, Languages, and Computation*, Addison-Wesley Publ. Company

Ide N., Veronis, J. (1995) Knowledge Extraction from Machine-Readable Dictionaries: An Evaluation, *Machine Translation and the Lexicon, Proc. of the 3rd International EAMT Workshop*, Heidelberg, Germany, Steffens P. (ed), pp. 19-34, Springer-Verlag

Johansson Kokkinakis, S., Kokkinakis, D. (1996) *Rule-Based Tagging in Språkbanken*, Research Reports from the Department of Swedish, Göteborg University, GU-ISS-96-5

Järborg, J. (1988) Towards a Formalized Lexicon of Swedish, *Studies in Computer-Aided Lexicology*, Data Linguistica 18, Dept. of Swedish Language, Göteborg University, pp. 140-158

Järborg, J. (1990) *Användning av SynTag*, Research Reports from the Department of Swedish, Göteborg University, Feb. 1990, (In Swedish)

Järborg, J. (1997) *Semantiska Roller i Definitioner*, Research Reports from the Department of Swedish, Göteborg University, Nov. 1997 (Under Preparation, In Swedish)

Kokkinakis, D., Johansson Kokkinakis, S. (1997) *A Robust, Modularized Lemmatizer/Tagger for Swedish Based on Large Lexical Resources*, Research Reports from the Department of Swedish, Göteborg University, GU-ISS-97-1

Leech, G., Wilson, A. (1996) EAGLES Recommendations for the Morphosyntactic Annotation of Corpora, *Document: EAG-TCWG-MAC/R*

Magerman, D.M., Marcus, M. (1990) Parsing a Natural Language Using Mutual Information Statistics, *Proc. of the 8th National Conference on Artificial Intelligence*, pp. 984-989, Boston, Massachusetts, AAAI Press/The MIT Press

Manning, C.D. (1993) Automatic Acquisition of a Large Subcategorization Dictionary from Corpora, *Proc. of the 31st ACL*

Marcus, M., Kim, G., Marcinkiewicz, M.A., MacIntyre, R., Bies, A., Ferguson, M., Katz, K., Schasberger, M. (1994) The Penn Treebank: Annotating Predicate Argument Structure, ARPA '94

Miller, G., Beckwith, R., Fellbaum, C., Gross D., Miller, K. (1993) *Introduction to WordNet: An On-line Lexical Database*, Cognitive Science Laboratory, Princeton University, `http://www.cogsci.princeton.edu/ ~wn/w3wn.html`. Site visited 17/09/97

Pollard, C., Sag, I.A. (1994) *Head-Driven Phrase Structure Grammar*, Studies in Contemporary Linguistics, The Univ. of Chicago Press

Poznański, V., Sanfilippo, A. (1996) Detecting Dependencies between Semantic Verb Subclasses and Subcategorization Frames in Text Corpora, In *Corpus Processing for Language Acquisition*, Boguraev, B. and Pustejovsky, J. (eds), pp. 175-190, Bradford MIT Press

Sanfilippo, A. (1996) Word Knowledge Acquisition, Lexicon Construction and Dictionary Compilation, *Proc. COLING 94*, Kyoto, Japan, Vol. I, pp. 273-277

Sinclair, J.M. (1987) Grammar in the Dictionary, Chapter 5, In *Looking Up - An Account of the COBUILD Project in Lexical Computing*, Collins Publishers

Toporowska Gronostaj, M. (1991) *Mot ett formaliserat verbvalenslexikon*, PhL Report, Språkdata, Göteborg University, (In Swedish)

Yarowsky, D. (1992) Word-Sense Disambiguation Using Statistical Models of Roget's Categories Trained on Large Corpora, *Proc. COLING-92*, Nantes, France, Vol. 2, pp. 454-460

Zernik, U., Jacobs, P. (1990) Tagging for Learning: Collecting Thematic Relations from Corpus, *Proc. COLING 90*, Vol. 1, pp. 34-39

Appendix: Cumulative Results for the Argument Structure of the Presented Verbs

Lemma/Inflection	Sense	Subcat. Pattern and Frame
BLÄDDRA (79/48)	-	
bläddra	-	ø att VPx ARGP(PPx(i))
bläddrar	-	SUBJ(NPx) VPx ARGP(PPx(i))
bläddrade	-	SUBJ(NPx) VPx MODx ARGP(PPx(i))
bläddrat	-	ø VPx ARGP(PPx(i))
bläddrades	-	SUBJ(NPx) VPx ARGP(PPx(genom))
FLYGA (32/19)		
flyga	1	SUBJ(NPx(PRON)) VPx
flyger	1	MODx VPx SUBJ(NPx(PRON))
flyger	2	SUBJ(NPx) VPx ARGP(PPx(ur))
flög	1	SUBJ(NPx) VPx MODx
flög	2	SUBJ(NPx) VPx MODx
flygs	1	ø VPx MODx MODx
flögs	1	MODx VPx SUBJ(NPx) MODx
TRYCKA (27/15)		
trycka	1	SUBJ(NPx) VPx ARGO(PPX(upp)) ARGF(PPx(i))
trycka	3	SUBJ(NPx) VPx ARGO(PPX(upp)) ARGS(PPx(om))
tryckas	3	ø ARGO(NPx) VPx ARGL(PPx(i))
trycker	1	SUBJ(NPx) VPx ARGO(PPx(på))
tryckte	4	VPx SUBJ(NPx) ARGO(PPx(till))
tryckt	1	SUBJ(NPx) VPx ARGO(PPx(på)) ARGF(PPx(i))
tryckt	3	SUBJ(NPx) VPx ARGO(PPx(i)) ARGS(PPx(om))
VÄXA (62/27)		
växa	1	SUBJ(NPx) VPx MODx
växa	2	SUBJ(NPx) VPx MODx ARGL(PPx(i))
växer	1	SUBJ(NPx) VPx MODx
växer	2	FSUBJ(det) VPx SUBJ(NPx) MODx
växt	1	SUBJ(NPx) VPx MODx
vuxit	1	SUBJ(NPx) VPx MODx
VÄXA_IFRÅN (21/16)		
växa_ifrån	-	SUBJ(NPx(PRON)) ARGO(NPx(PRON))
växer_ifrån	-	SUBJ(NPx) ARGO(NPx)
växt_ifrån	-	SUBJ(NPx) ARGO(NPx)
vuxit_ifrån	-	SUBJ(NPx) MODx ARGO(NPx(PRON))

CAPTURING MOTION VERB GENERALIZATIONS IN SYNCHRONOUS TREE ADJOINING GRAMMARS

M. PALMER, J. ROSENZWEIG AND W. SCHULER
Department of Computer Science
and Institute for Research in Cognitive Science
University of Pennsylvania
Philadelphia PA 19104-6389

Abstract. This chapter describes the use of verb class memberships expressed as semantic features as a means of capturing generalizations about manner-of-motion verbs in Synchronous Tree Adjoining Grammars, STAGs, (Shieber and Schabes, 1990; Shieber and Schabes, 1991), which characterize transductions between Lexicalized Tree Adjoining Grammars, LTAG, (Joshi, 1985; Schabes et al., 1988). Our approach allows STAGs, which are essentially transfer-based, to take advantage of the same types of generalizations which are generally thought of as wholly the domain of interlingua systems, without giving up any of the lexical specificity unique to transfer-based systems.

1. Introduction

We begin with a discussion of a particular class of verbs, namely motion verbs and other path-denoting expressions, and their defining characteristics. We go on to discuss their representation in Lexicalized Tree-Adjoining Grammars and how this representation can provide the basis for translation by Synchronous Tree Adjoining Grammars, STAGs, (Shieber and Schabes, 1990; Shieber and Schabes, 1991). The capability of lexical trees for the representation of lexical semantic information captured by Lexical Conceptual Structures is presented and a comparison is made to the ACQUILEX MT framework, which uses typed feature structures and unification to represent a categorial grammar.

P. Saint-Dizier (ed.), Predicative Forms in Natural Language and in Lexical Knowledge Bases, 229–256.

2. Motion verbs and other path-denoting expressions

One of the most difficult questions in lexical semantics is whether or not a prepositional phrase constitutes an argument to a verb or an adjunct. For instance, a verb's semantics may require that a path be specified, to complete the description of the eventuality the verb refers to. The prepositional phrase, if it denotes a path, may then be interpreted as the argument filling this slot. In other cases however, a path-denoting prepositional phrase is juxtaposed with a verb whose semantics does not seem to necessarily involve any paths, suggesting that the relationship between the semantic forms is more complicated. This situation is exemplified by sentences like:

(1) a. The bottle floated into the cave.
    b. The train roared through the station.

The verbs in these sentences can also occur without a syntactically related prepositional phrase, in a simple intransitive frame, where they do not seem to make reference to events involving a path:

(2) a. The bottle floated.
    b. The train roared.

This suggests that the semantic component contributed by the verbs in (1) does not contain a placeholder for a path, so it must be adjoined.

On the other hand, one may argue that the verbs in (2) are in fact different from the verbs in (1), despite their being spelled the same (Talmy, 1991; Levin and Rappaport-Hovav, 1995). That is, there are actually two verbs, *float* and *float′*, and two verbs *roar* and *roar′*. *Float* and *roar* take a prepositional phrase argument and have a semantic form with a slot that can be filled by a path. The other two verbs, *float′* and *roar′*, differ syntactically in taking no internal arguments, and semantically in not involving any paths.

If this *polysemy* argument is adopted, the verbs in (1) are standard models of predicate-argument composition and no special mechanisms are required to derive their semantics. However, one would still like to account for the strong similarity between the semantics of *float* and *roar* on the one hand, and *float′* and *roar′* on the other.

Indeed, the eventualities described in (1) seem to entail those in (2), suggesting that the semantic forms of the intransitive verbs are in some way contained in the semantic forms of the verbs which require a prepositional argument. If this is so, the situations described in (1) may be viewed as *event complexes*, since they comprise other, possibly simpler, events (Talmy, 1985).

Lexical rules, mapping from verbs like *float′* to verbs like *float*, have been proposed to formalize the relationship between the respective semantic forms, and capture the generalizations that would be lost if these verbs were treated as unrelated in the lexicon (Levin and Rappaport-Hovav, 1995).

Formulating such rules though poses many of the same problems as those which arise if the verbs *float* and *roar* are taken to be *monosemous*, that is, the same in both their intransitive and prepositional-phrase frames.

Under both hypotheses, the event involving a path must be formally related to the event not involving a path. If the verbs are viewed as polysemous, a lexical rule must operate on the semantic form of the verb sense that does not have a path argument, and which denotes the event not involving a path, yielding a semantic form that does take a path argument, and that denotes the event which does involve a path. If on the other hand the verbs are viewed as monosemous, then there must be an operation which combines the semantic form of the verb, which by itself evokes an event not involving a path, with the semantics of the path-denoting phrase, producing a composed semantic form that now does evoke an event in which a path is involved.

To achieve the goal of formally relating an event complex involving a path with the events that can be viewed as comprising it, it is necessary to look more closely both at the types of paths involved, and at the types of events that are evoked by the verbs which occur with path-denoting prepositional phrases in this way. For the purposes of this investigation, only intransitive verbs will be considered, although the principles of semantic composition which are appealed to for this set of verbs should be applicable to the transitivized form of the verbs as well (ex. "John floated the bottle" versus "John floated the bottle into the cove").

2.1. EXAMPLES OF VERBS WITH PATH-DENOTING PHRASES

There are a large number of verbs which can co-occur syntactically with prepositional phrases denoting paths. Even when these verbs are organized into broad semantic classes, in the style of (Levin, 1993), a variety of different relationships between classes and path-denoting phrases can still be identified. Even within a single such class, it may be possible to combine the semantics of a verb and an evoked path in more than one way.

For example, Levin and Rapaport Hovav discuss a set of verbs, which they call "sound emission" verbs, which under certain circumstances can occur in construction with a path-denoting prepositional phrase. In the absence of such a phrase, the verbs tend to denote eventualities in which their subject produces a sound. When the path-denoting phrase is present, the production of sound is still implied, but the traversal of a path is also predicated of the subject:

(3) a. The train roared into the station.
    b. The train roared.

(4) a. The truck rumbled into the driveway.

b. The truck rumbled.

In some cases, the absence of the prepositional phrase is odd:

(5) a. The bullet whistled through the air.
b. ? The bullet whistled.

Levin and Rapaport Hovav argue that the sound emission verbs are polysemous; each has, in addition to its base sense, an additional "directed motion" sense which is derived by a lexical rule. However, this lexical rule has context-dependent applicability conditions. The rule can only be applied to a sound-emission verb in a context where its subject is also traversing a path, in addition to emitting a sound, and where, moreover, the emission of the sound is a "necessary concomitant" to the traversal of the path. If these conditions are met, then the rule can be applied, and the second sense of the verb, whose semantic form contains a slot for a path, becomes available. The "necessary concomitant" constraint, which is never formally described, is meant to rule out cases like

(6) ? The dog barked down the street.

That is, it does not suffice for the two actions predicated of the subject, namely, path traversal and sound emission, to be occurring simultaneously; they must be more substantively related.

2.2. LEVIN CLASSIFICATIONS

The semantic verb classification by Levin makes reference to several other classes whose members can enter into construction with path-denoting prepositional phrases. Levin, in fact, describes this phenomenon as a "diathesis alternation": the "directional phrases with nondirected motion verbs" alternation. In addition to the above-mentioned "sound emission" verbs, four other classes are listed as having members which exhibit this syntactic behavior. These are the "*push/pull* verbs" ("verbs of exerting force"), "verbs of body-internal motion", and two subclasses of the "verbs of motion", "*waltz* verbs" and "*run* verbs".

The "*push/pull* verbs" typically require a direct object in order for a path-denoting phrase to be present. For the time being, therefore, these verbs will not be discussed, to keep the focus narrowed in on the intransitive verbs of this type.

According to Levin, the "verbs of body-internal motion" denote a movement of the body; displacement of the body is not entailed in the absence of the additional path phrase.

(7) a. The fish wriggled off the hook.
b. The fish wriggled.

A subset of the verbs in this class are cross-listed in Levin's taxonomy as also belonging to another semantic class, the "verbs of modes of being involved in motion". These verbs are said to describe states of motion of inanimate entities; the type of motion characterizes the existence of the entities. The verbs in the intersection of these two classes are *rock*, *sway*, *teeter*, *totter* and *wobble*.

The "*waltz* verbs" predicate of their subject that it performs the dance named by the verb; a direction of motion is not specified by the verb itself, but can be supplied by an additional phrase syntactically associated with the verb.

(8) a. The couple waltzed onto the balcony.
b. The couple waltzed.

The "*run* verbs", according to Levin, describe a manner in which an animate entity can move, or the displacement of an object in a particular manner or by a particular means. The direction of this movement is said to be unspecified in the absence of the path phrase. This is a heterogenous class, and many of its member verbs are cross-listed in other semantic classes as well. The largest cross-listed subsets are the "*roll* verbs" subset, the "verbs of modes of being involved in motion" subset, the "*swarm* verbs" subset, the "*meander* verbs" subset, and the "verbs of rushing" subset. These subsets may be viewed as an implicit subclassification within the Levin taxonomy, alongside the explicitly enumerated verb classes and subclasses (Palmer et al., 1997).

The "*roll* verbs" are said to specify a manner of motion characteristic of inanimate entities, often non-volitional. Verbs in the intersection of the *run* and *roll* classes are *bounce*, *float*, *glide*, *roll* and *slide*.

(9) a. I (glided / slid) down the hill.
b. I (glided / slid).

In the intersection with the "verbs of modes of being involved in motion" there are the verbs *creep*, *drift* and *float*. The "*swarm* verbs" subset includes *crawl*, *hop* and *swim*; they entail that their subject is a group or aggregate. (The verb *file*, though also a member of the *run* class with this entailment, is for some reason omitted from the cross-listing.) The "*meander* verbs" include *meander*, *sweep* and *wander*. Levin calls these verbs "pseudo-motional locatives", using motion to describe the location of a long continuous object such as a road or a river. Hence these verbs have the motion sense as their basic sense, and presumably their membership in the *meander* class is by virtue of a lexical rule.

(10) a. I meandered across the valley.
b. I meandered.

Finally, there is the "verbs of rushing" subset, comprised of the verbs *hasten*, *hurry* and *rush*, about which Levin only says that they denote an action that is performed quickly.

(11) a. I hurried down the street.
b. I hurried.

2.3. INHERENT PATHS

The majority of verbs in the *run* class already imply the traversal of a path by the subject, even in the absence of an additional phrase. When verbs like *climb* or *zigzag* (both members of the *run* class) are seen in the presence of an additional associated path description, this must be interpreted as further constraining the (underspecified) path inherent in the verb's semantic form. With *climb*, for instance, the inherent path must have some extent in the vertical dimension (notwithstanding examples of climbing along horizontal cracks in a cliff, which however must be part of a larger path that goes up or down). An additional path-denoting phrase may constrain the starting point, end point, direction, or other features of this path.

(12) a. I climbed to the top.
b. I climbed.

(13) a. I zigzagged across the field.
b. I zigzagged.

Formally, this combination of path descriptions may be modeled by a unification of semantic representations. The process may be similar to the semantic composition of a verb like *eat* with its direct object. Since *eat* can also occur intransitively, its direct object may be viewed as supplying optional additional descriptive constraints on an entity that is already implicit in the semantics of the verb by itself (since the intransitive *eat* implies that something edible is eaten). Semantically, this type of composition is in principle no different from the traditional predicate-argument model; the distinction is only in the syntax, insofar as traditionally the arguments of predicates are taken to require syntactic realization, while in these cases their realization is optional (although some of the verbs in this class, such as *hurtle*, seem odd when used without a prepositional phrase). Even when a verb such as *devour*, unlike *eat*, requires an explicit object, the semantic form of the object is still presumably only further refining an already existing partial description, inherent in the verb's semantics, of the expected argument. This inherent description corresponds to the semantic selectional restrictions of the verb. (This corresponds to the notion, discussed in Palmer, et. al., (Palmer et al., 1993), of semantically essential arguments, which sometimes may, and sometimes may not be syntactically obligatory.)

2.4. ADDITIONAL CHARACTERISTICS

Many of the verbs in the *run* class describe movements of the legs: for example, *hobble*, *hop*, *jog*, *limp*, *lope*, *lumber*, *march*, *mince*, *skip*, *stomp*, *stumble*, *trot*, *walk*, etc. It would be hard to apply any of these verbs to a subject that did not have legs to move. This semantic feature is shared also by the *waltz* verbs. Typically these verbs denote events which are iterated processes, although some of them, such as *jump* and *leap*, have related nominals which refer to a single iteration of the leg movement (unlike, for instance, *walk* or *dance*, where the nominal must refer to a set of multiple iterations).

In these cases, the additional contribution of a path-denoting phrase is also interpreted as predicating that the entity denoted by the subject traverses the path in question. In the case of a verb like *hop*, an iterated path is lexically specified for the legs, and its relative velocity and general shape are subject to certain constraints. This path is characterized only with respect to its local characteristics (within one iteration) and abstracts away from global path parameters involving endpoint and orientation with respect to other objects. An adjunct phrase co-occurring with *hop* may then specify such global parameters, augmenting the path description. However, such verb-adjunct pairs have a further entailment, since it is not the case that "I hop into the room" if I am hopping up and down on a cart which is being pushed into the room by someone else, even though the path followed by my legs may conform both to the constraints imposed by *hop* and those imposed by the path-denoting phrase simultaneously. Evidently, there is an additional requirement that the traversal of the entire path be accomplished by means of a single agent.

This entailment may perhaps be generalized to account for other restrictions on the co-occurrence of verbs and path phrases. For example, verbs like *bicycle* and *paddle* may freely accept path phrases as clausemates in such sentences as

(14) I bicycled into the town.

(15) I paddled across the river.

The interpretation of these sentences may be correctly inferred with the help of the "single-agency" entailment. Suppose that a verb such as *bicycle* predicates that the subject entity causes, by the application of force, subparts of an implicit instrument, a bicycle, to follow certain paths, again specified locally without any constraints on their global characteristics. The path phrase is again interpreted as further constraining this path, and the final computed path is consistent with both the assertions of the verb in isolation, and those of the path phrase.[1]

Though it is not mentioned in Levin's book, the verb *squeeze* may also occur with path-denoting phrases under some circumstances, but not others:

(16) I squeezed through the hole.

(17) ? I squeezed across the field.

In these sentences, the verb itself may be viewed as contributing the entailment that an object is contracting, which may in turn be decomposed into a simultaneous set of path traversals by subparts of the object towards the center of the object.[2] In (16), subparts of my body are both being squeezed in and are moving globally with respect to the space delineated by *the hole*. Moreover, this entire complex motion of the parts is being caused by a single agency; by moving *through the hole* I indirectly cause the squeezing if the edges of the hole do not yield, hence forcing my body to compress. In example (17), however, the same body parts may describe a path which satisfies the constraints imposed by *squeeze* and *across the field*. However, there is no readily available interpretation in which a single agency accounts for both motions. Indeed, there is hardly any possible way of accomplishing this simultaneously with two agencies, unless I am in some sort of vehicle that is moving across the field and at the same time applying pressure to my body with some sort of instrument, causing it to compress in a way consistent with the entailment of *squeeze*.

The above-mentioned "necessary concomitant" constraint suggested by Rapaport-Hovav and Levin may perhaps also follow from a generalization of the entailment that both the predicated motion described by the verb and the predicated motion described by the path phrase are caused by the same agency. In this case, however, there are two different paths, each followed by a different entity. In (1) above, *through the station* defines a path which the subject entity, *the train*, traverses. However, the verb itself predicates

[1]Similarly, with *paddle*, and in general with verbs derived from instrumental nominals, there is an entailment that the subject entity manipulates the incorporated instrument in a prototypical way. In the case of *paddle* this is again an entailment about the caused motion of the instrument. If moreover the same agency which causes this locally described motion can also bring about a global displacement of the paddle (and what's attached to it) by virtue of, for example, the resistance of water in a river, then a sentence such as (15) is acceptable and interpreted in the same general way as the others discussed so far. If however the conditions are such that a single agency cannot be responsible for the complex path traversed by the paddle (if for example the river is frozen, and I manipulate the paddle in the air while propelling myself across on ice skates), then (15) is out, even if the path of the paddle satisfies both predications.

[2]Note that *squeeze* when not used with a path phrase does not predicate contraction of its subject, but rather of its object.

motion of a sound wave, a *roar*, outward from its source, which in general is the subject entity, even when *roar* occurs intransitively. In cases where both entities are launched on their paths by the same agency, the single-agency constraint is satisfied and the sentence is interpretable. In the case of (2), the interpretation is that the same agency causing the locomotion of the train through the station, also causes the motion of the sound waves through the air around the train – either in parallel (e.g., if the same strokes of engine pistons which turn the wheels also displace a lot of air violently), or sequentially (e.g., if the propulsion of the train causes the train itself to displace the air[3]). Hence, the "single-agency" constraint may be viewed as a specification of the "necessary concomitant" constraint which applies to a number of other cases of verbs that appear with path-denoting phrases, and moreover also spells out more precisely the conditions under which the constraint may or may not be satisfied.

In section 3, we introduce Lexicalized Tree-Adjoining Grammars, and use them to represent motion verbs and sound emission verbs. Some of the motion verbs require *path* prepositional phrases as internal verb arguments, and some of them allow the *path* prepositional phrases to be adjoined on, as do the sound emission verbs. We give an account for how the adjunction of a *path* prepositional phrase can coerce a sound emission event into a *directed-motion* event. In section 4 we discuss the similarity between this treatment and construction grammar, and in section 5 we describe how LTAGs can be used in Synchronous Tree Adjoining Grammar systems to translate between English and French.

3. Lexicalized Tree-Adjoining Grammars

In the rest of the chapter we discuss how the semantic characteristics of verbs that take path-denoting expressions can be captured in concrete representations, in particular in Lexicalized Tree-Adjoining Grammars. Lexicalized Tree-Adjoining Grammars (LTAGs) are tree rewriting systems (Joshi et al., 1975), (Schabes et al., 1988; Schabes, 1990). The primitive elements of LTAGs are called elementary trees and are of two types: initial trees and auxiliary trees. The minimal, non-recursive linguistic structures of a language, such as a verb and its complements, are captured by initial trees. Recursive structures of a language, such as prepositional modifiers which result in syntactically embedded NPs, are represented by auxiliary trees.

[3]This is more the case with example (5), where the agency causing the bullet to move, say an explosion in the chamber of a gun, does not also cause the sound waves corresponding to the whistling directly. Rather, these are the indirect result of the motion of the bullet, which causes selective displacement of air molecules.

Elementary trees are combined by the operations of substitution and adjunction. Every tree is associated with a lexical item (or non-compositional phrasal form) of the language, called the *anchor* of the tree. The tree represents the domain over which the lexical item can directly specify syntactic constraints, such as subject-verb number agreement, or semantic constraints, such as selectional restrictions, all of which are implemented as features in *Feature-Based* LTAG, FB-LTAG (Vijay-Shanker and Joshi, 1991). Alternative syntactic realizations of a lexical item are grouped together into *tree families*, and the semantic constraints automatically apply to the same arguments in the alternative trees.

There are critical benefits to lexical semantics that are provided by the extended domain of locality of the lexicalized trees. Each lexical entry corresponds to a tree that will be used to build the parse of the sentence that that item appears in. If the lexical item is a verb, the corresponding tree is a skeleton for an entire sentence with the verb already present, anchoring the tree as a terminal symbol. The other parts of the sentence, the noun phrases that occur as subject and object, will be substituted or adjoined in at appropriate places in the skeleton tree in the course of the derivation. The composition of trees during parsing is recorded in a *derivation tree*. The derivation tree nodes correspond to lexically anchored initial trees, and the arcs are labeled with information about how these trees were combined to produce the parse. Since each lexically anchored initial tree corresponds to a semantic unit, the derivation tree closely resembles a semantic-dependency representation such as those used in the MT system of (Kittredge and Polguere, 1991).

Since a unique tree structure is associated with each lexical item, verb-specific idiosyncrasies can also be captured at the lexical level. Idioms and light verb constructions, such as *kick the bucket* and *take advantage of*, appear in the lexicon as single trees, with multiple lexical items anchoring the tree. However, the tree composition operations of the LTAG formalism allow appropriate lexical items to be inserted between words in such a phrase when necessary, permitting the handling of variations such as *kick the proverbial bucket* or *take unfair advantage of*.

3.1. TREE FAMILIES AND VERB FAMILIES

In the theoretical formulation of LTAG, there is a set of unique trees for each word or phrase. Even if many words anchor the identical tree structure, this structure must appear multiple times in the grammar, once for each such word. In practice, it is useful to generalize over *classes of words*, for instance, all verbs of a particular subcategorization frame, and associate trees with these classes, rather than with individual lexical items. The trees may be represented as uninstantiated lexical templates, with an open variable in the position where the lexical anchor occurs. The variable is constrained to take on only particular lexical values, for example, verbs from a particular class. This constraint may also be implemented by feature-structure unification, where each verb in the lexicon has semantic-class features associated with it.

Conversely, a particular word may appear as the anchor to many different syntactic trees, yet these trees may all share some information. For example, a verb such as *eat* may anchor a transitive tree as well as an intransitive tree. However, the same syntactic constraints between subject and verb apply in both cases, and there are also many commonalities in the meaning representation and predicate-argument structure associated with *eat* in these two frames. To capture such resemblances between related trees in the grammar, it is useful to define the notion of a *tree family*, or set of related (lexically uninstantiated) trees. Information common to all the trees can be stored once for the entire family; information special to particular trees in the family can be stored in that tree. The information common to all trees may be *parameterized*, with individual trees specifying the values of the parameters.

For example, in the case of the related transitive and intransitive trees for verbs such as *break*, the common information for the tree family may specify that the verb has two semantic arguments, say, *agent* and *theme*. However, the syntactic linking of these arguments is a parameter that is set differently by each tree in the family. The transitive tree in the family is specified as having its subject NP linked to the *agent* argument and its object to the *theme* argument. For the intransitive tree, a different linking for the *theme* is specified, and the *agent* argument is left unrealized. This linking may be specified by *coindexing* among the feature structures on the nodes of the tree (see Figure 1).

We are currently investigating the types of tree families that can be useful for an English grammar, building on work done by Becker (Becker, 1993) and Candito (Candito, 1996) on incorporating such generalizations into a TAG. We have implemented an explicit hierarchy for the LTAG English grammar that could be used to capture the inter-relationships between syntactic/semantic frames (Doran et al., 1997).

3.2. ADDING SEMANTIC FEATURES TO LTAGS

A primary task of lexical semantics is to find correct correspondences between the underlying semantic representation of a verb and its alternative syntactic realizations. Given any particular syntactic realization, the appropriate semantic representation is then available. A lexical semantic repre-

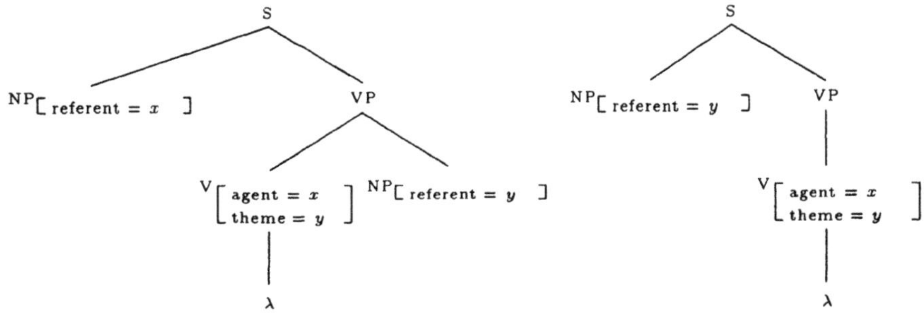

Figure 1. Linking rules in a tree family

sentation formalism should encode the constraints governing these syntax-semantics correspondences transparently and efficiently, enabling applications such as machine translation systems to retrieve semantic information from syntactic structure as quickly and directly as possible.

Selectional restrictions on a verb's arguments can be stated once for all the trees anchored by the verb, irrespective of where the arguments are realized syntactically. The individual trees, which reflect the verb's syntactic variability, each specify, by means of feature-structure co-indexing, where the selectional restrictions are enforced. For example, if one sense of *break* requires its *theme* to have the physical property *+brittle*, this can be noted as a constraint on the feature structure associated with the verb in the lexicon: [*theme* = [*brittle* = +]]. Because of the co-indexing on the *theme* feature structure in the verb tree, this constraint will be propagated to the syntactic position corresponding to the *theme* argument in any tree anchored by *break*, including both the transitive and intransitive trees in Figure 1.

Moreover, two different senses of *break*, separate-into-pieces *break* and non-functional *break*, may be differentiated by appropriate selectional restrictions on the verb object: *+brittle* and *+mechanical-device*, respectively (see Figure 2). If the sentence is "John broke the window," *window* will only be able to substitute into the verb tree with the *+brittle* feature, thus selecting the appropriate sense of *break*. It is often the case that a particular language will distinguish senses based on subtle shadings in meaning that are not relevant to other languages. For instance, Mandarin has different expressions for *breaking* events depending on the characteristics of the objects involved. *Da-duan* is used for objects shaped like line segments, *da-po* for objects that *break* into irregularly shaped pieces, and *da-sui* for objects that break into many small pieces, (Palmer and Wu, 1995). These features can be associated with the correct expressions in the Mandarin FB-LTAG,

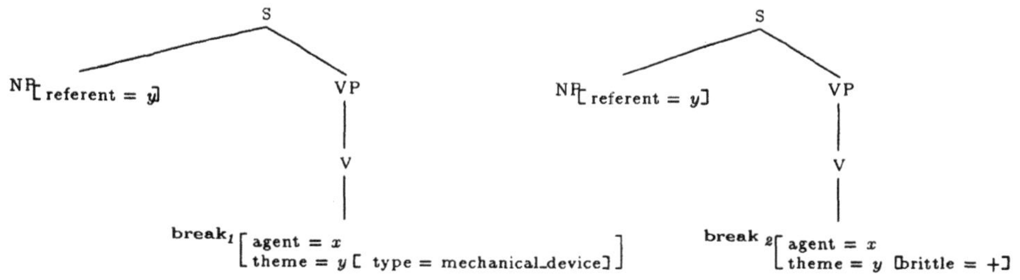

Figure 2. Selectional restrictions for two senses of *break*

without needing to make the same distinctions in other languages. When attempting translations, it is only necessary to pinpoint the range of possible translation, and the selectional restriction features will ensure that inappropriate combinations are not produced (Egedi et al., 1994), (Han et al., 1996).

3.3. EXAMPLES OF MOTION VERBS

Semantic features may also be used to represent verb-class membership. For instance, *change-of-state* verbs such as *dry* can be marked by a feature in the lexicon.[4] Selectional restrictions may be generalized over entire verb classes by specifying feature-structure equations using the verb-class membership features. Also, such features may be used to constrain or facilitate adjunctions in trees anchored by verbs in a particular class. For instance, intransitive *manner-of-motion* verbs such as *walk* and *float* can readily take *path* prepositional phrases such as *into the cave*, creating a directed motion event. Features can be used to associate the appropriate verb class with a lexical item, thus constraining the types of adjunctions that would be most suitable.

In Figures 3 and 4 we have illustrated a *manner-of-motion* verb such as *walk* with a *path* as an obligatory argument. If the prepositional phrase is not present in the syntactic realization, as in, "On his journey to his uncle's farm, the boy walked for hours before finding a friendly mule skinner who offered him a ride," the *path* is considered to be dropped, and the assumption is that this information should be recoverable from the context. It is always a *directed-motion* event.
This is in contrast with verbs like *float*, see Figure 5, which do not require a *path* phrase, although they may easily adjoin on, as in Figure 6. The feature

[4]The feature may be merely an atomic value, or it may be a more complicated, compositional structure depicting the structure of events which the verb denotes.

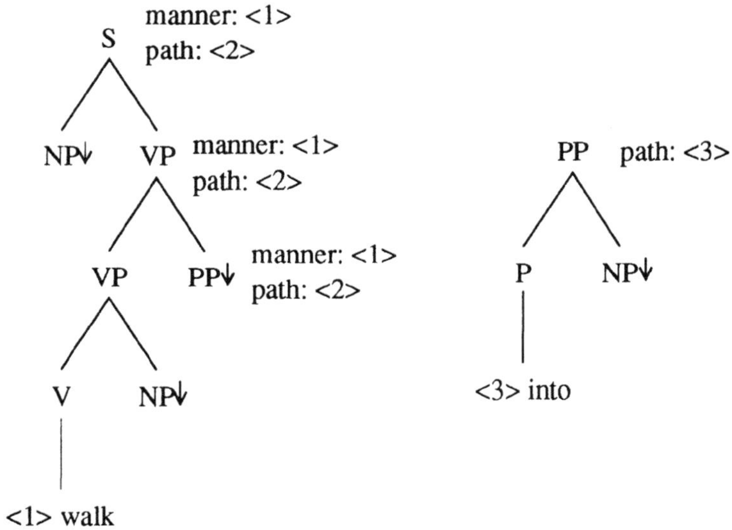

Figure 3. Elementary trees for *walk* and *into*

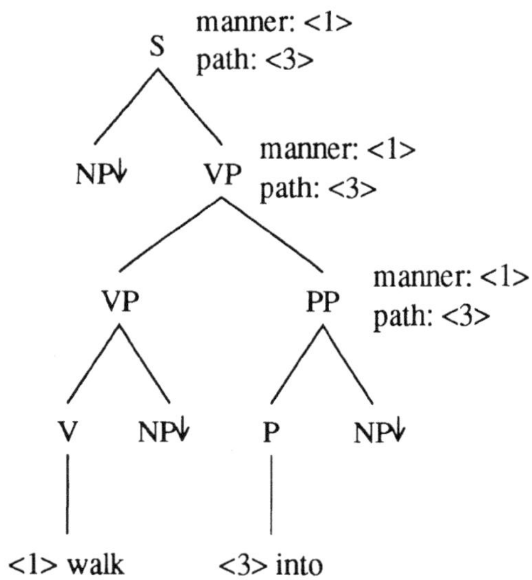

Figure 4. Tree for *walk* with *into* substituted in

unification has to allow *path* prepositional phrases which also contain features indicating that they comprise *directed-motion* events to adjoin onto *manner-of-motion* verb phrases, even when they are not specifically marked as *directed-motion* events. If the adjunction does not occur, the default *event* type should be *manner-of-motion* which can be handled through the use of top and bottom features (Vijay-Shanker and Joshi, 1991). In there is no explicit *path* prepositional phrase, and in the absence of a clear contextual path, it is assumed that the object is simply *floating* in place, without any change of location.

Notice that the auxiliary tree which allows prepositional phrases to be adjoined onto verb phrases already contains a distinct VP node (actually two). The claim we are making here is that, if the prepositional phrase is typed as a *path* prepositional phrase, then that VP in the auxiliary tree already carries the *directed-motion* event type with it.

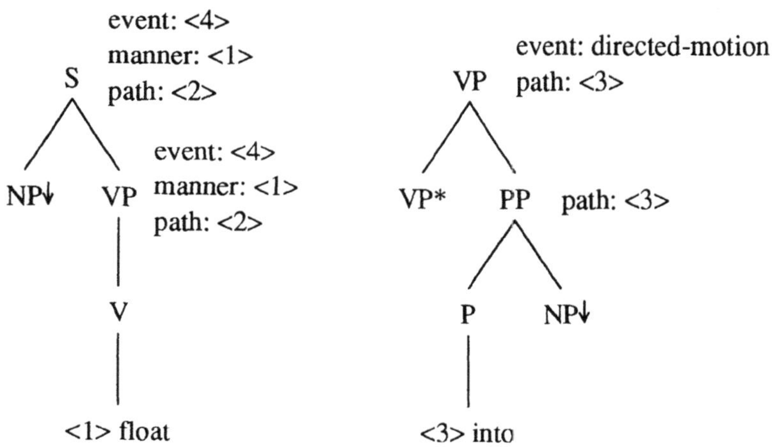

Figure 5. Trees for *float* and *into*

The same approach can then apply to *sound-emission* verb phrases. The feature unification simply has to allow *path* prepositional phrases with the *directed-motion* feature to adjoin on, and then the events described by the *sound-emission* verbs can also be considered *directed-motion* events, as in Figure 7.

4. Trees as constructions

In the previous section we described an adjunction of *path* prepositional phrases to both *manner-of-motion* verbs and *sound emission* verbs. By associating the *directed-motion* event type with the *path* prepositional phrase, and allowing this feature to be passed up to the S node during the adjunction onto the intransitive verb tree, we add this information to the *manner-of-motion* event type that is already associated with the intransitive verb. This approach effectively treats the path prepositional phrase as a *construction* in the sense of *Construction Grammar*, Construction Grammar,

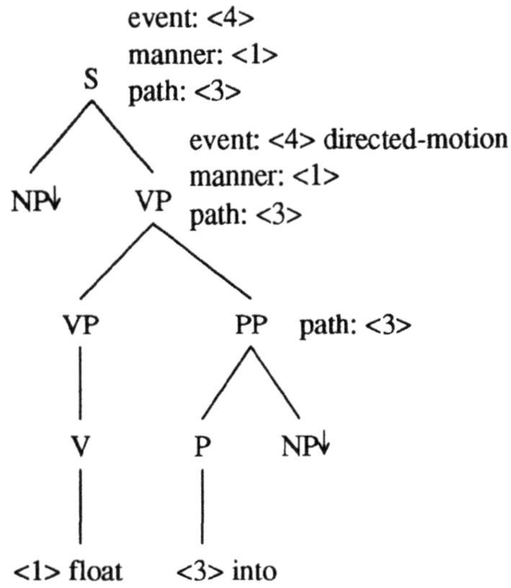

Figure 6. Trees for *float* and *into* adjoined together

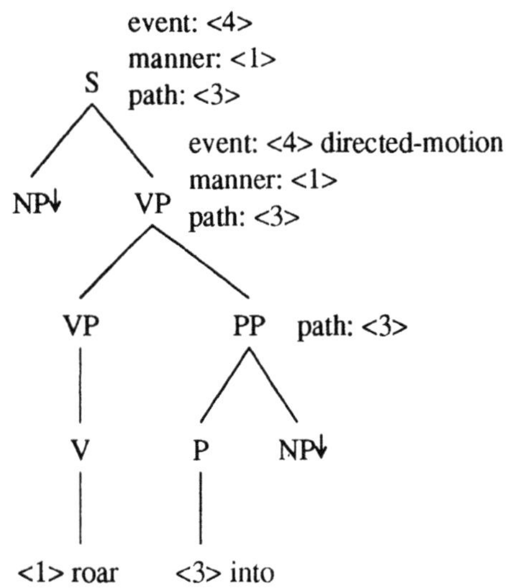

Figure 7. Trees for *roar* and *into* adjoined together

(C. J. Fillmore and O'Connor, 1988), (Goldberg, 1995). The claim of Construction Grammar is that there are atomic, non-decomposable meaning elements at higher levels of grammatical organization — in particular, at the level of phrase and sentence schemas. For example, Construction Grammar is able to directly associate meaning with a construction for ditransitive sentences, rather than having to specify how the implied meanings in such sentences are derived compositionally from the verb involved and its arguments. Construction Grammar simply adds a ditransitive syntactic schema to its lexicon.

Since many FB-LTAG trees are designed as phrase and sentence schemas, they are a natural vehicle for capturing the semantic constraints specified in Construction Grammar. Feature unification provides a mechanism for composing these constraints, where the violation of a constraint is signalled by the failure to find a unifier. The Construction Grammar constraints on composition may be formulated in a feature-based event-algebra framework, which can be implemented in an augmented FB-LTAG which makes use of a larger set of features and event annotations than is used in the current system.[5]

In the Construction Grammar approach, as discussed in Aske (Aske, 1989), the verb *swim* in the English sentence "John swam across the lake" is not analyzed as a predicate which conflates two events, as suggested by Talmy (see Section 5). Instead the sentence's meaning is viewed as the composition of a verb denoting a simple event, such as *swim*, with a syntactic schema which includes the path-denoting prepositional phrase, and which denotes a second simple event, the "directed-motion" event. The complex meaning is derived not by lexical incorporation, but by a compositional semantics of constructions.

An alternative approach to using FB-LTAGs as a vehicle for Construction Grammar is to treat each manner-of-motion verb as polysemous with two senses, one that is a simple intransitive and indicates a type of motion but no change of location, and a second sense that takes the path preposition and indicates directed motion, i.e., change of location, in addition, as discussed in Section 2. The relationship between the two intransitive senses can be captured by a productive lexical rule allowing the second sense to be generated automatically for all of the relevant verbs during a pre-processing stage, as was implemented by (Voss, 1996). This type of lexical rule is intended to be compatible with Levin and Rappaport's approach to lexical representation (Levin and Rappaport-Hovav, 1995), as discussed in Section 2. The distinction between these two approaches can be characterized from an implementation point of view as a choice between precompiling the pairs in the lexicon or producing them dynamically during run-time.

[5]Some of the constraints make reference to AI notions such as preconditions, which would have to be encoded in the semantics of the events described by verbs and trees.

5. Capturing cross-linguistic generalizations

The effectiveness of using semantic features in an FB-LTAG is most clearly illustrated by looking at translation examples which heretofore seemed to necessitate interlingua solutions.

5.1. SYNCHRONOUS TREE-ADJOINING GRAMMARS

FB-LTAGs can be used for translation based on the Synchronous Tree-Adjoining Grammar (STAG) formalism, (Abeillé and Schabes, 1989), (Abeillé, 1990), (Shieber and Schabes, 1990; Shieber and Schabes, 1991), (Palmer et al., 1995), which is uniquely suited to combining the strengths of both the transfer method and the interlingua method. In STAGs, LTAGs are developed independently for source and target languages. Lexicalized trees of the source language grammar are put in bi-directional correspondence with target language trees. The substitution and adjunction sites on the source trees are linked to substitution and adjunction sites on the target trees. By linking the source-language syntactic site for a verb's argument with the target-language site for the same argument, variations between languages in syntactic argument realization can be handled perspicuously.

In addition to the simple linkings, which pair one source-language tree with one target-language tree, *complex* linkings may also be defined. For example, entire tree families in the source grammar may be linked to target-grammar tree families, capturing generalizations about ranges of related verb frames in both languages. Another important type of complex linking pairs a *composition* of several source-language trees with a composition of target-language trees, as described below.

After a derivation tree for a source sentence has been produced, a derivation tree for the target sentence can be created by following the links and performing the indicated substitutions and adjunctions with the appropriate items from the transfer lexicon. Synchronous TAGs can readily handle the task of both recognizing an idiom and mapping from it to its natural translation in another language.

A common objection to this approach is that it is essentially transfer based, and loses the inherent efficiency of an interlingua approach. However, by enriching the lexicalized tree structures with verb class membership information, many of the advantages of an interlingua representation can be preserved. Having already discussed the classification of English motion verbs, we will now look at how they are translated.

5.2. STRUCTURAL DIVERGENCES

For example, a classic problem is the translation of motion verbs from English to French. In English the manner of motion can be incorporated into the matrix verb, with the direction of the motion being adjoined on by a prepositional phrase, as in "John swam across the lake." In many cases, this is not allowed in French, where the direction becomes incorporated into the matrix verb, and the manner is adjoined on as an adverbial or a prepositional phrase, as in "Jean a traversé le lac à la nage."

A theoretical explanation of this phenomena has been provided by Talmy, (Talmy, 1991), who differentiates between verb-framed languages (ex: English) and satellite-framed languages (ex: French): the former map the core schema into the main verb, while the latter map the core schema into the satellite[6].

This substantial difference in how the languages incorporate information is especially troublesome for transfer based systems, since there are many possible direction prepositions that can be adjoined on, and even more different manners of motion. Each possible unique combination of manner and direction must be spelled out explicitly and associated with its appropriate French equivalent.

This type of translation mismatch has been termed *structural divergence*, and is often cited as strong evidence for the advantages of an interlingual representation (Dorr, 1993). Dorr uses Jackendoff's Lexical Conceptual Structures (LCS), described below, as an interlingua which handles these types of divergence problems.

5.2.1. *Lexical Conceptual Structures*

Lexical Conceptual Structures (LCS), Jackendoff's seminal contribution to the representation of lexical semantic information, (Jackendoff, 1972; Jackendoff, 1990), provide an independent level of semantic representation that can be *linked* to a syntactic parse. The LCS approach views semantic representation as a subset of conceptual structure, i.e., the language of mental representation. It abstracts away from syntax just far enough to enable language independent encoding, while retaining enough structure to be sensitive to the requirements for multilingual processing. Jackendoff's approach includes *types* such as Event and State, which are specialized into *primitives* such as GO, STAY, BE, GO-EXT, and ORIENT.

This approach to semantic representation has been utilized in a number of natural language processing systems. One application of the LCS formal-

[6](Talmy, 1991) gives definitions for terms such as *core schema, satellite*. The *core schema* is the schematic core of the framing event. The *satellite to the verb* is the grammatical category of any constituent other than a nominal complement that is in a sister relation to the verb root.

ism, the PUNDIT/KERNEL text processing system (Palmer et al., 1993), makes use of LCS as a way of encoding the predicate-argument structure and selectional preferences of verbs. LCS allows verbs to be expressed as complex predicates which select for typed arguments. Linking rules are used to map to alternative syntactic realizations. In PUNDIT, the explicit syntactic realization of an AGENT or an INSTRUMENT results in substantial changes to the semantic representation that is produced.

They have also been used by the Unitran machine-translation system (Dorr, 1993), as an *interlingua*. To translate a sentence, this system first converts it into a language-independent structure (an interlingua based on the LCS representation). This structure is then used to generate the target sentence in the other language. The *linking rules* of LCS, which relate argument positions in a semantic predicate with syntactic positions in a parse tree, provide a framework for representing syntax-semantics correspondences that can vary from language to language. For instance, the same representation that was given above for "I like Mary," can also be used to generate the Spanish translation, "María me gusta," i.e., "Mary pleases me", with the syntactic distinction (i.e., the subject-object reversal) being captured by the linking rules for Spanish.

5.2.2. *An example*

Based on the assumption that motion and manner of motion are conflated in a matrix verb like *swim*, the use of LCS allows Dorr to tease apart the three separate concepts of motion, direction, and manner of motion (Dorr, 1993) in the example sentence. Each one is represented separately in the interlingua representation, as GO, PATH and MANNER, respectively.

John swam across the lake.

$[_{\text{Event}}$ CAUSE $([_{\text{Person}}$ JOHN],
$[_{\text{Event}}$ GO $([_{\text{Person}}$ JOHN],
$[_{\text{Path}}$ ACROSS $([_{\text{Position}}$ AT $([_{\text{Thing}}$ JOHN], $[_{\text{Thing}}$ LAKE])])])])]
$[_{\text{Manner}}$ SWIMMINGLY])]

Then the French linking rules can allow these components to be recombined in a different order, this time incorporating GO and DIRECTION into *traverser* and leaving MANNER by itself as an adjunct.

The STAG approach, though fundamentally a transfer approach, can still translate these classes of sentences in an efficient manner, but this time based on a different linguistic assumption: that the GO component, (or *directed-motion* feature) is not conflated with the matrix verb, but rather is contributed by the prepositional phrase when it adjoins. The end result is the same, namely that the event of *swimming across the lake* is characterized by three separate semantic components, represented as features in the STAG implementation, which must unify with the characterization in the target language. This allows us to capture the generalization that the *path* component of some events, which is described by *across* in English, is also expressed by the verb *traverser* in French (see Figure 8). To this degree, the STAG solution to this problem incorporates the advantages of interlingual semantic generalization. However, a separate level of representation with the full semantic decomposition of the source input is not required. Instead, the STAG transfer lexicon pairs an abstract representation of a generic directed-motion phrase from the source language with a similar abstract representation of a generic directed-motion phrase from the target language. The transfer lexicon thus abstracts away from individual pairs of lexical items to pairs of verb classes and co-indexed shared semantic features. This is possible because the lexical items in the two grammars are equipped with cross-linguistic verb-class membership features and certain other cross-linguistic semantic features such as directionality.

In the above example, if *swim* is categorized as a *manner-of-motion* verb, and *across* is recognized as supplying *path* information and a *directed-motion* event; a combination of the trees anchored by these two words (using the adjunction operation) results in a *directed motion* event with a specific *path* and *manner*. In the transfer lexicon, a *complex* linking specifies that the composition of an English *manner-of-motion* verb tree with an English *path* prepositional tree can be mapped to a French composition of *directed-motion* verb tree and *manner* preposition phrase tree (see Figure 9). The explicit lexical items on both sides remain unspecified, so that these two partial structures can represent all the possible combinations of manner of motion and path for their respective languages, as in Figure 9. However, the transfer algorithm ensures that the particular lexical items on the source side constrain the selection of correct corresponding lexical items on the target side, by enforcing co-indexation of the cross-linguistic shared semantic features, such as *path* and *manner* features. So for our example sentence only lexical items with features that unify with *path: across* and *manner: swim* can instantiate the source language tree.[7]

[7]We are currently implementing a new version of Synchronous TAGs in C in which mappings can be predicated between underspecified constructions, including partial derivations, that will facilitate this type of analysis. Crucially, this new implementation allows coreferences of features (such as *manner* and *path*) with lexical anchors.

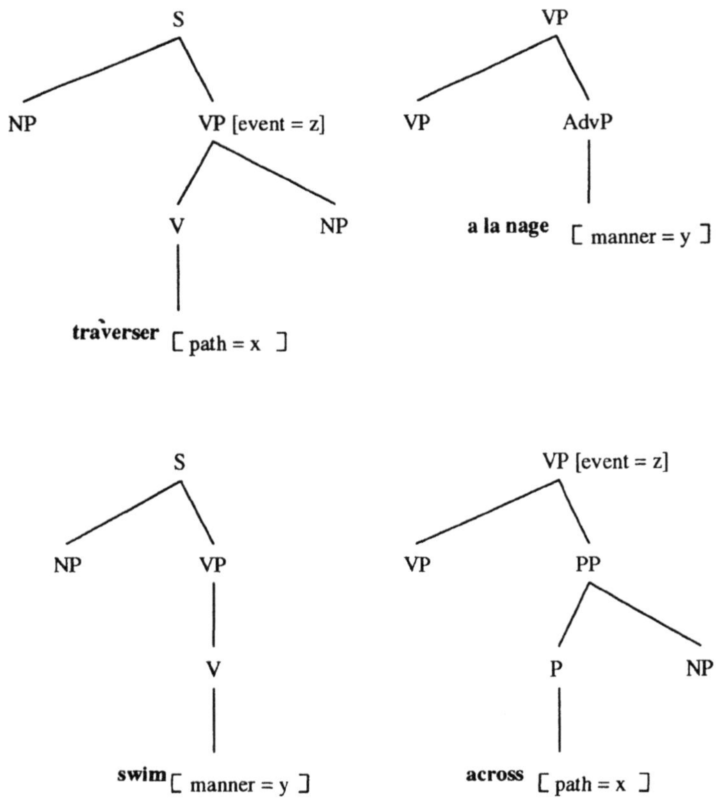

x = ACROSS and y = SWIM

Figure 8. Lexical entries for *swim*, *across*, *traverser* and *nager*

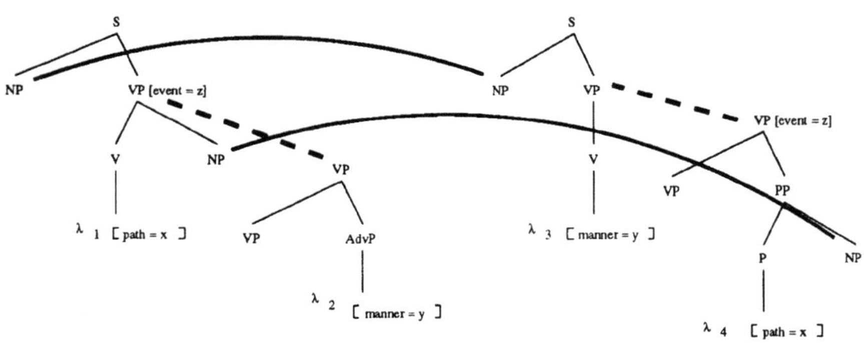

Figure 9. Transfer lexicon link between directed-motion schemas in English and French

6. Comparison to other MT approaches

In this section we compare our approach to the LCS interlingua approach described in Section 5.2.1 in terms of its ability to capture lexical semantic information that is relevant to making cross-linguistic generalizations. We also discuss the ACQUILEX system, a typed-feature structure framework that, like STAG, is essentially transfer-based. It uses a similar approach for structural divergence examples like the manner-of-motion verbs.

6.1. LCS DISCUSSION

To make use of the LCS, an implementation must include an independent syntactic parser which produces parse trees for its input. Then, a semantic interpreter which encodes the linking rules of the LCS component is run on the parse tree, resulting in the production of instantiated LCS predicate-argument structures. Our LTAGs allow the semantic generalizations captured by an LCS to be encoded directly in the syntactic structure of the lexical entry, allowing much closer interaction between syntactic and semantic processing. The application of syntactic and semantic constraints is performed in a single unification step during the parsing process, allowing simultaneous feedback from syntax and lexical semantics on possible linguistic structures.

The table below summarizes the mechanisms that are used by Lexical Conceptual Structures and Tree-Adjoining Grammars for *transparent* and *efficient* representation of the requisite lexical semantic information. The tree families provide an explicit representation of the alternative syntactic realizations which are implicit in the LCS linking rules. The predicate argument structure is captured explicitly by both methods, as the domain of locality of the TAG and the thematic relations of the LCS. Both methods can place arbitrary selectional restrictions on these arguments. The semantic primitives of the LCS are represented as the LCS predicates, whereas in the TAG approach they are expressed as features.

We have argued that the extended domain of locality of FB-LTAGs provides an especially suitable environment for representing a level of lexical semantic information that is equivalent in depth of coverage to LCS. FB-LTAGs have a significant advantage, though, in that they do not require two separate structures for syntax and semantics - the syntactic structures are the building blocks of the semantic representation.

Criteria	*LCS*	*FB-LTAG*
Levels of Representation	Two	One
Syntactic Realizations	Linking Rules	Tree Families
Predicate-Argument Structure	Lexical Conceptual Structure	Elementary Trees
Selectional Restrictions	Added to LCS as Constraints	Features in Trees
Semantic Components	Predicates of LCS	Features for Class Membership

6.2. ACQUILEX DISCUSSION

The ACQUILEX system (Copestake and Sanfilippo, 1993; Briscoe et al., 1994) provides a typed-feature structure framework for doing MT, based on a HPSG/categorial grammar formalism for the source and target languages. Like an LTAG-based MT approach, the ACQUILEX MT framework uses a set of bi-directional transfer rules, called *tlinks*, to pair up translation equivalents in the two languages. The *tlinks* pair feature structures from the source and target languages. A translation is performed by parsing an input sentence with a unification-based source-language parser, resulting in a source-language feature structure that is mapped to a target-language feature structure, which is used to generate the output.

Since the linked feature structures may be partially parameterized, the ACQUILEX system, like the LTAG-based system, is able to make generalizations about the translations of semantically similar sentences, such as the directed-motion sentences discussed above.

While the two approaches share many of the same underlying assumptions about lexical semantic representation, the LTAG approach offers some unique advantages stemming from its use of tree composition operations. Its efficient handling of non-compositional phrases, discussed below in section 7, is particularly important in the context of MT, where it is undesirable to treat such phrases too rigidly. For example, in translating the phrase *take unfair advantage of* from English, it should be possible to recognize that the target form is systematically related to the translation of the phrase *take advantage of.* An LTAG representation of these two English phrases highlights their similarity in such a way that it is easy to make reference to the common elements of both phrases when defining a transfer lexicon (Abeillé, 1990). This is not the case for the representation of these phrases in other grammars, such as the categorial grammars used by ACQUILEX.

7. Accommodating lexically specific translations

The combination of generalizations based on verb class memberships with an essentially lexicalized transfer-based approach offers a powerful basis for handling the range of phenomena presented by the machine translation task. The structure of FB-LTAGs allows syntactic dependencies, such as agreement and subcategorization, to be localized in the lexical items and their associated elementary trees. Since transfer rules are stated as correspondences between nodes of the elementary trees associated with lexical entries, we can define lexical transfer rules over a larger domain of locality. This provides an elegant method for translating the language-specific idioms and constructions which are not strictly compositional and which can pose problems for interlingua approaches. The fixed parts of an expression are built into the tree as lexical anchors to provide for non-compositional readings.

For example, verb particle constructions such as *lift up*, or *give away*, appear as trees, but here the verb and the particle both anchor the tree, and are already in the correct position. However, the flexibility of the parsing system allows noun phrases to be inserted in between the verb and its particle if necessary, so both *lift the baby up* and *lift up the baby* can be represented. In the same way, as mentioned above, idioms and light verb constructions, such as *kick the bucket* and *take advantage of*, appear in the lexicon as trees, with all of the essential lexical items anchoring the tree. Once again, these "frozen strings" are flexible enough, that where there is a need to treat them productively, where other lexical items need to be inserted, as in to *prendre une vraie veste* or *vraiment prendre une veste*, (to come a real cropper, or to really come a cropper),[8] these can also be accounted for (Abeillé and Schabes, 1989), (Abeillé, 1990). Since the anchor for the tree is the entire phrase, *prendre une veste*, the modifier, whether it is an adjective or an adverbial, adjoins on (albeit in different locations) and is dependent from the same node in the derivation structure. The transfer lexicon pairs the phrasal trees in one language with phrasal trees in another language, and then adjoins on modifiers appropriately. In this way, Synchronous TAGs can readily handle the task of both recognizing an idiom, and mapping from it to the natural translation in another language.

[8]This is a British idiom that means to run into trouble

8. Conclusion

In summary, we have presented an approach to machine translation based on Synchronous Tree Adjoining Grammars that is truly a hybrid approach – it seamlessly combines the advantages of transfer-based techniques with a capability for utilizing powerful generalizations that are usually only associated with interlingua approaches. This unique combination is possible because the lexicalized style of our grammar allows a range of syntactic and semantic features to be stored with each lexical item. The semantic features can be used to associate particular event types, such as *directed-motion*, with the auxiliary trees which allow *path* prepositional phrases to adjoin on. Then, when a *path* prepositional phrase is adjoined onto a suitable type of verb, (and that type can range from *manner-of-motion* verbs to *sound-emission* verbs), then the event associated with the entire verb phrase becomes a *directed-motion* event. The features associated with the verb phrase, which in the case of "John swam across the lake" would be *event: directed-motion, manner: swim, path: across*, are co-indexed in the transfer lexicon with the features for a French partial derivation. Once the features of the French partial derivation have been instantiated, then only French lexical items whose own *manner* and *path* features can also unify with *manner:swim* and *path:across* can be substituted into the derivation. This provides an elegant and efficient method of associating together in the transfer lexicon entire classes of lexical items that are structurally divergent in the source language and target language, yet in a predictable fashion. The lexicalized approach also allows a fine-grained treatment of frozen and semi-frozen expressions such as idioms.

Much work remains to be done on finding optimal verb classification schemes and appropriate typed feature structure systems for encoding them. We will continue to look at Levin classes as well as the ACQUILEX framework for the most appropriate ways to organize our tree families without losing any of the linguistic desirability of the Tree-Adjoining Grammar formalism.

Acknowledgements

We gratefully acknowledge the gentle wisdom and guidance of Aravind Joshi and the members of the XTAG group, as well as Bonnie Dorr, Patrick St. Dizier, Laura Wagner and our reviewers. This work has been supported in part by NSF grant SBR 8920230 and Darpa grant #N66001-94C-6043.

References

Abeillé, A. (1990). Lexical Constraints on Syntactics Rules in a Tree Adjoining Grammar.

In 28th *Meeting of the Association for Computational Linguistics.*

Abeillé, A. and Schabes, Y. (1989). Parsing Idioms in Lexicalized TAGs. In *Proceedings of EACL '89*, pages 161–65.

Aske, J. (1989). Path predicates in english and spanish: A closer look. In *Proceedings of the Berkeley Linguistics Societ*, Berkeley, California.

Becker, T. (1993). *HyTAG: A new Type of Tree Adjoining Grammars for Hybrid Syntactic Representation of Free Word Order Languages.* PhD thesis, Universität des Saarlandes.

Briscoe, E. J., Copestake, A., and de Paiva, V. (1994). *Inheritance, defaults and the lexicon.* Cambridge University Press.

C. J. Fillmore, P. K. and O'Connor, M. C. (1988). Regularity and idiomaticity in grammatical constructions: The case of *let alone*. *Language*, 64(3).

Candito, M.-H. (1996). A principle-based hierarchical representation of ltags. In *Proceedings of COLING-96*, Copenhagen, Denmark.

Copestake, A. and Sanfilippo, A. (1993). Multilingual lexical representation. In *Proceedings of the AAAI Spring Symposium: Building Lexicons for Machine Translation*, Stanford, California.

Doran, Hockey, Hopely, Nasr, Rambow, Rosenzwieg, Sarkar, Srinivas, and Xia (1997). Maintaining the forest and burning out the underbrush in xtag. In *Computational Environments For Grammar Development And Linguistic Engineering, ACL-97 Workshop.*

Dorr, B. (1993). *Machine Translation: A View from the Lexicon.* MIT Press, Boston, Mass.

Egedi, D., Palmer, M., Park, H., and Joshi, A. (1994). Korean to english translation using synchronous tags. In *First Conference of the Association for Machine Translation in the Americas*, Columbus, MD.

Goldberg, A. E. (1995). *Constructions. A Construction Grammar Approach to Argument Structure.* University of Chicago Press, Chicago, Ill.

Han, C., Xia, F., Palmer, M., and Rosenzweig, J. (1996). Capturing language specific constraints on lexical selection with feature-based lexicalized tree-adjoining grammars. In *Proceedings of the International Conference on Chinese Computing.*

Jackendoff, R. (1972). *Semantic Interpretation in Generative Grammar.* MIT Press, Cambridge, Massachusetts.

Jackendoff, R. (1990). *Semantic Structures.* MIT Press, Boston, Mass.

Joshi, A. K. (1985). Tree Adjoining Grammars: How much context Sensitivity is required to provide a reasonable structural description. In Dowty, D., Karttunen, I., and Zwicky, A., editors, *Natural Language Parsing*, pages 206–250. Cambridge University Press, Cambridge, U.K.

Joshi, A. K., Levy, L., and Takahashi, M. (1975). Tree Adjunct Grammars. *Journal of Computer and System Sciences.*

Kittredge, R. and Polguere, A. (1991). Dependency grammars for bilingual text generation: Inside fog's stratificational models. In *Proceedings the International Conference on Current Issues in Computational Linguistics*, Penang, Malaysia.

Levin, B. (1993). *English Verb Classes and Alternation, A Preliminary Investigation.* The University of Chicago Press.

Levin, B. and Rappaport-Hovav, M. (1995). *Unaccusativity. At the syntax-semantics interface.* MIT Press.

Palmer, M., Park, H., and Egedi, D. (1995). The application of korean-english machine translation to a military message domain. In *Fifth Annual IEEE Dual-Use Technologies and Applications Conference.*

Palmer, M., Rosenzweig, J., Dang, H., and Xia, F. (1997). Capturing syntactic/semantic generalizations in a lexicalized grammar. presentation in working session of Semantic Tagging Workshop, ANLP-97.

Palmer, M., Weir, C., Passonneau, R., and Finin, T. (1993). The kernel text understanding system. *Artificial Intelligence*, 63:17–68. Special Issue on Text Understanding.

Palmer, M. and Wu, Z. (1995). Verb semantics for english-chinese translation. *Machine Translation*, 10:59–92.

Schabes, Y. (1990). *Mathematical and Computational Aspects of Lexicalized Grammars.* PhD thesis, Computer Science Department, University of Pennsylvania.

Schabes, Y., Abeillé, A., and Joshi, A. K. (1988). Parsing strategies with 'lexicalized' grammars: Application to Tree Adjoining Grammars. In *Proceedings of the* 12th *International Conference on Computational Linguistics (COLING'88)*, Budapest, Hungary.

Shieber, S. and Schabes, Y. (1990). Synchronous Tree Adjoining Grammars. In *Proceedings of the* 13th *International Conference on Computational Linguistics (COLING'90)*, Helsinki, Finland.

Shieber, S. and Schabes, Y. (1991). Generation and Synchronous Tree Adjoining Grammars. In *Proceedings of the Fifth International Workshop on Natural Language Generation*, Pittsburgh, Pennsylvania.

Talmy, L. (1985). Lexicalization patterns: Semantic structure in lexical forms. In *Language Typology and Syntactic Description*, volume 3. Cambridge University Press.

Talmy, L. (1991). Path to realization: A typology of event conflation. In *Proceedings of the Berkeley Linguistics Society Parasession*, Berkeley, CA.

Vijay-Shanker, K. and Joshi, A. K. (1991). *Unification-based Grammars*, chapter Unification Based Tree Adjoining Grammar. MIT Press.

Voss, C. (1996). *Interlingua-based Machine Translation of Spatial Expressions.* PhD thesis, University of Maryland, Department of Computer Science.

SOME SYNTACTIC CONSEQUENCES OF ARGUMENT STRUCTURE DIMENSIONS

CHARLES JONES
English Department, 3E4
George Mason University
Fairfax, VA 22030, U.S.A.

Abstract. This paper discusses the nature and scope of the relation between morphology and (phrasal) syntax. I propose two dimensions along which the "space" of lexical argument structure might be partitioned, and then I illustrate how these dimensions can be exploited to address puzzles presented by various morphosyntactic phenomena in English. The direction of the argument will be toward support for a conception of the relation between morphology and syntax in which the rules of one differ in interesting ways from the rules of the other—a "lexicalist" conception—and away from a "syntactic" conception of that relation, in which the rules of morphology *are* the rules of syntax.

In this paper I propose a couple of formal dimensions along which the mechanisms of θ-role assignment may be aligned. The configurations of θ-properties admitted by this novel realignment (along with previously proposed alignments) make sense of various morphosyntactic phenomena.

(Pustejovsky 95: 5) notes the increasing interpenetration of linguistic research and computational natural language processing (NLP) research: "Positioned at the center of this synthesis is the study of word meaning, lexical semantics," and he goes on to observe that, "without an appreciation of the syntactic structure of a language, the study of lexical semantics is bound to fail. There is no way in which meaning can be completely divorced from the structure that carries it." This paper is located toward the theoretical linguistic end of a dimension that stretches from "pure" linguistics to "pure" computational problems of NLP.

Within the broad outlines of a linguistic theory like Government and Binding (GB: Chomsky 81), one might expect an enrichment of one component of the grammar to yield a corresponding simplification in some other component. The two components of central relevance to

P. Saint-Dizier (ed.), Predicative Forms in Natural Language and in Lexical Knowledge Bases, 257–284.

morphosyntactic phenomena are, of course, morphology and syntax: the theories of word structure and phrase structure. Let us agree with (Selkirk 82) that both domains of structure have a kind of "syntax". The question remains what the nature of the relation is between the two syntaxes. In one conception of the relation, call it the Morphology = Syntax (M=S) conception, the rules of morphology are (perhaps a subset of) the rules of syntax. This conception of the relation between morphology and syntax informs many studies, including (Baker 85, 88), (Lieber 92), and (Hale & Keyser 93). At another extreme, a (strong) Lexicalist Hypothesis would maintain that, while the two syntaxes may share a certain vocabulary (category labels, etc.), the nature of the rules of each differ from the other. (DiSciullo & Williams 86) is a recent comprehensive exposition of such a position.

The enrichment of argument structure proposed in the present paper is intended to support the Lexicalist Hypothesis; hence it bears some comment at the outset about what kind of evidence might be seen as advantageous for one theory over the other. Proponents of the M=S theory point out that, to the extent that the syntactic mechanisms invoked for morphological phenomena are independently justified in the theory of phrasal syntax, their extension into the realm of morphology involves no net complication to the theory as a whole. It is the point of view of this paper that the syntactic mechanisms employed in the M=S theories are usually a crucially distinctive subset of the syntactic mechanisms available to phrasal syntax: they generally involve only "head-movement" mechanisms. Head-movement mechanisms differ from phrase-movement mechanisms by their extreme locality constraints: they are constrained by the ECP to a strictly uninterrupted chain of antecedent government from antecedent to trace. As (Williams 87, etc.) has pointed out in various places, this extreme locality is characteristic of θ-role assignment. On this conception, the rules of morphology are no accidental subset of the rules of phrasal syntax, they are a separable and distinct set of rules. The coherence and locality of the morphological rules indicate to the lexicalist that the rules of word formation can be isolated in an interesting way from those of phrasal syntax.

Evidence in favor of a theory A over a theory B typically takes the form of a demonstration that the properties of a morphosyntactic phenomenon follow naturally from the principles of theory A in a way that can only be covered by theory B with the addition of construction specific or ad hoc principles. In this regard, the M=S theory has at its disposal the entire technicalia of modern syntactic research, while the Lexicalist theory is typically supposed to labor within the strict confines of classical 1981 GB theory, where θ-roles are mere place holders for the number of a lexical item's arguments, and the θ-criterion requires only that each be assigned to a unique NP.

One familiar way of beefing up the theory of θ-roles is to suppose that they have "thematic content", like "actor", "theme", etc., which can be referred to in syntactic rules. We will not pursue this course here. An alternative enrichment of θ-roles would be to assign some structure to the space they assume within the θ-properties of a lexical entry. It is this kind of enrichment that the θ-dimensions proposed here are intended to articulate. The idea underlying the present proposal is that θ-properties that can be readily discerned from the primary linguistic input may interact to determine certain morphosyntactic phenomena along lines that are neither obviously syntactic nor obviously related to "thematic content".

In section 1, I outline and clarify various modes of θ-role licensing, and I locate the present proposal within the particular theory of predication initiated in Williams 80, refined most relevantly in Williams 87 and developed in subsequent work (e.g., Di Sciullo & Williams 87, Williams 89, 94). In section 2, I lay out the two dimensions of argument structure that form the central hypothesis of this paper and then in section 3 show how these dimensions make sense of various V → A derivations. In sections 4 and 5, I show how a lexicalist account of some morphosyntactic phenomena compares with syntactic accounts that have been proposed in the literature.

1. θ-role Assignment

The essential idiomatic argument array of a lexical item is contained in its list of θ–roles. The Projection Principle and the θ-criterion in general require that each θ-role of a lexical item be assigned to a unique NP. The θ-theoretical mechanisms by which these θ-role assignments are accomplished have an extremely local character. Of the three mechanisms we consider here, two crucially work within the domain of government by a lexical head. The remaining mechanism, predication, is not so locally constrained. In section 2 we return to the matter of constraints on predication.

1.1. INTERNAL ARGUMENTS

There are various modes of thematic discharge. Take, for example, the assignment, under government, of a θ-role to an internal argument, as in (1).

(1)

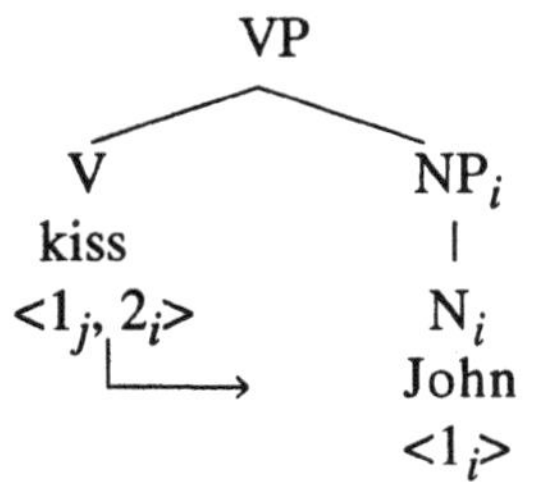

In (1), *kiss* has an internal θ–role, subscript-*i*, to assign to NP. The head of NP itself projects some θ–role to its maximal projection. Here I follow (Williams 81) in supposing that the characteristic external θ-role associated with N is "R", mnemonic for "referential". The discharge of the *i*-role of *kiss* consists of assigning the *i*-index to the NP. The discharge of the projected R θ-role of N consists in having its projected θ–role indexed by the *i*-argument of *kiss*. Hence, as (Higginbotham 83) argues, θ–discharge is asymmetrical. As indicated by the directionality of the arrow in (1), the tail corresponds to the independent assigner, the head points toward the dependent assignee. (In the following discussion, I assume the typographically simple notation of coindexing in cases where this asymmetry is not at issue.)

1.2. MODIFICATION

Consider the structure of prenominal modification. For (Higginbotham 85), the central thematic mechanism of modification is *θ-identification.* In a case of ordinary prenominal adjectival modification, like *smart dog,* the θ–role projected by *smart*, presumably corresponding to the "subject" of the predicate *smart,* must be identified with the projected θ–role of *dog,* so that the cumulative referent will be something smart and something that is a dog.[1]

(2)

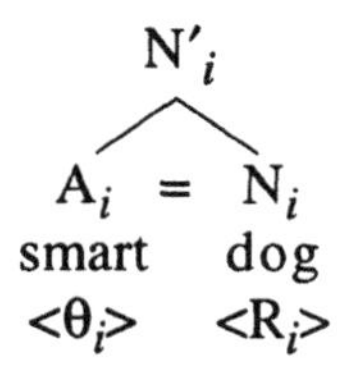

Note that in θ–identification, as in θ–assignment, the θ–discharge is

[1] We will not address here the question of how an adjective becomes relativized to a context; e.g., a smart dog is a dog that is smart relative to other dogs. I assume, following (Higginbotham 85), that this relativization is itself a form of thematic discharge, of a second argument of the adjective. Hence, an adjective like *smart* has one argument for θ-identification with the element it modifies, and another argument place for the property relative to which the adjective is to be evaluated.

accomplished in a configuration of government.

1.3. PREDICATION

Contrast the above two forms of θ–discharge with an ordinary case of predication, in which a θ–role is projected to the maximal projection of its head, and is assigned outside of that projection.

(3)

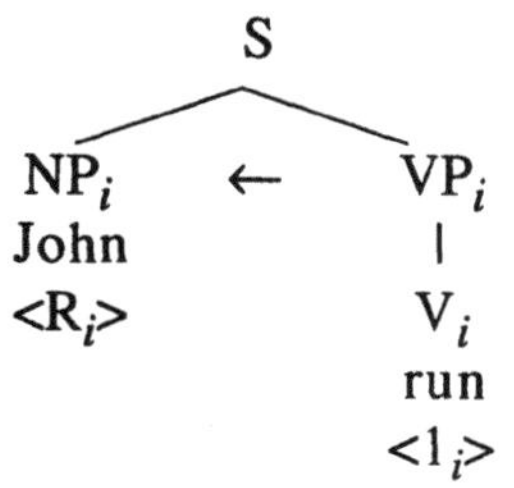

In (3) NP is the dependent element, whose projected θ–role is discharged by being assigned an index. However, the independent assigner is not a lexical head, and the assignment is outside of the government domain of the head with which the assigned θ–role is associated.

The mechanism by which the *run* θ-role in (3) projects to VP we call *vertical binding,* after (Williams 87). We suppose that the projection of a lexical head can carry at most one index of a θ-role, and it is via the discharge (= predication) of the index of that θ-role that the phrasal projection of the lexical head lawfully enters into the syntax.

1.4. θ-THEORY OF RAISING, PASSIVE, AND AUX

The present conception of argument structure is most conveniently located within a theory in which argument structure takes part in the syntax in certain ways. Specifically, I adopt (Williams 87)'s theory of raising and passive as operations on argument structures.

Briefly, where the standard GB theory of raising involves syntactic movement from embedded subject to matrix subject position, as in (4a), the argument structure theory supposes that it is the projection of the external θ–role of the embedded predicate "through" the projection of the raising predicate, as in (4b).

(4) a. John$_i$ seems [t$_i$ to run]
 b. John$_i$ [$_{VP}$ seems$_ø$ [$_{IP}$ e$_i$ [$_{I'}$ to [$_{VP}$ run$_i$]$_i$]$_i$]$_i$]$_i$

In terms we adopted above, the index of the external argument of *run* in (4b) is vertically bound by all of the phrasal nodes between *run* and *John.*

This kind of vertical binding is possible only because neither the raising predicate itself nor the embedded clause projects an external θ–role of its own. In fact, this absence of vertical binding of an independent θ–role is the only condition under which such long distance projection of an embedded θ-role is allowed.

Likewise, in the case of passive, the object assumes the position of subject not from syntactic movement, as in the standard analysis in (5a), but rather from the projection of some internal θ–role out of the VP and through the inflectional system to the θ–empty subject position, as in (5b).

(5) a. $John_i$ was kissed t_i
b. $John_i$ $[_{IP}$ was $[_{VP}$ kissed $e_i]_i]_i$

In general, we suppose that the auxiliary system is transparent with respect to the vertical binding of external arguments—the elements of the auxiliary system exist only to mediate the projection of the external argument of the matrix V.

(6) $They_i$ $[could_\emptyset$ $[have_\emptyset$ $[been_\emptyset$ $[being_\emptyset$ [beaten $e_i]_i]_i]_i]_i$

2. Verbal θ-Mechanics

Let us now "look under the hood" of θ-role assignment to consider various ways in which the different modes outlined above vary. In this section we concentrate on verbal θ-role assignment because it has the robust property of generally being obligatory. That is, nouns and adjectives in general are not required to phonetically realize their complements in the syntax.

(7) a. $[_{VP}$ feared *(John)]
b. $[_{NP}$ fear (of John)]
c. $[_{AP}$ afraid (of John)]

While, in GB-theoretical terms, the Projection Principle requires that "lexical structure...be represented categorially at every syntactic level" (Chomsky 86: 84), the examples in (7) indicate that only with V is that representation strictly required to be phonetically realized. It will be useful to have a term to distinguish these apparently different phonetic requirements on the realization of argument structure. When phonetic realization of lexical structure is required, as in V, I will say that that structure is "visible to the Projection Principle", and I adopt a [± vis] notation to distinguish two modes of θ-assignment. Within this notation, the lexical structure of V is [+ vis], that of N and A are [- vis].[2]

2 I generally ignore P in the discussion, though see fn. 9.

2.1 THE MARK OF THE EXTERNAL ARGUMENT

Underlying the standard theory of predication is a natural and economical assumption about argument structure. This assumption can be broken into two parts.

(8) a. The argument structure of a lexical item is maximally structured by marking one and only one argument as a distinguished argument. Let us designate this significant marking with a lowercase Greek sigma: σ-marking.

b. The invariant choice of the "one and only one distinguished argument" is the "external argument", that argument that is characteristically associated with "subject position", external to the maximal phrasal projection of the relevant lexical item.[3] Let us call this special θ-role the *characteristic external argument* of the verb.

It does not follow from the assumptions in (8) that the characteristic external argument always projects externally to VP. In fact, it must not follow, because different outcomes, for example passive, are possible. So, in a sense, any distinctive marking of the characteristic external argument is irrelevant to the actual choice of the θ-role that ultimately gets projected externally. Let us reserve the term "external θ-role" for a θ-role that actually projects externally from its head's phrasal projection.

Suppose that the irrelevant marking of the characteristic external argument does not exist. That is, suppose that the characteristic external argument is a featureless cipher, unmarked in any way by its lexical head. It will be convenient to refer to the absence of any special marking on a characteristic external argument. Let us call it "Ø-marking".

If the characteristic external argument receives no marking from its V, we may ask where the subject (of a typical verb) gets its typical thematic connotations of "actor". An idea with a long tradition, one we will assume here, is that the thematic content of the subject is determined compositionally by the verb phrase as a whole.[4]

2.2 THE FUNCTIONAL MECHANICS OF PREDICATION

We may now inquire how this Ø-marked θ–role comes to participate in the

3 This assumption was implicit in (Williams 80), was made explicit in (Williams 81), and has been tacitly assumed in many places since.

4 See, for example, (Williams 74) and (Marantz 84). Also, (Hale & Keyser 93) set out an explicit theory of how this "actor" connotation follows from the configurational properties associated with being external to the VP. See the discussion of Hale & Keyser below in section 5.

ubiquitous, powerful, non-governed relation of predication. Our appeal will be to a form of the canonical requirement on non-governed relations, the ECP.

A "non-governed relation" is in general a binding relation that spans one or more domains of government. The paradigmatic relation of this kind in phrasal syntax is the relation between a *wh*-moved element and its trace. The relation between an externally assigned θ–role and its lexical head is also in this general sense a non-governed relation.

In its general form, the ECP requires the bound element in a non-governed relation to be "kept track of" in some way. Let us call this "keeping track" *licensing*. The syntactic ECP allows the bound element to be licensed in two ways: it could either be directly governed by a lexical item, *lexical government*, or connected with its antecedent through a strictly locally connected chain of c-commanded dependencies, *antecedent government.*

Consider now an ordinary case of predication. Assuming (Chomsky 86)'s generalized X′ theory of S = IP, predication is ordinarily "across" an inflection relation mediated by Infl.[5]

(9)

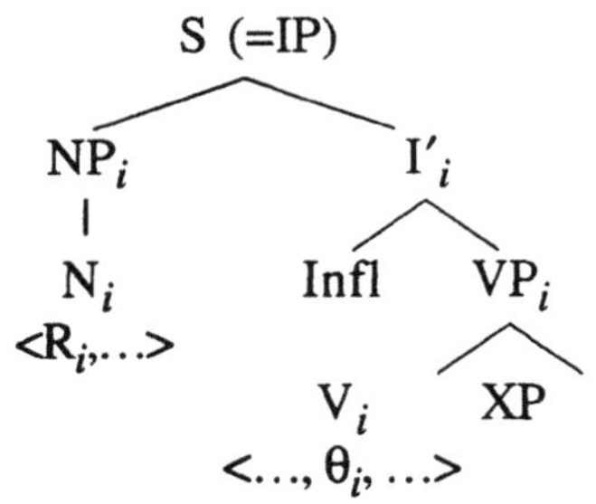

Let us assume that this government by a functional head like Infl is not simply a concommitant characteristic of predication, but rather is one of its crucial components. In ECP terms, let us suppose that government by a functional head is how an externally projecting θ–role can be licensed.

(10) *External θ –role projection principle* (XPP)
External θ–roles must be θ-licensed.

(11) A θ-role is θ-licensed if
θ is governed by a functional head.

In the next section we consider an additional mode of θ-licensing.

[5] In recent theory, Infl has "exploded" into a number of functional categories, each with its own X′ phrasal projection. I will assume these phrasal projections are "transparent" to vertical binding, like the auxiliaries in (6).

2.3 THE *DIRECT* ARGUMENT

A lexical item with multiple θ–roles must be capable of marking its internal arguments in various ways. For example, a three-argument verb like *put* must determine that the argument receiving the "theme" interpretation must receive accusative Case, that it is an obligatory argument, and perhaps other things. Likewise, it must mark the argument that receives the "location" interpretation that it must be governed by the appropriate preposition, that it also is an obligatory argument, etc.

(Levin & Rappaport 86), following (Marantz 84), propose a special kind of argument marking that proves to be especially useful in the analysis of adjectival passives. They propose that, in addition to the above mentioned kinds of marking, a lexical item can mark one of its internal arguments as the *direct* argument. We can incorporate this special marking into the present proposal, and at an advantage. Because the characteristic external argument is no longer considered to be σ-marked in any way, we can consider internal *direct* argument marking to be the "one and only one" kind of σ-marking considered in (8a). The maximal structure of an argument structure of a verb, then, would be as in (12).

(12)

< external,	direct,	internals ..>
|	|	|
Ø	σ	non-σ
	direct Case	P-governers
	(optionality)	(optionality)
	etc.	etc.

One of the themes of some of the sections to follow is that σ-marked arguments are θ-identified, in the XPP sense above in (10). Hence the θ-licensing conditions begun in (11) are completed here in (13).

(13) A θ–role is θ-licensed if

i. θ is governed by a functional head, or
ii. θ is a σ-marked argument of a lexical head.

2.4 NONDEFAULT Θ-ROLE ASSIGNMENT

What has been outlined so far constitutes the "unmarked" version of θ-role assignment, which I refer to as *default*. It is possible that a lexical item may not θ-mark its arguments in this default way. Consider the verb *move*. *Move* has a "physical" sense that is fully in keeping with default θ-role assignment. In (14a), the subject argument has the default "actor" connotations.

(14) a. John moved the piano.
b. The concert moved John.

In (14b), on the other hand, *move* has a more "emotive" sense, in which the subject does not have "actor" connotations. A typical way of distinguishing between the two modes of θ-role assignment would be to assign to *move* two different lists of thematically contentful θ-roles, "Actor/Patient" for the *move* in (14a) and "Impact/Experiencer" for (14b). This kind of solution is not in the spirit of the present proposal. Instead, I propose that the difference between the two modes of θ-role assignment for *move* consists simply of the difference between the *default* θ-role assignment of (14a), in which nothing out of the ordinary happens, and a *nondefault* θ-role assignment in (14b), in which the lexical head dictates the special properties of each of its arguments. We need not worry about the characterization of the specific thematic contents of the arguments in the nondefault θ-role assignment. All we are supposing to be significant is the formal difference between default and nondefault θ-role assignment.

For concreteness, we will suppose that *move,* then, has two modes of θ-role assignment, default and nondefault. In the default mode, only the direct argument is σ-marked by the lexical head. In the nondefault mode, on the other hand, where every argument potentially differs from its default counterpart, each argument is σ-marked by the lexical head. The effect, in the marked nondefault case, is that the "one and only one" limitation on σ-marking is abandoned.

The picture of θ-role assignment that emerges from the foregoing considerations has the following details. There are various kinds of external θ-roles: R for nouns, Ø-marked for default verbs. Adjectives and their arguments will be discussed in the following section.

As for internal θ-roles, they are σ-marked as direct, or not σ-marked. The dimension of default versus nondefault θ-role assignment (in which all arguments are σ-marked) is orthogonal to these differences.

3. Adjectives

The properties of adjectives of various sorts can begin to show how the argument structure dimensions outlined above can be put to some use. I begin in section 3.1 with adjectives that have no apparent derivational history—lexical adjectives. In section 3.2 I proceed to consider three different kinds of morphologically derived adjectives.

3.1 LEXICAL ADJECTIVES

Non-clausal predicates embedded under verbs of appearance, for example, *seem, appear, look, sound,* etc., present a blunt paradigm.[6]

(15) a. John seems [$_{AP}$ smart].
b.* John seems [$_{NP}$ a doctor].
c.* John seems [$_{PP}$ in the shed].
d.* John seems [$_{VP}$ running].

The problem posed by (15) is familiar from the very earliest work in transformational grammar (Chomsky 1955, 1957). The judgments in (15) are not particularly subtle, so we would like to locate their explanation in a salient property of adjectives. Chomsky's original treatment of this phenomenon exploited a categorial difference between adjectives and verbs—here we will exploit a difference in argument structure that sets adjectives apart from other categories.

A particularly common property of the English adjective is its function as a prenominal modifier. In the structure of prenominal modification, (2), repeated here as (16), the adjective does not externally project any θ-role; rather, it has (one of) its θ-roles discharged by θ-identification.

(16) N'_i
$A_i = N_i$
smart dog
$<\theta_i>$ $<R_i>$

So a perfectly ordinary use of an adjective is consistent with, and even suggestive of, the absence of the kind of characterstic external argument typical of V. Rather, let us suppose that the characteristic θ-role of an adjective is a governed internal θ-role. In fact, the paradigm in (15) indicates that the θ-role of A is σ-marked, like a direct argument, as in (17).

(17) ..., A, < *direct*,... >, ...

An θ-licensing kind of account of the paradigm in (15) can now be based on the distinction just drawn. External arguments require θ-licensing. In (15b-d) the maximal projections of N, P, and V are governed by lexical *seem,* which is not a functional head. Hence the external arguments of these categories can project no further, not being θ-licensed. In (15a), on the other hand, the projected argument of *smart* is θ-licensed as the direct argument,

6 A "small clause" analysis of the phrases in (15), in which they have PRO "subjects", does nothing in itself to solve the problem.

hence it is able to project through the argument structure of *seem* to the subject position.

On the assumption that copular *be* is an inflectional element, the paradigm in (18) follows as a direct consequence.

(18) John is smart_{AP} / a doctor_{NP} / in the shed_{PP} / running_{VP}.

Further, in (19) (as well as in raising structures in general), the external argument of the embedded predicate projects through, and is θ-licensed by, the functional inflectional system of the embedded clause.

(19) John_i seems $[_{IP}\ e_i$ to be smart_{AP} / a doctor_{NP} / in the shed_{PP} / $\text{running}_{VP}]$

3.2 SOME DERIVED ADJECTIVES

In this section we consider three V → A derivations. In each case the V that feeds the derivation is accompanied by verbal inflection. That is, the morphemes that inflect the root verbs productively derive verbs. I assume in each case that the V → A derivation is true zero-derivation from an inflected verb, as in (20).

(20) $[_V\ [\ V\ ]$ + inflectional morpheme$]$ →
$[_A\ [_V\ [\ V\ ]$ + inflectional morpheme$]]$

Thus, the V → A derivations we consider here are all headless morphological constructions, in the sense of (DiSciullo & Williams 87). That is, on the assumption that derivational morphemes are the functional heads of the words they derive, zero-derivations by definition have no such heads. We have, then, in these V → A derivations, verbal argument structures embedded in morphological structures without governing functional heads. We will see that the argument structures of the V roots of these V → A derivations determine how the derived As fare in the syntax. Specifically, we will see that some arguments are able to project out of these headless structures and some are not. It will be convenient to have a term to refer to the property of being able to project externally from a construction. I will call a θ-role with this property *Xprojectable*.

3.2.1. *V+ing*

The *running* of (15d) may also, of course, be used as a prenominal adjective: *the running man*. Why then does it not pattern with *smart*$_A$ in the *seem* paradigm (15)?

We can find the answer in argument structure. The argument projected by the verb *run* is the default Ø-marked characteristic external argument of a typical V. We would expect that such an undistinguished argument could not project externally from a construction without some assistance from a functional head of some sort. In the terminology adopted above, the external argument of *run* is, by itself, not Xprojectable.

I assume that *-ing* does nothing to alter the argument structure of *run*, hence [$_V$ *running*] has the same argument structure as *run*. When [$_A$ *running*] is zero-derived from [$_V$ *running*], we are left with a lexically derived category with no Xprojectable argument. Hence its failure, even as an adjective, with *seem*.

Even with no Xprojectable θ-role, [$_A$ *running*] can still function as a prenominal adjective, because in that configuration external projection is not necessary. In prenominal position an adjectival θ-role is discharged by θ-identification.

When the root verb of the adjectival *ing*-derivation has nondefault θ-assignment, the σ-marking of its characteristic external argument allows θ-licensing through the headless zero-derivation. Note that only the nondefault, "emotive" reading of *move* in (22) is Xprojectable.

(21) a. John moved his foot.
b.* John's foot seemed moving.

(22) a. The concert moved John.
b. The concert seemed moving.

Likewise, so-called "psych" predicates, long noted for their noncanonical θ-role assignments, pattern with the nondefault properties of *move*.

(23) a. The idea interested John.
b. The idea seemed interesting.

3.2.2. *Passives*

Not all deverbal adjectives carry with them their characteristic external arguments. Adopting (Levin & Rappaport 86)'s theory of adjectival passives as essentially correct, these derived adjectives have had their characteristic external argument removed by the passive *-en* morphology. Instead, they project their direct argument.

(24) a. The car was sold.
b. The customer was sold a car.

(25) a. an unsold car
b.* an unsold customer

The syntactic passives (24) show that either one of the internal arguments can project externally. This is due to the functional auxiliary head *be*: external projection of either internal argument can be θ-licensed by the passive auxiliary.

The restriction to direct argument assignment in adjectival passives in (25) calls for explanation. We begin by distinguishing between the two English past participles (PstPrt). While the perfective PstPrt does not alter the argument structure of its host V, the passive PstPrt has the well-known property of suppressing its characteristic external argument. Hence, unlike the perfective PstPrt (and *-ing*), the passive PstPrt performs a function, which qualifies it as a functional head for the derived adjective. For reasons that will become clear below, I will characterize this property of the passive PstPrt as the "zero external role" property (ØX). Observe now that only the ØX PstPrt allows V → A.[7]

(26) a. Our agent has purchased these goods
b.* our purchased agent
c. These goods were purchased.
d. these purchased goods

We can adopt Levin & Rappaport's story about why only the direct argument is allowed to externalize from the derived A, illustrated in (25). Briefly, the direct argument must be directly marked by its lexical head. This property is inherited by the derived adjectival passive, and is passed on to the projected AP. If the direct argument is not externalized, this θ-marking will go unaccomplished, in violation of the θ-criterion.

The direct argument of the adjectival passive is thus θ-licensed in two ways. It is σ-marked by its lexical head, and its external projection is mediated by a functional head. As we would expect from a lexical item with an Xprojectable argument, it can be embedded beneath the verbs of appearance.

(27) The car seemed unsold.

3.2.3. *Intransitives*

Levin & Rappaport note (Bresnan 82)'s observation that the past participles

[7] Apparently *-ing*, rather than perfective PstPrt, takes over the general task of non-ØX V → A derivation:

i. * our purchased agent (= 26b)
ii. our puchasing agent

of some intransitive verbs may become adjectives. ((28)-(29) = Levin & Rappaport's (101)-(102).)

(28) wilted lettuce, fallen leaf, escaped convict, collapsed tent, burst pipes, rotted railings, sprouted wheat, swollen feet, rusted screen, vanished civilizations, expired passport, failed bank

(29) *run man, *coughed patient, *swum contestant,*flown pilot, *cried child, *exercised athlete, *sung artist, *yawned student, *laughed clown

Levin & Rappaport's characterization of the relevant difference between the two sets of verbs is that the verbs in (28) can independently be characterized as unaccusative, having a single internal argument, while those in (29) are unergative, having a single external argument. Within Levin & Rappaport's framework, this distinction can remain only an observation—there is available no explanation for the difference, only the vocabulary of its description. Within the present theory, the unaccusative/unergative distinction lends itself to a more insightful account.

I assume that the external argument of the unergative is simply the Ø-marked characteristic external verbal argument. With respect to the internal argument of the unaccusative, it does not have a property that often accompanies direct lexical marking: it has no Case. It is reasonable to assume, then, that the internal argument of the unaccusative has no special property other than being subcategorized.

Recall that, of the two PstPrts, only the ØX PstPrt allows V → A. The array of data in (28)-(29) would follow if the PstPrts in (28) were all adjectivalizing ØX PstPrts, while those in (29) were non-adjectivalizing perfect PstPrts. However, we have no need to stipulate this difference. Unaccusatives have just the property that the ØX PstPrt is sensitive to: they have no external arguments. If we assume that ØX PstPrt must check for this property, then it again performs a function, and serves as a functional head.

In the case of intransitive ØX PstPrt, the internal argument of the unaccusative is θ-licensed only by the ØX-functional head. It appears that this θ-licensing allows the derived intransitive adjectives to be embedded beneath the verbs of appearance.[8]

8 Some of the derived intransitive adjectives in (28) function neither as complements of verbs like *seem* nor as predicates at all.

i. * The convict seemed escaped.
ii. * The convict was escaped.

The derived adjectives that do function in these ways appear to be derived from "middles" of verbs that have direct objects.

(30) The lettuce seemed wilted.

The unaccusative roots in (28) have an addtional way to adjectivalize. Recall that adjectives can be derived from *ing*-inflected verbs. Presumably *ing* inflects a verb without checking for any special property. Hence we would expect that *ing* does not function as a functional head, entailing that adjectives derived from *ing*-unaccusatives cannot be embedded beneath verbs of appearance.

(31) * The lettuce seemed wilting.

Example (31) forms a kind of minimal pair with (30), demonstrating the difference between a functional morpheme that checks for a property (31) and one that does not (30).

4. Derived Categories and -*ing*

(Milsark 88) argues that a single -*ing* is responsible for all V+*ing* → X derivations. That is, he proposes that, unique among English morphemes, V+*ing* can take on the properties of any category. Milsark's generalization of the properties of the single *ing* is appealing. However, implicit in the generalization is a kind of limitation. Milsark supposes that *ing*-inflected verbs can take on the properties of clausal Infl (more about which below) and of any lexical category: V, N, A, and, in very rare instances, P.[9] The implicit limitation in Milsark's theory is that *ing* assumes properties of already existing categories.

Within the present theory, in which θ-assigning properties are located along dimensions that are separable from lexical category, there exists the possibility of a kind of overgeneralization of the properties of *ing*. This overgeneralization would be that *ing* can assume not only the θ-assigning properties of pre-existing categories but in fact can assume any coherent (that is, non-canceling) set of θ-properties. This overgeneralization allows novel lexical accounts of a couple of constructions that usually receive syntactic analyses.

iii. The sun wilted the lettuce.
iv. * The plan/friends escaped the convict.

Perhaps not surprisingly in the present context, it is the "middles" that are related to verbs that σ-mark the "middle" argument.

9 The non-productiviy of V+*ing* → P is no doubt due to the fact that the class of prepositions is relatively closed. Whatever the reason, there is no reasonable analysis other than V+*ing* → P (like, say, reduced relative) for examples like (i) (= (Milsark 88)'s (5c)).

i. We spoke with John [*regarding* his application].

4.1. NOMINAL GERUNDS

The so-called nominal gerund is a syntactic sphinx: one kind of animal from the neck up, another from the neck down.

(32) [Mary's kicking him] disturbed us.

As is well-known, the nominal gerund has the distribution of NP, and the genitive Case on the determiner also is characteristic of N. The accusative Case on the object, however, requires the head of the construction to be a Case assigner like V, indicating a clausal structure. Further evidence of clausal structure is the availability of the aspectual auxiliaries and adverbial modification in (33a), in contrast to its unavailability with non-Case-assigning N in (33b). Example (33c) illustrates that true N-headed gerunds allow only adjectival modification.

(33) a. [$_{XP}$ John's having kicked Bill maliciously] disturbed us.
b.* [$_{NP}$ John's having kicked of Bill maliciously] disturbed us.
c. [$_{NP}$ John's malicious kicking of Bill] disturbed us.

Milsark assumes a transformational account like that outlined by (Baker 85), in which *ing* is an inflectional element that, like *ing* in general, can recategorize its host V to any other category. In this case of course we are interested in recategorizing to N. The requisite recategorization follows movement of the *ing*-Infl to the V, as in (34 (≈ Milsark's (4)).

(34) DS: [$_S$ PRO [$_I$ ing] [$_{VP}$ eat salmon]] is delightful
(internal assignment of Case to *salmon* by *eat*) ⇒
SS: [$_{NP}$ PRO [[$_N$ eating] salmon]] is delightful

In Milsark's scheme, the possibility of genitive Case assignment to the position occupied by PRO in (34) arises as a consequence of the category change to N.

Within the present theory, we can find an argument structure account that does not involve transformation. We begin by determining what kinds of argument structure properties the head of a nominal gerund would have to have in order to project its properties. Its NP distribution would follow if it had the characteristic external argument of an N, that is, R. If we associate the NP's ability to assign genitive Case with the presence of R (as is implicit, though not in these terms, in Milsark's account), then we have one half of the required properties. The ability to assign Case to internal arguments is a property of V. As pointed out at the outset of section 1, V differs from A and N in that its arguments are "visible" to the θ-criterion. Suppose that the

[+ vis] requirement on V that (obligatory) arguments be syntactically realized is always accompanied by the ability to effect that syntactic realization: Case assignment. If the head of the nominal gerund has the [+ vis] property we have the other half of the required properties. That is, a [+R, +vis] head would project a phrase with the properties of the nominal gerund.

Note that the properties associated with [+R] and [+ vis] do not overlap, so they do not cancel. Assuming that *ing* freely assumes non-canceling θ-properties, a [+R, + vis] head would project exactly the structure of the nominal gerund.

4.2. RETROACTIVE NOMINALS

(Clark 90) proposes a transformational account of a restricted kind of nominal construction, which we follow (Hantson 84) in calling a "retroactive nominal", like that in (35).

(35) That room needs [a good cleaning]

The nominal determiner *a* and the adjectival *good* modification in (35) indicate that the bracketed constituent is an NP. On the other hand, there appears to be a relation of obligatory control between the subject *room* NP and the empty object of *clean.* Clark proposes a "'Move NP' inside NP" analysis, in which a "Control PRO" is derived via syntactic movement to a leftward NP position.

(36) That room needs [$_{NP}$ PRO$_i$ a good cleaning t_i].

There are several problems with this structure. For example, the left-of-determiner "landing site" of the PRO movement corresponds to no familiar NP (or DP) Specifier. (Jones 91) discusses these problems at some length, and proposes an alternative analysis in which the *ing* inflection of the retroactive nominal has the external argument suppressing property of passive *en,* the property we have characterized in the present context as ØX. Again, if we suppose that *ing* can freely assume any non-canceling set of θ-properties, we have available a non-transformational account of the retroactive nominal, along the following lines. Verbs like *need* that idiosyncratically subcategorize for retroactive nominals select for the ØX property of the head of their retroactive complements. The complement satisfies the θ-criterion by projecting the argument of the non-suppressed internal θ-role, to be satisfied by θ-marking by *need,* as in (37) (in which we ignore, for the time being, the exact category of the retroactive nominal).

(37) That room$_k$ needs [$_{XP}$ a good [$_X$ clean+ing$_i$] e_j]$_j$.

needs	clean+ing$_i$
[__ XP$_{\emptyset X\text{-}ing}$]	[__ NP]
<1$_k$, 2$_j$>	<1$_i$, 2$_j$>
k controls	
(external arg of) *j*	

As noted above, the retroactive XP in (37) has some properties of NP. The analysis in (37) would require, however, that XP project the θ-role of an internal argument, rather than the R θ-role characteristic of N heads. There are a couple of pieces of evidence that suggest that the retroactive XP is not a full-fledged NP. First, it does not allow genitive Case in determiner position, as in (38a), a property that we have above associated with the R θ-role. Second, in some American dialects (for example, western Pennsylvania) the *ing*-retroactive nominal is replaced with what appears to be a purely verbal passive, complete with *en*-morphology, as in (38b) (in which the "%" marking simply indicates dialectical variation).

(38) a.* That room needs [John's cleaning].
b.% That room needs [cleaned].

Short of identifying the retroactive XP as an NP, it appears that the retroactive nominal presents another case of a synthetic category whose properties follow from a novel combination of θ-properties, a combination associated with no lexical category.

5. Denominal Verbs

(Hale & Keyser 93) present an analysis of denominal verbs like *laugh* and *shelve* (*the books*) that makes extensive use of syntactic verb movement. In this section I briefly present the mechanics of their theory and then I present an outline of an alternative lexical account. It is beyond the scope of the present paper to fully address the full range of data and phenomena that Hale & Keyser treat, however I hope that the brief proposed analysis here is clear enough to indicate along which lines further work might be fruitful.

5.1. THE VERB MOVEMENT ANALYSIS OF DENOMINAL VERBS

Hale & Keyser primarily treat two different kinds of denominal verbs, unergatives like *laugh* and verbs of "putting" like *shelve*. In both kinds of derivation the noun from which the verb is derived is associated with a syntactic argument structure, or *lexical relational structure* (LRS) of a certain kind of verb.[10]

5.1.1. *Laugh*

The LRS of a verb like *laugh,* as in (39a), is essentially that of verb-complement structure, as in (39c), reflecting an underlying θ-structure like (39b).

(39) a. Bill laughed.

b. Bill (had/did/made) a laugh

c.

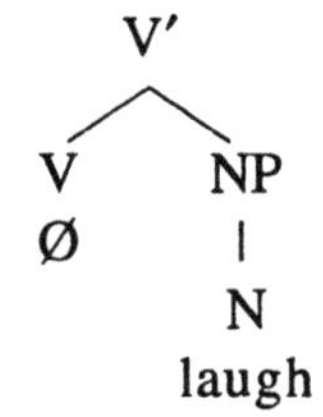

Head movement of the N to adjoin to the V head that governs it yields a structure like that in (40).

(40)

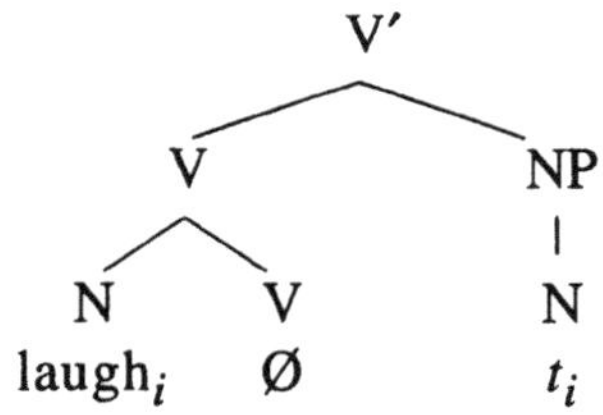

On the assumption that adjunction nodes do not count as branching nodes for c-command, the V node dominating N in (40) is not a branching node. Only the V′ node counts as a branching node, and so *laugh* c-commands its trace, satisfying the ECP.

5.1.2. *Shelve*

A more complex example is provided by the LRS of a verb like *shelve*. *Shelve*'s LRS is essentially that of a *put* structure, as in (39c), reflecting an underlying θ-structure along the lines of (41b).

10 Hale & Keyser's own characterization of the LRS below explicitly emphasizes its essentially syntactic nature: "During the course of our investigations, we have become persuaded that the proper representation of predicate argument structure is itself a syntax. That is to say, as a matter of strictly lexical representation, each lexical head projects its category to a phrasal level and determines within that projection an unambiguous system of structural relations holding between the head, its categorial projections, and its arguments (specifier, if present, and complement). We will refer to these projections somtimes as *lexical argument structures* and sometimes as *lexical relational structures* (LRSs), and we will use the now conventional tree diagrams to represent them in our discussions here."

(41) a. Bill shelved the books.
b. Bill (put) the books (on the) shelf

c.
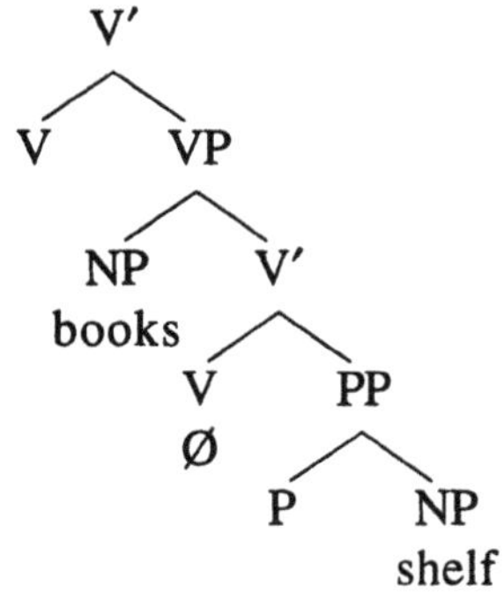

Successive head movements of the N *shelf* to adjoin first to P, then to the (null) V, and finally to the V head of V′ that governs it yields a structure like that in (42).

(42)
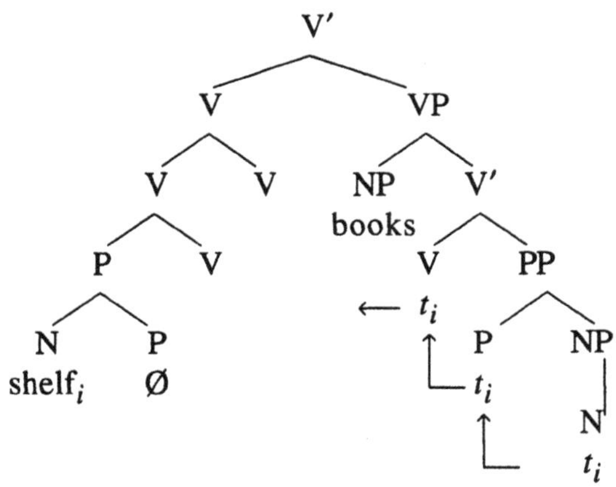

Again, because all of the adjunction nodes (on the left branch of the dominating V′) are "transparent" to c-command, *shelf* c-commands its trace, satisfying the ECP.

5.2. A LEXICAL ALTERNATIVE

The crucial step that sets up Hale & Keyser's derivations is the association with the target noun of an argument structure of a particular type of verb. The idea behind this seems to be the reasonable idea that nouns become verbs by virtue of the manner in which they typically participate in certain relations. If we avail ourselves of the same kind of possibility within the present set of assumptions, then the resulting lexical items have the requisite properties.

Before turning to the mechanisms of the lexical alternative in section 5.2.2, we briefly deal in section 5.2.1 with an (important) matter that will not be at issue here.

5.2.1. *X′ structure and thematic content*

Within the framework of their syntactic assumptions, Hale & Keyser present a quite detailed theory of how thematic connotations are linked to various X′ configurations. For example, they argue that the "actor" connotations arise from being associated with a second "cause" VP "shell". For our purposes, we have simply been assuming that, for instance, "actor" connotations arise compositionally from various properties of the elements of VP. I leave it to future argument whether Hale & Keyser's articulation of X′ structure is necessary to achieve the kinds of thematic distinctions Hale & Keyser arrive at. In what follows, I assume minimally necessary syntactic constituent structure.

5.2.2. *Denominal verbalizations*

Let us begin with the matter of the morphology that accompanies denominal verbalizations. Recall the zero-derivation that accomplished the V → A derivations in section 3.2. These zero-derivations crucially did not affect the argument structure of the root V. With denominal verbalizations, on the other hand, we verbalize the root N by giving it the argument structure of a V. For concreteness, let us associate this change in argument structure with a (generally) phonetically null morpheme "μ".

Hale & Keyser associate *laugh* with the argument structure of an abstract, typical (in our conception, default) transitive V. Suppose we follow Hale & Keyser and hypothesize that the verbalizing morpheme μ assigns a verbal argument structure like that in (43a) to *laugh,* in which *laugh*$_N$ itself is assigned the θ-role of the internal argument. A lexical item like (43a) would project into the syntax as in (43b).

(43) a. *laugh*: $[_V [_N \text{laugh}_i] + \mu]$,
|
$<1, 2_i>$
2 θ-binds R of root N

b. Mary$_j$ $[_{VP}$ laugh$_i$ ed$]_j$
|
$<1_j, 2_i>$

The same kind of account can be given for the verbs of location like *shelve,* which Hale & Keyser associate with the argument structue of an abstract verb like *put*. If we allow the verbalizing morpheme μ to θ-bind the indirect argument of a *put* θ-structure, then we obtain the derived θ-structure in (44a), which will project into the syntax as in (44b).

(44) a. *shelve*: $[_V [_N \text{shelf}_i] + \mu]$,
$\mid$
$<1, 2, 3_i\ (= \text{Loc})>$
3 θ-binds R of root N

b. $\text{Mary}_j\ [_{VP}\ [_V\ \text{shelv}_i\ \text{ed}]_j\ [\text{the books}]_k]_j$
$\mid$
$<1_j, 2_k, 3_i>$

5.3. NONEXISTENT VERBS AND THEIR EXPLANATIONS

So far we have simply coöpted Hale & Keyser's fundamental mechanism, that of associating verbal argument structures with nouns, and put it to work within argument structure. To the extent that the mechanism is the same, we might expect that the empirical coverage would be the same. However, if the underlying mechanisms of syntax and argument structure differ, then we might reasonably look for instances that distinguish one kind of account from the other. Below in section 5.4 I point out just such a distinction, one which appears to favor the argument structure account over the syntactic. Before considering this distinction, however, let us consider a kind of explanation that does not appear to distinguish between the two theories.

Hale & Keyser consider several kinds of non-existent denominal verbs and show how their syntactic theory rules them out. In some of these cases, the present theory has access to considerations that are analogous to those that the syntactic theory brings to bear. For example, consider argument structures like those in the (a) examples of (45) and (46). They derive denominal verbs like those in (b), but not those in (c).

(45) a. A cow had a calf.
b. A cow calved.
c.* It cowed a calf

(46) a. A machine got the wine into bottles.
b. A machine bottled the wine.
c.* It machined the wine into bottles.

Hale & Keyser note that the (c) examples would involve incorporating a subject rather than of a complement, something that is not allowed across incorporation phenomena. However, they go further to point out that a conclusion that they will arrive at later—that the subject of verbs like those in (45) and (46) are not present at all in LSR—will formally rule out any possibility of such incorporation. Hale & Keyser's conclusion about the external arguments in question is quite similar to the supposition in the present theory that they are featureless ciphers at the lexical level, receiving Case and thematic content only outside of the phrasal projection. Why do subjects not incorporate? Because, having no lexical thematic properties, they cannot satisfy any lexical properties of their governing head.

5.4. SYNTAX VS. MORPHOLOGY

A phenomenon for which Hale & Keyser crucially invoke syntactic ECP involves the non-existence of examples like (47a), with readings analogous to those of (47b), on the assumption that they share the LRS in (47c).

(47) a.* They housed the paint.
b. They painted the house

c.

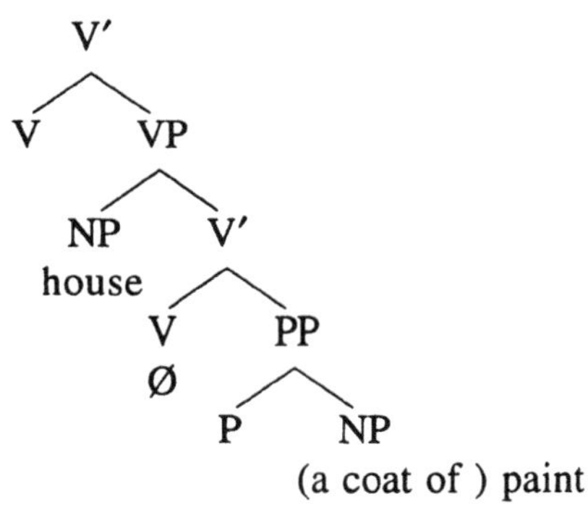

Hale & Keyser suppose that the LSR in (47c) corresponds to the LSR of a verb like *provide,* as in *provide a house with a coat of paint.* In Hale & Keyser's incorporation theory, it follows as a consequence of the syntactic movement involved that incorporation of the *house* NP results in an ECP violation: if *house* "lowers" to incorporate with the embedded (null) V, its trace will not be c-commanded, hence not properly governed. On the other hand, as illustrated above in (42), incorporation of the PP object like *paint* does not violate the ECP.

Within the present theory we cannot appeal to principles of syntactic movement. So the question boils down to why we can allow a θ-structure like that in (48a) but not that in (48b).

(48) a. *paint*: $[_V [_N \text{paint}_i] + \mu]$

$$\begin{array}{c} \mu \\ | \\ \langle 1, 2, (with)\ 3_i \rangle \\ \text{3 } \theta\text{-binds R of root N} \end{array}$$

b. **house*: $[_V [_N \text{house}_i] + \mu]$

$$\begin{array}{c} \mu \\ | \\ \langle 1, 2_i, (with)\ 3 \rangle \\ \text{2 } \theta\text{-binds R of root N} \end{array}$$

First, let us consider what it is that allows verbal structures (which we have been associating with the verbalizing morpheme μ) to be assciated with nouns. Consider, for example, the *shelf* → *shelve* derivation, in which *shelf* is associated with the locative argument of an abstract verb like *put*. Apparently it is some property of *shelf* that allows this association, because not every noun that can function as the locative argument of a verb like *put* has an associated verb.

(49) a.* I rugged it. (I put it under the rug.)
b.* I sinked it. (I put it in the sink.)

Apparently, the extent to which a noun is typically associated with a particular kind of containment, or location, is the extent to which it can associate with the argument structure of an abstract *put*-like verb that subcategorizes for such an argument. Similar qualifications seem to be in order for nouns typically associated with particular kinds of application. For example, *paint* is meant to be applied to a surface, a *pantry* is not. Hence we get (50a), but not (50b).

(50) a. We painted the house. (We provided the house with a coat of paint.)
b.* We pantried the house. (We provided the house with a pantry.)

Note that the *with* preposition governing *paint* must crucially invoke application. If it is the *with* of possession with respect to the object of *provide*, then the verbal association is impossible.

(51) * We painted Bill. (We provided Bill with paint (for his house))

Relturning to the ungrammatical (47a) above, the relevant question to ask is how the LSR in (47c) could come to be associated with *house* in the first place, since *house* presumably is not primarily associated with being a location for paint. If we associate *house* with its more familiar use as a location for people, it verbalizes naturally.

(52) a. Mary (put) the people (in the) house

b. *house*: $[_V [_N \text{house}_i] + \mu]$,
$\quad\quad\quad\quad\quad\quad\quad\quad |$
$\langle 1, 2, 3_i\ (= \text{Loc})\rangle$
3 θ-binds R of root N

c. $\text{Mary}_j\ [_{VP} [_V \text{hous}_i\ \text{ed}]_j\ [\text{the people}]_k]_j$
$|$
$\langle 1_j, 2_k, 3_i\rangle$

So if on independent grounds *house* cannot be associated as the second argument of a *provide*-type LSR, then the ECP never need be invoked to rule out (47a).

It may be supposed that the argument structure theory has merely dodged a bullet that the syntactic theory is able to take full on and deal with within its syntactic architecture. In particular, there is nothing within the argument structure theory that necessarily precludes a structure like that in (48b), repeated and generalized in (53), in which the argument structure of an verb marks its 2-argument with the appropriate "location" or "application" connotations, while Hale & Keyser's syntactic theory rules them out in principle, by the ECP.

(53) $[_V [_N X_i] + \mu]$
$|$
$\langle 1, 2_i, 3\rangle$
2 θ-binds R of root N

In fact, within the present conception of argument structure, properties of the *provide*-type argument structure indicate that a structure like (53) must be possible.

Let us assume (with Hale & Keyser) that *provide*-type LSRs are structurally similar to the *put*-type LSRs that derive verbs like *shelve*. Recall that, for the *shelve* LSR to work, the indirect argument of the associated abstract *put* LSR had to be the incorporating argument. Incorporation of the higher direct argument resulted in an ECP violation. If the *provide*-type LSR that derives *paint* is to work analogously, then, as in (54), *paint* would have to be the indirect argument of the associated abstract *provide* LSR.

(54)

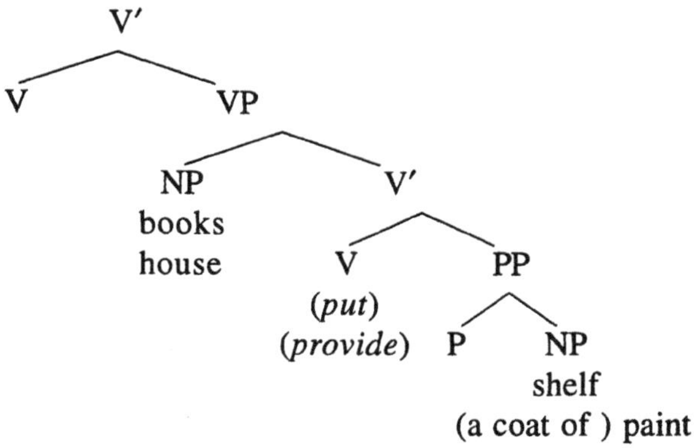

However, by Levin & Rappaport's tests for the direct argument, namely the obligatory argument test in (54) and adjectival passive test in (55), *paint* cannot be the indirect argument of a *provide*-type LSR. *Paint* must be the direct argument.

(55) a.* We provided the house. (*on the reading in (54))
b. We provided the paint.

(56) a.* the provided house (*on the reading in (54))
b. the provided paint

If *paint* is underlyingly the direct argument of a *provide*-type LSR, then its lexical position would be in the upper VP-internal "subject" position, occupied by *books* and *house* in (54). By Hale & Keyser's own syntactic ECP argument, incorporation from this position should be ungrammatical.

5. Conclusion

The mechanisms of argument structure differ enough from those of (phrasal) syntactic structure that even Hale & Keyser adopt different terms for them: *l-syntax* for argument structure and *s-syntax* for phrasal syntax. It has been the argument here that this difference is not merely terminological, but is real, with interesting possibilities for morphosyntactic explanation. L-syntactic mechanisms are by their very nature constrained to operate within severely local domains of argument structure and θ-role assignment. I have proposed here that the single case of non-local θ-role assignment, external argument projection, is mediated by an ECP-type requirement: that externally projecting θ-roles be licensed either by a functional head or by some special lexical marking.

It is possible to expend some effort explaining certain morphosyntactic phenomena in terms of the mechanisms available in s-syntax. What I have tried to demonstrate here is that, while s-syntactic mechanisms may be sufficient to characterize certain morphosyntactic phenomena, they are not necessary for them. An appropriately articulated θ-space allows l-syntax, with its severely local mechanisms, just enough room to get the required work done.

References

Baker, M., The Mirror Principle and morphosyntactic explanation, *Linguistic Inquiry* 16, 373-415, 1985.

Baker, M., *Incorporation: A theory of grammatical function changing,* University of Chicago Press, Chicago, IL, 1988.

Bresnan, J., The passive in lexical theory, in Bresnan (ed.), *The mental representation of grammatical relations*, MIT Press, Cambridge, MA, 3-86, 1982.

Chomsky, N., *The logical structure of lintuistic theory,* University of Chicago Press, Chicago (1955/75).

Chomsky, N., *Syntactic structures,* Mouton, The Hague, 1957.

Chomsky, N., *Lectures on government and binding,* Foris, Dordrecht, 1981.

Chomsky, N., *Barriers,* MIT Press, Cambridge, MA, 1986.

Clark, R., *Thematic theory in syntax and interpretation,* Croom Helm, London, 1990.

Di Sciullo, A.-M., Williams, E., *On the definition of word,* MIT Press, Cambridge, MA, 1987.

Hale, K., Keyser, S.J., On argument structure and the lexical expression of syntactic relations, in Hale & Keyser (eds.), *The view from Building 20,* MIT Press, Cambridge, MA., 53-109, 1993.

Hantson, A., Towards an analysis of retroactive nominals, inW. de Geest & Y Putseys (eds.), *Sentential complementation,* Foris, Dordrecht, 1984.

Higginbotham, J., Logical Form, binding, and nominals, *Linguistic Inquiry* 14, 395-420, 1983.

Higginbotham, J., On semantics, *Linguistic Inquiry* 16, 547-593, 1985.

Jones, C., 'Anti-internalization': Suppression and projection of external θ-roles, in G. Westphal, J. Dai, & B. Ao (eds.), *ESCOl '91 Proceedings,* Linguistics Department, Ohio State University, Columbus, OH, 1991.

Lieber, R., *Deconstructing morphology: Word formation in syntactic theory,* University of Chicago Press, Chicago, IL, 1992.

Levin, B., Rappaport, M., The formation of the adjectival passive, *Linguistic Inquiry* 17, 623-661, 1986.

Marantz, A., *On the nature of grammatical relations.* MIT Press, Cambridge, Mass, 1984.

Milsark, G. L. Singl-*ing*. *Linguistic Inquiry* 19, 611-634, 1988.

Pustejovsky, J., *The generative lexicon,* MIT Press Cambridge, Mass., 1995

Selkirk, E., *The syntax of words,* MIT Press, Cambridge, Mass., 1982.

Williams, E., *Rule ordering in syntax,* Doctoral dissertation, MIT, Cambridge, Mass., 1974.

Williams, E., Predication, *Linguistic Inquiry* 11, 203-238, 1980.

Williams, E., Argument structure and morphology, *Linguistic Review* 1, 81-114, 1981.

Williams, E., Implicit arguments, the Binding Theory, and Control, *Natural Language & Linguistic Theory* 5, 151-180, 1987.

Williams, E., The anaphoric nature of θ-roles, *Linguistic Inquiry* 20, 425-456, 1989.

Williams, E., *Thematic structure in syntax,* MIT Press, Cambridge, Mass., 1994.

PRAGMATIC CONNECTIVES AS PREDICATES. THE CASE OF INFERENTIAL CONNECTIVES

JACQUES JAYEZ
EHESS–CELITH, Paris
54, Boulevard Raspail
75006, Paris, FRANCE

AND

CORINNE ROSSARI
Université de Genève
Département de Linguistique Française
1211 Genève 4, SWITZERLAND

Abstract. This chapter investigates the linguistic description and formal representation of some pragmatic *inferential* connectives in French. We show that connectives expressing consequence, opposition or reformulation (like *anyway* in English) presuppose an abstract relation between propositional arguments of certain semantic types. We first contrast inferential and non–inferential connectives, then we turn to the semantic types of the propositional arguments to lay down some basic distinctions. Finally we substantiate the relations themselves. We use a version of generalized quantification over proofs to describe the inferential constraints which define the various relations presupposed by the connectives. The very possibility of such a description suggests that inferential connectives have a genuine (presupposed) predicative content.

1. Introduction

Although pragmatic particles (or 'discourse markers') such as *then*, *therefore*, *but* or *OK* for English, have been studied in different languages from

P. Saint-Dizier (ed.), Predicative Forms in Natural Language and in Lexical Knowledge Bases, 285–319.

a variety of points of view,[1] there is at this time no standard framework in which these analyses can be combined and compared. Syntactically, discourse markers fall into five main classes: sentential adverbs like *therefore*, subordination conjunctions like *because* or *if*, prepositions like *owing to*, parentheticals like *you know*, and interjection–like expressions like *yes!* or *good!*. Semantically, the first three classes impose a semantic connection between propositions or, more generally, intensional entities. That is why we will call members of these classes pragmatic *connectives* or PCs.

In this paper, we will be interested in some members of the first class of PCs in French,[2] that of sentential adverbs. By and large, they fall into three categories: inferential PCs like *however* exploit an inferential relation between propositions, type–based PCs like *even*, *also*, *for instance*, *first* or *next* point to various typing relations between sentences or discourse fragments, temporal PCs like *then* connect eventualities.

We claim that PCs are predicates or relations bearing on intensional arguments of various types. The difficulty is to substantiate the constraints which license such relations. While there has been some suggestive work in this direction for type–based and temporal PCs (see Bonomi & Casalegno 1993, Asher 1993, Glasbey 1993 for recent examples),[3] the treatment of inferential PCs has been less uniform, and we will focus on them in this paper. We show that they impose precise, although abstract and underspecified, requirements on their intensional arguments. These requirements can be divided into two categories: the nature of the relation between the arguments, which corresponds to the question 'how are the arguments linked?', and the typing of the arguments which corresponds to the question 'What are the semantic types of the arguments?'. More precisely, we propose to analyze inferential PCs as generalized quantifiers on sets of intensional entities, such as propositions or modal formulas. These sets are linked by the existence of certain *proofs*, which capture some aspects of the inferential properties exhibited by inferential PCs.

We begin by explaining intuitively the difference between inferential PCs and the other two classes (section 2). Turning to inferential PCs in

[1]See for example (Reichmann 1985; Schiffrin 1987; Knott 1996) for English, (Lang 1984; Grote & Lenke 1995) for English and German, (Roulet *et al.* 1985; Jayez 1988; Moeschler 1989) for French, (Rossari 1994) for French and Italian, (Ferrari 1995) for Italian, (Gil 1995; Fernandez 1996) for Romance languages. Note that the terminology is not fixed: in the literature, one can find *connectives*, *connectors*, *particles*, *markers* to designate the same lexical items. Also, some authors prefer *utterance* to *discourse* (cf. Fernandez 1996).

[2]To ease understanding, we will provide English examples and write the French connective in small capitals. We sometimes provide rough English equivalents for French connectives. Readers interested in translation issues should consult (Grieve 1996).

[3]See also (König 1981) for the German so–called *scalar* particles.

section 3, we first describe the essential properties of the inferential classes we study, consequence, oppositive, and reevaluation PCs (sections 3.1–3.3). Then, in section 3.4, we examine the semantic typing of arguments or semantic *scope*. Next, in section 4, we analyze the linking relation and distinguish between abductive and deductive constraints. Finally, we consider more sophisticated examples of generalized quantification behavior for some consequence and reevaluation PCs.

2. Basics

Most PCs belong to one of the classes illustrated in (1)–(8).

(1) If you don't understand such a simple point, then, you should drop linguistics

(2) He has dropped linguistics. Yet, he was really gifted

(3) He has dropped linguistics. Anyway, he could not understand even simple points

(4) He has dropped linguistics. EN EFFET he could not understand even simple points

(5) John fixed the vacuum–cleaner. Then, he started repairing the mower

(6) He has dropped linguistics. On the other hand, Mary keeps on studying formal semantics

(7) I won't go for a walk. The weather is not so fine; besides, I'm tired

(8) Some executives of the company do not consider favorably the new project. For instance, John is sure it will be a flop

For simplicity, we have kept a structure XCY, where C is the PC, throughout the examples.[4] (1) contains a consequence connective: X brings about Y or allows one to conclude that Y. In (2), we find an oppositive PC: X and Y are somehow incompatible. In (3), the relation is that of *rephrasing* or *reevaluation*: Y installs a new setting for evaluating X. (4) exemplifies an explanatory relation: Y is supposed to account for X. (5) is a typical example of a temporal relation (of sequencing in this case). (6) exploits a contrast between two states of affairs. (7) shows how arguments can be enumerated in intuitive reasonings. Finally, (8) illustrates exemplification. We did not attempt to classify *all* PCs. Some of them are hard to categorize. E.g. *et encore*[5] and *sinon* in French do not seem to fit naturally into one of the present classes. Moreover, borderline cases remain (some PCs belong

[4] We will refer to dialogue uses only marginally. For studying dialogue uses, extra parameters are required, but the bundle of basic parameters about linking and argument types would remain unchanged.

[5] See (Nemo 1992) on this PC.

to several classes). This was to be expected since what really classifies the PCs in our perspective is the set of precise semantic constraints assigned to them, while intuition-based classes reflect these constraints only partially.[6]

Granted that individual PCs are often difficult to classify because they can exhibit hybrid behavior, there are nevertheless some broad differences between temporal, type-based, and inferential PCs, which we explain briefly in the next two sections.

2.1. TEMPORAL PCs

Temporal PCs introduce constraints on temporal relations, promoting relations such as succession (*then*, *next*), simultaneity (*at the same time*, *meanwhile*), or anteriority (*previously*, *before that*). Their predicative character is obvious: the temporal relation bears on eventualities, which can be for instance viewed as events, or intervals in Allen style (Allen 1981). Note that temporal relations are independent from other relations but must be compatible with them. For instance, a causal relation of form x CAUSE y tends to be expressed temporally as a succession x *then* y. In this paper, we will not be interested in these PCs, since their predicative character is not open to question.

2.2. TYPE-BASED PCs

The behavior of items in this class is less clear. Such PCs exhibit the following basic property: they associate or contrast pieces of information sharing some informational type(s). As pointed out by Asher (1993), contrastive or parallel PCs, like *but*[7] or *too*, can be analyzed as morphisms between the propositional contents of the sentences they connect. For instance, a contrast like (6) above would get a representation similar to (6').[8]

[6] See König (1986) who says "There are many cases of overlap and neutralization so that a watertight system of classification and analysis does not seem to be possible." (Knott 1996) contains an interesting discussion of the classification problem and proposes an empirical method to address it.

[7] Note that there is more to *but* than just the idea of a contrast.

[8] We use x, y etc. as variables for non-temporal entities, e and s for events and states, *now* for the time of utterance, o for temporal overlapping. We stick uncritically to Asher's DRT notation and assume that events and states are comparable entities.

(6')

$$\begin{array}{|l|}\hline \\ \hline \mathit{K1} = \begin{array}{|l|}\hline x\ y\ e \\ \hline e < now \\ person(x) \\ linguistics(y) \\ e\text{-}drop(x,y) \\ \\ \hline \end{array} \quad \mathit{K2} = \begin{array}{|l|}\hline z\ u\ s \\ \hline s \circ now \\ named(z, Mary) \\ formal\text{-}semantics(u) \\ s\text{-}keeps\_study(z,u) \\ \\ \hline \end{array} \\ \\ contrast(K1,K2) \\ \\ \hline \end{array}$$

In the DRT–based representation of (6'), the two DRSs *K1* and *K2* are opposed with respect to the two pairs of conditions[9] $\{linguistics(y), e\text{–}drop(x,y)\}$ and $\{formal\text{–}semantics(u), s\text{–}keeps\_study(z,u)\}$. This opposition is licensed by the information contained in a background knowledge base, from which we can extract the fact that the two pairs of conditions have opposite *polarities.* Polarity opposition extends the traditional concept of Boolean contradiction, by allowing more pragmatic oppositions to be taken into account. For instance, oppositions between verbs like *assert* and *prove* (Ducrot 1980), *say* and *do*, or modifiers like *clever* and *very clever*, can be used in a contrast relation. Parallelism relations are defined on sets of conditions which exhibit the same polarity. Roughly speaking, for two–sentence connections of the form we examine in this paper, Asher's model predicts that any contrast or parallelism relation has to be licensed by identifying a subset of conditions in the DRT representation of each sentence, and observing that these two subsets have opposite or similar polarities. Then, clearly, the treatment offered is relational: connectives like *too* and *but*, or the like, are relations (contrast and parallelism) on subsets of conditions. Two points should be noted to clarify the status of such connectives.

First, in a more detailed study of this class, one would have to substantiate the *admissible* subsets of conditions which can constitute the arguments of the contrast/parallelism relation. Consider for example temporal adverbs like *today* and *yesterday.* While they can provide a basis for oppositive polarities, as in (9), with the French connective *par contre*, they are unable to do so with the stronger connective *au contraire*, as evidenced by (10).

(9) Mary went skiing yesterday, PAR CONTRE Paul went skiing today
(10) ??Mary went skiing yesterday, AU CONTRAIRE Paul went skiing today

[9] Remember that, in DRT, a predicative assertion of form $R(x_1 \ldots x_n)$ is a condition.

Second, the inferential status of these connectives is open to discussion: in some cases the contrast or parallelism relation makes use of an inference which provides the missing term of the relation. Consider (11) and (12).

(11) Mary had her dinner, but Paul is hungry
(12) Mary is hungry, Paul did not have his dinner either

In (11), we can oppose two kinds of themes. One is the hunger: if Mary had her dinner she is (normally) not hungry, in contrast with Paul. The other is 'having had his/her dinner': if Paul is hungry, he probably had no dinner, in contrast with Mary. A similar remark holds for (12). In both cases, the contrast/parallelism relies on propositions which are *inferred* from the explicit sentences rather than asserted by the sentences themselves. In this respect, inferencing is crucial for interpreting the connection. However, the connectives we study here are not inferential in this weak, derivative, sense, but in a stronger one: inferential connectives do not only occasionally make use of inference, they give information on how to combine which kind of inference to obtain interpretations.

Exemplification PCs (*for instance*, *in particular*, etc.) also make use of inferential relations in a limited way. They cannot access the whole range of inferences. Basically, a PC like *for instance* selects a member from a set: when a sentence describes a situation which can be realized in several ways, or a property which can manifest itself in several ways, it can be connected by *for instance* to a sentence which expresses one of these ways, as in (13) or (14).

(13) Give him a newspaper, the Herald for instance (from Manzotti 1995, ex. 2/30)
(14) Mary is grumpy, for instance she did not even have a look at Paul this morning

In (13), an instance of a given class of entities (newspapers) is selected and presented as an illustration. In (14), an instance of a general behavior or disposition is similarly selected and used as an illustration. In the two cases, there is a type or subsort relation between the class/property and its instance: the instance is a particular case subsumed by the class/property. This instantiation relation is subject to many restrictions examined in detail for Italian in (Manzotti 1995), but its general sense is clear and seems rather stable across, for instance, French, English and Italian: an exemplification relation is an instantiation and remains quite distinct from an inferential link. Due to the possibility for instantiation of mentioning the symptoms or manifestations of a given state/situation, there is some crossover in causal cases. E.g., (14) can be perceived as causally relating a state of mind to one of its consequences. Similarly, (15) conveys a causal relation.

(15) It was very hot here in summer. For instance, the tar melted on the roads, even in the morning

However, the causal relation is not essential to the instantiation relation, as shown by (13) and by the impossibility of cashing the exemplification PCs on causal relations when they do not support instantiation, as in (16).

(16) ?? It was very hot here in summer. For instance, we decided to move to less stuffing climates

The last two classes of PCs, contrast/parallelism and exemplification, rely on the fact that the connected sentences share an informational type: for contrast/parallelism relations, it must be a *theme*, that is, minimally, a *skeleton* in Asher's sense, or, more generally, some structured subset of conditions; for exemplification, the type is defined as the set of conditions of the sentence to be illustrated. Let $cond(X)$ be the set of conditions associated with X in a sentential form $X\ C_{exempl}\ Y$, where C_{exempl} is an exemplification PC. Possible sets of conditions $cond(Y)$ for Y include the following (see Manzotti 1995 for a more detailed and better motivated analysis):

- if $cond(X)$ conveys a non–assertive propositional attitude of the speaker with respect to some propositional content p, $cond(Y)$ conveys the same attitude of the speaker with respect to a propositional content q which is a subsort of p.
- If $cond(X)$ conveys an assertive propositional attitude of the speaker with respect to some propositional content p, $cond(Y)$ conveys the same attitude of the speaker with respect to some propositional content q which is a symptom or manifestation of q.

An example of the first case is (13), represented as (13').

(13')

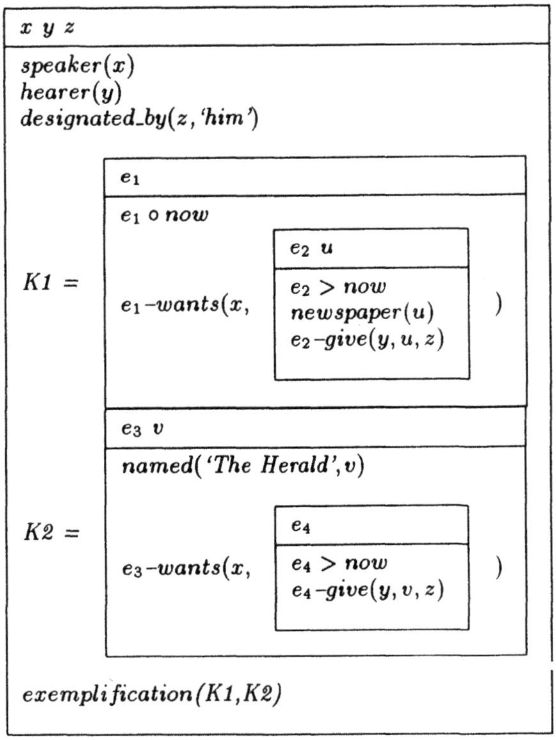

In (13'), the two structures *K1* and *K2* are parallel: they represent the propositional attitude of the speaker ($wants(x, \_)$) concerning two propositional structures, the second being a subsort of the first under the assumption that the thing named 'The Herald' is a newspaper. In addition to providing a simple case of exemplification, (13) and (13') feature the central scope distinction between propositional attitude and propositional content to be used below in section 3.4.

The last category of non–inferential PCs, enumeration PCs (*first(ly)*, *on one hand*, etc.), exhibits a pretty different behavior than contrast/parallelism and exemplification PCs. Enumeration PCs 'live on' the discourse relations that are available from the discourse structure and/or from other PCs. Consider the next contrast.

(17) a. I won't go to the party. First, I am tired, second, I can't stand the hectic bunch of people who are likely to come

b. ??I won't go to the party. First I am tired, second I worked too hard yesterday

c. I have just two things to say. First, it's true that Janice sings very well, but, second, we really don't need a contralto in the choir by now

In (17a), the two sentences introduced by *first* and *second* are perceived as explanations or justifications of the first sentence. In (17b), no such connection is available: the third proposition sounds as an explanation for the second one. Although there is an explanatory connection between the two sentences, there is no connection between those sentences and a third, different, one. (17c) merges the two preceding cases: there is an oppositive connection between the two sentences introduced by *first* and *second*, but there is also an elaborative connection between those sentences and the first one. Such observations suggest the following constraint on enumeration PCs: for a form $C_{enum}X, C'_{enum}Y$ to be admissible, it is necessary that X and Y share a common discourse function with respect to some third sentence Z. This condition might not be sufficient: for instance, some discourse functions might be more appropriate than others to license enumeration PCs, or might differ according to Z's position (before or after X and Y). But we will not discuss these problems further, since we will not be interested here in this kind of PCs.

As for contrast/parallelism and exemplification PCs, we note that enumeration PCs do not rely crucially on inference for setting up a connection. They connect sentences with a common discourse function type. So, the three classes of PCs reviewed in this section require that the sentences they connect be somehow *comparable*: they must share an informational type or a discourse function type. In this respect, the corresponding PCs do not hierarchize sentences but rather associate them on the basis of an information or discourse–based similarity.

3. Inferential PCs

We will consider here only three classes of inferential PCs: consequence PCs, like *so* or *therefore* in English, and *alors* or *donc* in French, oppositive PCs, like *but* or *yet* in English and *mais* or *pourtant* in French, and reevaluation PCs, like *anyway* in English and *de toute façon* or *en tous cas* in French.

3.1. CONSEQUENCE PCs

Consequence PCs[10] can introduce causal (18a) or causally abductive relations (18b).[11]

(18) a. John had forgotten his passport. So he was delayed at the frontier

[10]See (Hybertie 1995) for an introduction to their descriptive properties in French.

[11]We will call a relation ***causally abductive*** when it amounts to deduce some causal explanation of a state of affairs.

b. John was delayed at the frontier. So, he must have forgotten his passport

In the two cases, there is a consequence relation, from a cause to its effect in (18a), and from a belief to an entailed belief in (18b). Usually, the explicit reasoning is enthymematic and the speaker expects the hearer to provide for missing additional premises. Interestingly, this abduction process is not unconstrained: in (Jayez & Rossari 1996) it is noted that the French *donc* ($\approx$ *therefore*) may not link a conclusion to a conditional sentence, while *alors* ($\approx$ *then*) may.

(19) a. ?? If the weather is fine, DONC I will go out

b. If the weather is fine, ALORS I will go out

This contrast is not predictable if consequence connectives are insufficiently parameterized, which is typically the case when consequence is simply equated with a form of necessary entailment (Iatridou 1994). We will propose a more complex analysis in section 5.1.

3.2. OPPOSITIVE PCs

The use of oppositive PCs[12] revolves around three types of properties.
(a) The geometry of opposition. In a form $X\ C_{opp}\ Y$, X and Y can be directly opposed, as in (20a), or point towards opposite conclusions ('indirect opposition') as in (20b).

(20) a. It is raining, however, nobody in the street has an umbrella

b. It is raining, but I need a walk

Note that, in (20b), there is no clash of intuition between the two facts. Rather, they favor different opposite conclusions ('I will take a walk' vs 'I will not take a walk').
In addition to the direct–indirect distinction, there is a difference between 'backward' and 'forward' connectives in the case of direct opposition, as illustrated by:

(21) a. John was late, yet he did not specially hurry up

b. John was late, but he did not specially hurry up

(21a) can be reversed ('John did not specially hurry up, yet he was late'), while (21b) cannot in general.[13]
(b) The strength of opposition. In French, an oppositive PC like *pourtant*

[12] See (Morel 1996) for an introduction to their descriptive properties in French.
[13] The reverse form is possible when it stands for an answer to a dialogue opponent who somehow denies that John was late. This and similar observations show that backward opposition must be contextually primed in the case of *mais*.

is direct oppositive and presents the terms it associates as strongly incompatible.
(c) Scalar opposition. Some indirect oppositive PCs[14] allow for restriction moves on a scale, in the style of (22).

(22) John is intelligent, but he is not a genius.

3.3. REEVALUATION PCs

The prototypical member of this class[15] in English is *anyway*. These PCs have quite a few properties, of which we will mention only the following two.
(a) The ability to weaken or cancel the effect of the first proposition. In (23), the second proposition Y makes a potential conclusion (such as 'there was no meeting this afternoon') true, no matter whether X is true or not.

(23) (X) John did not feel like going to a meeting this afternoon. Anyway, (Y) the idea was abandoned

(b) The possibility of a scalar connection, on the model of (24).

(24) John is probably a genius, anyway he is extremely clever

3.4. SEMANTIC SCOPE

Inferential PCs highlight inferential relations between intensional entities. Consequence PCs point to causal or deductive relations between beliefs or speech acts, oppositive PCs to the potential clash between some proposition and the negation of some of its consequences, reevaluation PCs rely on inferential updating, i.e. the operation of drawing various conclusions from some new information. Intuitively, the 'inferences' we allude to here can be viewed as some form of logical deduction of a given (set of) conclusion(s) from a set of premises, but the status of the intensional entities which are used as premises or conclusion(s) is less accessible to intuition.

It has been generally recognized in the pragmatic (Searle & Vanderveken 1985) or semantic tradition (Bierwisch 1980) that the basic elements of discourse units are: (i) illocutionary forces, such as assertion or question, (ii) propositional attitudes, such as belief, desire, etc., and propositional content, which corresponds to the basic predicative structure of a sentence.

[14]The category 'indirect oppositive' will be assigned to PCs which tolerate indirect oppositions. Some of them also allow for direct opposition, while 'direct oppositive' PCs, like *pourtant* in French, are not compatible with indirect opposition.

[15]See (Rossari 1994) for an introduction to their descriptive properties in French, and (König 1986) for preliminary remarks on *anyway*.

Analogously, Halliday and Hasan (1976) have proposed a distinction between *external* and *internal* conjunction, which seems to correspond in part to the illocutionary vs (attitudinal + propositional) distinction. Knott discusses various presentations of this distinction in his 1996 thesis (sections 6.2.1 and 6.2.9).

More recently, elaborating on (Bierwisch 1980), Ferrari (1995) proposes that any utterance has a structure:[16]

$$IF(ATT(pc))$$

where IF is the illocutionary force, ATT the attitude, and pc the propositional content. If X is a surface form, we will note $IF(X)$ (resp. $ATT(X)$, $pc(X)$) its illocutionary force (resp. attitudinal content, propositional content). Starting from a different tradition, Knott (1996) discusses the semantic/pragmatic distinctions

There is no general agreement as to which entities (force, attitude, propositional content) are necessary to describe sentential adverbs or PCs. We show below that the distinction between force/attitude, on one side, and propositional content on the other provides a sufficient approximation of a number of phenomena. Yet, we agree with Knott (1996) that it is necessary to move to a more complex position to explain certain other observations. Knott proposes to take into account perlocutionary effects, in (Rossari & Jayez 1997) we describe attitudes as transitions between information states in a dynamic semantic framework. However, a dynamic treatment of attitudes complicates substantially the inferential system and we will abstract here from the dynamic dimension for that reason.

In the static treatment to follow, PCs may connect semantic objects like forces, attitudes and propositional content. We will use the term *semantic scope* or *scope* to denote the entities connected by PCs. In a form X C Y, the term *left scope* will refer to the entities associated with X in the connection, and similarly for *right scope* and Y.

3.4.1. *Basic properties*

The basic observation is that, while all PCs can connect assertions, many of them cannot felicitously connect other illocutionary forces in monologues. There are many variations and borderline cases, especially when dialogue is taken into account, but we will limit ourselves to the following three clear–cut structures.

In a form X C Y, some PCs prefer assertions (or equivalent illocutionary

[16] We change and simplify the notation of Ferrari for intuitive readability: Ferrari would probably consider the illocutionary force as a communicative elaboration over a more elementary cognitive attitude (called *Satzmodus* after Pasch 1989). Doing justice to these distinctions would require working out a complex system of semantic and pragmatic features, a task we will not undertake here.

forces) in X and Y. This is typically the case for the consequence PCs *du coup* and *de ce fait.*

(25) This machine is dangerous. DU COUP/DE CE FAIT (you must not touch it vs ?? don't touch it)

Other PCs have a mixed behavior: they accept illocutionary forces different from assertion in Y, but not always in X. The consequence PCs *donc* and *alors* illustrate this case.

(26) a. This machine is dangerous. DONC/ALORS don't touch it

b. ??Don't touch this machine. DONC/ALORS you must obey me

Last, some PCs do not seem to impose restrictions on the illocutionary force of the terms they relate. The reevaluation PC *de toute façon* illustrates this case.

(27) a. This machine is dangerous. DE TOUTE FAÇON don't touch it

b. Don't touch this machine. DE TOUTE FAÇON it does not work

Interpreting observations of this kind is in general difficult. We will not carry out a detailed discussion here, but will instead summarize the two factors which affect scoping.

[1] Scoping is 'hierarchical': a PC with scope on attitude bears also on propositional content, a PC with scope on illocutionary force bears also on attitude, and, therefore, on propositional content. E.g., one may not issue an order (illocutionary force) without communicating at the same time that the speaker desires (attitude) that some state of affairs (propositional content) hold.[17] To illustrate this point, consider *donc*. There are two possible cases: in a form X *donc* Y, either $pc(X)$ is the cause of $pc(Y)$ (see (18a)), or $pc(Y)$ is the cause of $pc(X)$, in which case the link between X and Y is causally abductive, as in (18b). In the abductive case, the connection between forces and attitudes is cashed on the causal relation from $pc(Y)$ to $pc(X)$. Considering (26b) from this perspective, we observe that the propositional content is not used by the epistemic and illocutionary dimensions: if there is any causal relation between propositional contents, it goes from the obligation (the hearer must not touch the machine) to the

[17]Of course, the link between force and attitude is probably non–monotonic, since, for instance, one might issue an order without any desire that the situation mentioned in the command obtain. This is not the case in most current situations, however, and, anyway, the command can be considered as satisfied only when the relevant situation is created or brought about by the hearer. This suggests that the modality of 'desire' is too strong and should be replaced by a more neutral one, expressing the conditions under which a command would be considered as obeyed, irrespective of the psychologically plausible correlates of the command.

situation (the hearer does not touch the machine). But this possible causal link does not provide a ground for desires, beliefs, command and assertion. More precisely, the fact that the obligation to do p can cause doing p for the hearer, has nothing to do with the fact that issuing the order that p in front of the hearer creates for her an obligation to do p, or with the fact that desiring that the hearer do p favors the belief that the hearer is under an obligation to do p. The importance of propositional content is also suggested by contrasts like the following.

(28) a. I was not at the meeting. DONC/ALORS I don't know whether John came

b. #I was not at the meeting. DONC/ALORS did John come ?

(28b) is acceptable (for some speakers) only if the speaker's absence at the meeting is perceived as influencing John's presence or absence at the same meeting. If the question is (literally) understood as a request for information, the connection is not natural. This is readily explained by noting that, in the latter interpretation, there is no causal link between the speaker's absence at the meeting and John's presence or absence. In contrast, (28a) is perfect because there is clearly a causal link between the speaker's absence at the meeting and his ignorance of John's presence or absence at the same meeting.

[2] In general, the intensional entity on which a given PC has scope must be 'true'. E.g., if a PC bears on some propositional content pc_i, pc_i must be 'true' in the world of the speaker. More precisely, there are some admissible degrees of likelihood for the intensional object, in a present or future state of the world of the speaker. We may not use an intensional object as the value of the scope if the object has not got one of these admissible degrees. This is illustrated by the following contrast.

(29) a. ??My car broke down, DU COUP call a taxi

b. My car broke down, DU COUP would you call a taxi?

Forces and attitudes determine degrees of likelihood in such cases. This does not mean that they constitute the scope, however, since, in this case, (29a) would be as normal as *My car broke down, so I want you to call a taxi* (scope on attitude). Imperatives tend to present the propositional contents on which they bear as false in any situation in which the addressee controls her own behavior (without any external influence). This is not the case for more polite requests, which do not presuppose that the addressee could be uncooperative, hostile, or unconcerned. While such distinctions could be represented in an appropriate modal setting, we will ignore them in the rest of this paper, and focus on the simple binary truth–conditional cases, in which an intensional object is or is not in (some temporal slice of) the world of the speaker.

To illustrate these two aspects, we will turn to the case of consequence PCs.

3.4.2. *A case study: the scope of consequence PCs*

There are three reasons for assuming that *donc* and *alors* have scope on attitudes or forces, while *du coup* and *de ce fait* have scope on propositional contents.

[1] While *donc* and *alors* can introduce imperatives and questions, this is not possible for *du coup* and *de ce fait*, except for questions which are covert or hedged assertions. This was illustrated by the contrast between (25) and (26a), which we repeat below, adding a variation for *du coup.*

(25) a. This machine is dangerous. DU COUP/DE CE FAIT (you must not touch it vs ?? don't touch it)

b. This machine is dangerous. DU COUP/DE CE FAIT wouldn't it be better to keep it in a locked room?

(26) a. This machine is dangerous. DONC don't touch it

We can explain the contrast between (25a), on one side, and (25b) and (26a) on the other, by noting that the intensional entities which form the right scope of the PC must be compatible with the truth–conditional 'effect' of the PC. For consequence PCs (and probably oppositive ones), the intensional entity is presented as *true.* This was to be expected since consequence PCs draw consequences from a given set of premises, that is, present them as *true* given the premises. Let $X\ C_{cons}\ Y$ be a consequence connection. When C_{cons} has right scope on force or attitude, that is on $IF(Y)$ or $ATT(Y)$, their exact nature is irrelevant. When it has scope on $pc(Y)$, the force is preferably (some form of) assertion, because imperatives and questions bearing on $pc(Y)$ imply at least that the speaker ignores whether $pc(Y)$ is true. On the other hand, in a sentence like *This machine is dangerous,* DU COUP *you must not touch it,* there is a relation between two states of affairs, that of the machine being dangerous and that of the obligation of not touching it, thus between a factual and a deontic state of affairs. In a sentence like (26a), we can represent the consequence relation as: 'the speaker believes that the machine is dangerous, then she desires that the addressee not touch it' (relation between attitudes), or 'the speaker believes that the machine is dangerous, then she orders that the addressee not touch it' (relation between attitude and force).

[2] The second reason for distinguishing the scope of *donc/alors* and *du coup/de ce fait* is the impossibility of causal abduction with the latter two

PCs, as shown by the contrast between variants of (18b).

(18) c. ??John has been delayed at the frontier. DU COUP/DE CE FAIT he must have forgotten his passport.

d. John has been delayed at the frontier. DONC/ALORS he must have forgotten his passport.

Such contrasts are easily explained if we assume that consequence PCs promote links between a premise and a conclusion. Under this hypothesis, (18c) is anomalous because there is no premise–conclusion directed link from the first propositional content to the second (the link has the reverse direction). In contrast, (18d) exploits a directed link between two beliefs (and their corresponding assertions).[18]

[3] The interaction between consequence (premise–conclusion directed link) and scope allows one to explain other impossibilities which are not cases of causal abduction, such as:

(30) ??Could you terminate this work very quickly, DONC I'm in a hurry

Assuming that *donc* has scope on attitudes, (30) has the following abnormal consequence structure: 'I want you to terminate this work very quickly, so I believe (know, etc.) I'm in a hurry'. Of course, one expects the reverse order, which indeed can host a *donc*.

So, given some plausible assumptions, observations are clearly in favor of a difference of scope. This does not imply, however, that *donc* and *alors* are insensitive to propositional content. In fact, PCs are all the more natural as they are licensed by connections at *all* the levels (attitude and propositional content, or force and propositional content) which are semantically relevant. This is why (28b) sounds strange: *donc* and *alors* cannot be supported only by attitudes or forces. Not unexpectedly, there is variation among speakers when the relation between propositional contents is not clear. For instance, consider (31).

(31) #I already told you to go to bed. I'm your father, DONC/ALORS turn this TV off

Speakers do not rate (31) uniformly. Under the present assumption, this can be explained by noting that (i) the force/attitude link is supported by (31) ('I know I am your father, so I order you to turn off the TV'), (ii) the propositional content link is supported only under the interpretation 'The fact that I am your father will cause the future event of turning off the TV', that is, under an interpretation which makes the social situation the

[18] See (Sweetser 1990) for a similar analysis.

cause of an event, as in *C'est son père qui le lui ordonne, du coup Jean va éteindre la télé* ('It is his father who ordered it, DU COUP John is going to turn the TV off'), when issued in the same context. Note that (31) would be quite odd with *du coup* or *alors*: this is predicted by the hypothesis introduced in (3.4.1,[**1**]) above, according to which the intensional entity on which consequence PCs have scope must be 'true', a condition which is not fulfilled by propositional contents introduced by imperatives.

Other differences[19] are not amenable to scope problems, but pertain to individual differences between consequence PCs, a topic which we will not consider further here. The reader will note that we have not tried to decide between force and attitude in scope assignment. Our observations tend to show that scope on attitudes is sufficient in many cases to account for possible or impossible linkings. However, a careful answer to this question would require a thorough investigation of the relation between forces and attitudes, which is well beyond the scope of this paper.

4. Inferential relations

In all the previous examples, interpretation relies on an inferential linking between a proposition–like entity, considered as a *premise*, and another entity which plays the part of the *conclusion*. This is obvious in the case of consequence PCs. For oppositive ones, opposition between X and Y depends on having X 'entailing' $\neg Y$ or Y 'entailing' $\neg X$ in some sense, or on having X 'entailing' some Z while Y 'entails' $\neg Z$. In most cases, the entailment presupposes implicit premises. We will note this situation $(\Sigma), X \vdash Y$, where Σ is a (possibly empty) set[20] of additional implicit premises, and $\vdash$ some consequence relation. Now, the question comes up naturally of which constraints we are to put on entailment to provide more precise representations.

4.1. ABDUCTIVE CONSTRAINTS

Interpolating additional premises is an *abductive* process, by which we postulate facts which could explain some other given fact *via* some inference rule. This phenomenon has been known for long and appears in the classical theory of syllogisms under the guise of so–called *enthymematic* reasoning. Its importance for natural language understanding has been recently underlined (Hobbs *et al.* 1993)

[19]Such as that which is revealed by the contrast *I happen to realize that you should not attend this meeting,* DONC/ALORS/DE CE FAIT vs ?? DU COUP *I ask you to leave.*

[20]In this paper, we consider only consistent sets of propositions. So any mention of a 'set' means 'consistent set'.

Our basic tool will be a consequence relation $\vdash$ between premises and conclusion. $\Sigma \vdash p$ means that p can be deduced from Σ modulo the rules of the logic which characterizes $\vdash$. We will not discuss the detailed properties of $\vdash$ in this paper, but we need to distinguish between two possible interpretations of $\Sigma \vdash p$. In the first interpretation, which we will adopt here, the formula means that any state of information in which we have only Σ and its consequences contains p. This interpretation, which is an unordered version of Veltman's validity$_1$ (Veltmann 1996) amounts to say, that, in a world, that is, in a set of propositions, we might have conjointly (i) Σ, (ii) $\Sigma \vdash p$, without having p. This is so because the interpretation in question only requires that p be deducible from Σ when there is no other information available, apart from Σ and its consequences. If some world contains additional information, it might block the derivation of q. The second interpretation is much stronger: it requires that p be true in any information state where Σ is true. This is typically the monotonic interpretation of the deduction symbol $\vdash$.

Let p, q be propositions and $\vdash$ a consequence relation, we define:
${}^{\vdash}p =_{df}$ the set of finite non–empty sets Σ such that $\Sigma \vdash p$.
We call ${}^{\vdash}p$ the set of *antecedents* of p.

Let us see how antecedents come into play by considering the following example.

(32) John has just gulped down two big glasses of beer, yet (POURTANT) he had refused to drink ten minutes ago

This example can be understood *via* the reconstruction of the following underlying logical structure, where $t \ll t'$ notes the fact that t is prior but near to t', t and t' are time variables, θ and *now* are time constants.
1a. John refuses to drink at θ,
1b. $\theta \ll now$,
2a. John is not thirsty at θ,
2. John is not thirsty at $t \vdash$ John refuses to drink at t,
3. John is not thirsty at t, $t \ll t' \vdash$ John is not very thirsty at t',
4. John is not very thirsty at $t \vdash$ John does not drink much at t,
5. John drinks two glasses of beer at $t \vdash$ John is very thirsty at t.
Note that the second sentence of (32) should be represented by the conjunction 1a $\wedge$ 1b. We have the following proof schema, where $\phi\{u = v\}$ designates the result of substituting v for u in ϕ:
from 2a, $2\{t = \theta\}$, 1b **deduce** 1a $\wedge$ 1b
from 2a, 1b, $3\{t = \theta, t' = now\}$, $4\{t = now\}$ **deduce** 'John does not drink much at *now*'.

This schema delineates a proof from 2a and 1b to a conclusion 'John does not drink much at *now*', which is contradictory to the explicit sentence

John has just gulped down two glasses of beer. The use of 2a as a premise shows that we have to take into account propositions obtained by causal abduction, that is, propositions which are causes of the explicit proposition(s) mentioned in the first sentence. More generally, a relation $POURTANT$ between p and q will be licensed by any situation which consists of (i) a set of propositions which allows one to derive p, (ii) a set of propositions which, conjointly with some subset of the previous set allows one to derive $\neg q$; that is, in any situation in which we have: a set of premises Σ which is a member of $^{\vdash}p$, a set of premises Σ' which, conjointly with some subset of Σ is a member of $^{\vdash}\neg q$. Intuitively, the speaker communicates that (i) she imagines some situation Σ in some world w which licenses p and (ii) in her world ($w_{speaker}$), this situation would entail $\neg q$. As we run mentally through situations imagined by the speaker, we obtain a more familiar paraphrase: *p pourtant q* is 'true' whenever we get $\neg q$ in every minimal reasonable situation where we would have p, from the speaker's point of view. In short, every p-situation is a $\neg q$–situation for the speaker.[21] *Pourtant* will also be licensed in the symmetric situation, where two sets of propositions entail respectively q and $\neg p$, as in *John refused to drink ten minutes ago, yet he has just gulped down two glasses of beer*. We will allow for sets Σ and Σ' from different situations or worlds, indexing them accordingly: the notation Σ_w will designate a set of propositions Σ which are true (in particular) at w.[22] The world $w_{speaker}$ refers to the set of propositions which are true from the speaker's point of view (the world 'of' the speaker). We can now define the relation $POURTANT_{p,q}$.

(33) $POURTANT_{p,q}(\Sigma_w, \Sigma'_{w_{speaker}})$ iff there exists a non–empty set $\Sigma'' \subseteq \Sigma$ such that:
(i) $\Sigma'' \cup \Sigma'$ is consistent, and
(ii) $\Sigma \vdash p$ and $\Sigma' \cup \Sigma'' \vdash \neg q$ or $\Sigma \vdash q$ and $\Sigma' \cup \Sigma'' \vdash \neg p$.

Three points should be noted. First, this definition covers the cases where p (with, possibly, some additional premises) entails $\neg q$ (or q entails $\neg p$). Let $\{p\} = \Sigma_w = \Sigma''_w$, $\Gamma, p \vdash \neg q$ and $\Sigma'_{w_{speaker}} = \Gamma$. Then $\Sigma \vdash p$, since $p \vdash p$, and $\Sigma' \cup \Sigma'' \vdash \neg q$. We can accommodate in this way examples such as *He worked very hard, yet he is not tired* or *He is not tired, yet he worked very hard*. Second, the definition does not say that p and $\neg q$ are true in the same world: it requires only that the speaker believe that p 'entails' $\neg q$

[21] This formulation is intended only to help intuition. In fact, *matching* readings (Rothstein 1995) are infelicitous with many PCs: ?? *Each time John is hungry,* POURTANT *he does not eat enough*, ?? *each time John is hungry,* ALORS *he eats much*, etc.

[22] As usual, we assume that sets of propositions extracted from the same world or situation are consistent. However, we do not reject the possibility of inconsistent worlds, that is, worlds containing two mutually inconsistent sets of propositions.

(viz. that some proof of p leads to a proof of $\neg q$, modulo the additional premises Σ''). This reflects the intuition that speakers endorse oppositions or entailments, that is, the consequence relation itself, while they not necessarily endorse the premises or the conclusions which are linked by this relation. This is particularly clear in *if ... then* sentences, where speakers are only responsible for the consequence relation from the *if*-sentence to the *then*-sentence. This can happen also in ironical sentences, where the speaker draws conclusions from a premise she obviously does not approve. Third, the definition does not say that the speaker has contradictory beliefs or that its world is contradictory, but that she can deduce $\neg q$ from p modulo additional premises (Σ''). This problem is more visible when the premise is an assertion. When a sentence is an assertion, we can assume that the world in which the proposition associated with it is interpreted is the world of the speaker, $w_{speaker}$.[23]

(34) If p is associated with X and $IF(X) = assertion$, then every set Σ_w which provides premises for p is such that $w = w_{speaker}$.

In an example like (32), we are in the speaker's world ($w = w_{speaker}$). So, (i) p and $\neg q$ might be true in $w_{speaker}$, and (ii) q might be true in $w_{speaker}$. If the relation $\vdash$ is monotonic, the speaker's world is contradictory: $p \vdash \neg q$ means that $\neg q$ is true in any situation where p is true, in particular in $w_{speaker}$. If $\vdash$ is non–monotonic, it is not necessarily so, because additional information in $w_{speaker}$ could block the derivation of $\neg q$. Recall that we chose a non–monotonic deduction relation. There is anyway a 'local' contradiction, defined with respect to Σ and Σ'. This is as it should be, since *pourtant* is considered as expressing surprise in French: this agrees with the speaker's belief that the two assertions connected by *pourtant* create a local inconsistency: somehow, every reasonable p–situation is a $\neg q$–situation in this part of $w_{speaker}$ which includes only Σ'.

This example introduces a recurring theme in the study of PCs: their *Generalized Quantification* behavior. Many determiners and adverbs have been analyzed as Generalized Quantifiers (GQs) in the past fifteen years by several authors elaborating on Barwise and Cooper (1981) initial insights and proposals (see Keenan & Westerståhl 1997 for a recent survey). In what follows, we will assume that a PC can be represented as a GQ of form:

$$R_{p^{type1}, q^{type2}}(\Sigma_1, \ldots, \Sigma_n),$$

where *type*1 and *type*2 are the intensional types (propositional content, attitude, force, or disjunction of such types) which are possible in view of the PC's semantic scope(s). The Σ_i are sets of propositions (usually indexed w.r.t. worlds). The relation $\vdash$ can associate propositions of various types

[23] This is a default rule, but we will not be concerned here with its non–monotonic properties.

(e.g. forces and attitudes). We will omit the type restrictors *type*1 and *type*2 for readability.

Under the form of the previous definition for $POURTANT_{p,q}$, *pourtant* is a GQ of type <1, 1> (Lindström 1966), that is a relation between sets. Although this representation is much simplified (it does not use particular properties of $\vdash$), it allows one to distinguish between oppositive PCs of various 'strengths'. It has been noted (Jayez 1988) that the French indirect oppositive PC *mais* ($\approx$ *but*) is in a sense weaker than *pourtant* ($\approx$ *yet*), even when it conveys a relation of direct opposition. Compare:

(35) a. John is intelligent and reliable, POURTANT Microsoft did not hire him

b. John is intelligent and reliable, MAIS Microsoft did not hire him

While these two pairs of sentences describe the same situation, (35a) would suggest to a French speaker that this situation is somewhat strange or surprising, in contrast with (35b) which just points to a direct opposition. By using (35a), a speaker implies that, as far as she knows, there is no clear explanation for the decision of Microsoft.

To account for this difference, we recast the analysis of Jayez (1988) in the present GQ-based approach, by defining a relation $MAIS_{dir.op.}$, for the direct oppositive use of *mais*.

(36) $MAIS_{dir.op.:p,q}(\Sigma_w, \Sigma'_{w_{speaker}})$ iff there exists a non-empty set $\Sigma'' \subseteq \Sigma$ and a set $\Omega \supseteq \Sigma'$ such that:
(i) $\Sigma'' \cup \Omega$ is consistent,
(ii) $\Sigma \vdash p$,
(iii) $\Omega \cup \Sigma'' \vdash \neg q$.

This definition selects pairs of sets in which the first member entails p while the second member can be extended to a set (Ω) which supports a proof of $\neg q$ modulo some additional premises (Σ''). In the case of *pourtant*, the second set Σ' allowed by itself the construction of a proof of $\neg q$. In the present case, Σ' has not necessarily these resources: this reflects the intuition that *mais* conveys the idea of a possible opposition (modulo some additional information), while *pourtant* conveys the idea of a necessary opposition. Intuitively, the definition says that, in every reasonable situation where p is true, the speaker can imagine some reasons (Ω) for having $\neg q$. In short, every p-situation is a potential $\neg q$-situation for the speaker. When *mais* connects assertions, as in (35b), p and q are in the speaker's world. However, $\neg q$ is not necessarily in the speaker's world, even in the part of this world limited to Σ', since the set Ω might contain propositions outside this world. In consequence, contrary to *pourtant*, there is no impression of contradiction. We will not consider the other uses of *mais* here.

Turning to consequence connectives, we can assign to *donc* and *par conséquent* the following GQ form.

(37) $DONC_{p,q}(\Sigma_w, \Sigma'_{w_{speaker}})$ iff there exists a non-empty set $\Sigma'' \subseteq \Sigma$ such that:
(i) $\Sigma'' \cup \Sigma'$ is consistent,
(ii) $\Sigma \vdash p$,
(iii) $\Sigma' \cup \Sigma'' \vdash q$.

So, a pair $<\Sigma_w, \Sigma'_{w_{speaker}}>$ licenses a consequence from p to q whenever Σ proves p and Σ' augmented with some elements of Σ proves q.

In French, a distinction similar to the difference between *pourtant* and *mais* exists for *donc* and *alors*. *Donc* is modally 'stronger' than *alors*, which is more 'circumstantial' (the consequence is not presented as necessary). Once again, this can be reflected by relaxing the condition on *alors* in the manner of that for *mais*.

(38) $ALORS_{p,q}(\Sigma_w, \Sigma'_{w_{speaker}})$ iff there exists a non-empty set $\Sigma'' \subseteq \Sigma$ and a set $\Omega \supseteq \Sigma'$ such that:
(i) $\Sigma'' \cup \Omega$ is consistent,
(ii) $\Sigma \vdash p$,
(iii) $\Omega \cup \Sigma'' \vdash q$.

Intuitively, the relation $ALORS$ says that every reasonable situation where p is true can be extended to a q-situation, from the speaker's point of view.

The present approach is very 'weak' because it does not require that there be a special form of connection between p and q: what it says is simply that we can distinguish two sets Σ_w and $\Sigma'_{w_{speaker}}$ which prove p and q or $\neg q$ respectively, when Σ' is enriched with some elements of Σ. This seems compatible with most approaches on conditionals and informational entailment (see Veltman 1986 for a discussion), but it remains to make more precise which sort of consequence relation $\vdash$ we have in mind, a topic which is the subject of the next section.

4.2. DEDUCTIVE CONSTRAINTS

4.2.1. *General relevance requirement*

It is well-known that material implication $\Rightarrow$ and its associated deduction relation is in some cases inadequate for capturing linguistic entailment, because it does not require that there be any special link between premises and conclusion. We could assume that $\vdash$ corresponds at least to *relevant entailment* $\rightarrow$ (Anderson & Belnap 1975), which requires that every premise be effectively used in a deduction. This would forbid situations $\Sigma, p \vdash q$ in which p is simply added to Σ but is completely dispensed with in the proof of q. Note that the relevance constraint is a general pragmatic one, and is not inherited from the set of constraints on PCs.

However, this hypothesis would lead us to reject so-called *disjunctive syllogism* as a valid form of reasoning (Anderson & Belnap 1975, pp. 164–166): relevance logics based on the rejection of weakening also block disjunctive syllogism.

$$\text{(Weakening) } A, B \vdash A$$

$$\text{(Disjunctive syllogism) } \neg A, A \vee B \vdash B$$

This is clearly unwanted, since many PCs use disjunctive syllogism: *It's you or Sam who wrote the letter. You tell me it's not you, so it's Sam.*

We observe that disjunctive syllogism is dangerous because it can be used in a proof of weakening, but this proof needs $\vee$–introduction as a crucial step:

$$(\vee\text{–introduction) } A \vdash A \vee B$$

$\vee$–introduction is notoriously problematic as it is not really 'relevant' in any intuitive sense (Schurz 1991). We could adopt here, at least in the case of $\vee$–introduction, the proposal of Sperber and Wilson (1986) that there be no introduction rule in natural inference systems. Forbidding $\vee$–introduction is sufficient to avoid the derivation of weakening when disjunctive syllogism is maintained, as can be shown by a proper Gentzenization of the system (Thistlewaite, McRobbie, Meyer 1988).

Accordingly, we will assume in the following that our deduction relation $\vdash$ does not obey weakening and does not support $\vee$–introduction. However, we will not take a firm position as to whether other introduction rules must also be dismissed, as suggested in relevance theory (Sperber & Wilson 1986), or a quite different notion of analytic deduction should be preferred (Tzouvaras 1996).

4.2.2. *The two forms of connection*

There are two main forms of connection used by PCs, that between two distinct propositions (e.g. (18a) or (20b)), which we will call *discrete connection*, and that between two propositions expressing different degrees of the same quality (e.g. (22)), which we will call *scalar connection.* Another form of scalar connection is shown in (39).

(39) Most students have registered, but not all of them yet

Are examples like (22) and (39) amenable to a case of *semantic opposition* (Lakoff 1971), as *Peter is tall but Bill is short*? Semantic opposition is a form of contrast in which two symmetric predicates (*tall* vs *short*) are predicated of two distinct entities (*Peter* and *Bill*). We could relax the condition on the distinctness of entities and hypothesize that semantic opposition is sufficient to account for (22) and (39). However, there are two reasons to reject this move. First, as noted by Lakoff, semantic opposition is not compatible with *although* (?? *Although Peter is tall, Bill is short*). But we have:

(40) Although John is intelligent, he is not a genius
(41) Although most students have registered, not all of them have done yet

Second, in some cases, X is used to concede something akin to Y, in a form X *but* Y.

(42) A– All students have probably registered by now
B– No. True, most of them have registered, but not all of them

So, even if one sees (42) as a case of semantic opposition, it has still to be explained in which sense a form *most* x P can be 'akin' to a form *all* x P. Therefore, we need to provide a representation for scalar connection, since we cannot simply eliminate it in favor of semantic opposition.[24] In the next section, we address the problem of defining the consequence relation for scalar connection.

4.2.3. *Scalar connection*

The basic idea in this section is that a judgment like 'John is intelligent' means that John's intelligence is superior to a certain intelligence degree that we call the intelligence *threshold*. This information does not determine the exact degree of John's intelligence, which could be specified by adding information. It is natural to represent intelligence degrees on a scale (total order), and to consider consistent information states which define possible positions on this scale. Let l be an arbitrary 'intelligence threshold': we consider any individual whose intelligence degree is greater than l to be intelligent. So, if x^i is the degree of intelligence of x, 'x is intelligent' will correspond to $x^i > l$. We assume also that there is a 'genius threshold' $L > l$. 'x is not a genius' will be reflected as $x^i < L$.
The initial information state s_0 is just the set $\{x^i > l, L > l\}$. By adding information stepwise, we obtain sequences of information states. A piece of information is any expression $t_1 > t_2$ or $t_1 < t_2$, where t_1, t_2 can be l, L, x^i, or any variable from a fixed domain.

[24] In a very interesting discussion, Blakemore (1987, pp. 125–141) argues in favor of a unified analysis of *but*, which would merge the contrast and opposition categories. In French, semantic opposition licenses the use of *par contre*, but *par contre* would be clumsy in the mentioned examples. So, we cannot extend Blakemore's proposal to French in a simple way.

A finite sequence is a *proof* of an expression if (i) its last member is consistent and (ii) one of its states allows for a proof of this expression. Every state in a sequence can be *revised*, in the sense of (Gärdenfors 1988). We will consider here what we call *proper* revisions: a proper revision amounts to replace a proposition by the contradictory proposition, here $x > y$ by $x < y$ and make all the necessary alterations to preserve consistency. In the revision process, some expressions can be *protected*, that is, may not be modified by the revision. The only expression we will protect is $L > l$: it expresses a default preference (the threshold of intelligence is most naturally conceived as inferior to the genius threshold). A proper revision must leave the protected expressions untouched.

We now define the notion of *admissibility* for a proof. A proof of an expression A is admissible whenever it satisfies the following constraint: if A becomes provable at s_{i+1} and $s_{i+1} = s_i \cup \{r\}$, r must not be consistent with any proper revision of s_i.

Our proof system will be a mini Gentzen–style system with just the two following axioms:

$x > y, y > z \vdash x > z$,

$x > y \dashv\vdash \neg(y > x)$.

The definition of proof and consistency are standard.

We now prove the following: no proof of $x^i < L$ ('John is not a genius') is admissible, while some proof(s) of $x^i > L$ ('John is a genius') is (are).

Suppose that we have some proof of $x^i < L$, $s_0 \ldots s_n$. This means that $s_0 \cup s_1 \cup \ldots \cup s_{n-1} \cup \{r\} \vdash x^i < L$. Assume that r is inconsistent with every revision of s_{n-1} where $x^i > l$ is replaced by $x^i < l$. Observe now that the set $\{x^i < l, l < L, r\}$ is consistent. In the contrary case, we would have that $l < L, r \vdash x^i > l$. This is only possible if $r = x^i > l$. So, our proof has actually the form $s_{n-1}, x^i > l \vdash x^i < L$. But it can be shown by induction on the length of the proof that every proof $\Sigma, x > y \vdash x < z$ can be reduced to a proof $\Sigma \vdash x < z$, which indicates that $x^i < L$ was already established in s_{n-1}. So, the set $\{x^i < l, l < L, r\}$ is consistent and can be extended to a consistent set including for each A of s_{n-1} either A or its negation. This amounts to say that we have constructed a proper revision of s_{n-1} including r.

On the other hand, consider the sequence:

$s_0 = \{x^i > l, l < L\}$,

$s_1 = s_0 \cup \{x^i > L\}$.

This sequence is an admissible proof (any attempt to construct a proper revision will lead to inconsistency).

This construction is motivated by the relevance requirement mentioned above: no previous information can be 'useless', in the sense that replacing it by contradictory information would not matter for the proof's result. In other words, only proofs of 'John is a genius' from 'John is intelligent' can obey relevance.[25] Proofs of 'John is not a genius' from 'John is intelligent' always imply producing and subsequently turning out some information.

The present treatment uses only indirect properties of the phenomenon of scalar inference. For space reasons, we will not discuss here other solutions, which link scalar inference to *argumentative* monotony. The interested reader is referred to (Anscombre & Ducrot 1983; Ducrot 1995; Elhadad 1993; Jayez 1988; Merin 1994) for further discussions.

4.2.4. *Discrete connection*

When no scale is implied, the situation is simpler. Viewed as sequences of information states, chains of premise sets are automatically admissible in the family of relevance logics. Suppose increasing sets of premises $\{p_1\}$, $\{p_1, p_2\}$, ..., $\{p_1 \ldots p_n\}$. If adding p_{n+1} allows one to prove q and p_{n+1} is consistent with a proper revision of $\{p_1 \ldots p_n\}$, then either the proof is not relevant or the information state is inconsistent. As an illustration consider the following strange variant of (18a).

(43) ?? John had forgotten his passport. So they did not delay him at the frontier.

No classical theory of enthymematic reasoning can explain why (43) is odd: one has only to add a premise such as 'the custom officers did not notice John' to get the unwanted conclusion. Relevance simply says in this case that the first premise 'John had forgotten his passport' is no longer used in the reasoning, which is then not admissible (although correct in classical logic).

Inferential PCs differ as to whether they support scalar connection: consequence PCs do not, while some opposition and reevaluation PCs do, within some limits. For space reasons, we will not discuss their detailed behavior nor propose a systematic account of these variations. We simply assume that the deduction relation $\vdash$, which was parameterized for discrete connection has a sibling $\vdash_{><}$ for scalar inference. E.g., the definition for *mais* given in section 4.1 is now changed to the following.

$MAIS_{DOPP:p,q}(\Sigma_w, \Sigma'_{w_{speaker}})$ iff $MAIS_{dir.op.:p,q}(\Sigma_w, \Sigma'_{w_{speaker}})$ or, $\Sigma_w = \{p\}$, $\Sigma'_{w_{speaker}} = \{q\}$ and $p \vdash_{><} q$.

[25]There are several kindred senses of the term *relevance*. An admissible proof of form $x^{\iota} > l, l < L, x^{\iota} > L \vdash x^{\iota} > L$ is 'relevant' because the unprotected expression $x^{\iota} > l$ cannot be negated. However, this proof is not 'relevant' on relevance logic's account, since the two premises $x^{\iota} > l$ and $l < L$, are not used in the proof, which is just an instance of the axiom $A \vdash A$. It seems that the basic intuitive notion behind all these cases is that of information's usefulness.

5. More GQ behavior

In this section, we will refine and extend the model presented in section 4.1, by considering first consequence PCs, then reevaluation PCs.

5.1. CONSEQUENCE PCs AND *SI*-SENTENCES

It has been proposed that forms *if* X, Y introduce a very weak form of implication (Veltman 1986). *Si* has been analyzed along similar lines by de Cornulier (1985) who equates *si* X, Y with 'in a situation where X, Y'. If consequence PCs such as *then* presuppose only such weak conditions, it explains their well-known compatibility with conditional sentences. However, it does not explain why *donc* is incompatible with such sentences, as evidenced by the contrast in (19), which we repeat below.

(19) a. ?? If the weather is fine, DONC I will go out

b. If the weather is fine, ALORS/DU COUP/? DE CE FAIT I will go out

At first sight, we just have to assume that *donc*, in contrast with other consequence PCs, must introduce an 'absolute' attitude or force, that is, an attitude or force which holds independently of the protasis, whence an incompatibility with *si*. In other terms, while consequence PCs in general require that the intensional entity associated with Y be true whenever that associated with X is, *donc* requires that the intensional entity associated with Y be true in the speaker's world $w_{speaker}$, or, at least, in the restriction of the speaker's world to the premises Σ'' and Σ' of the definition (37). However, in a conditional structure *if* X, Y, Y is not necessarily true in the speaker's world, since its truth depends on that of the protasis X, which is only presented as possible.

While this seems reasonably simple and plausible, it raises a technical problem. Up to now, the approach we followed is essentially *static*: definitions are built over a deduction relation $\vdash$ between premises and conclusions(s). If we just say, for capturing *donc*, that the conclusion must be true in any world containing the necessary premises (Σ'' and Σ' in the definition 37) and their consequences, we do not add anything to the definition (37). What we must require is that the conclusion q be true whenever the material used in the proof of p (i.e. Σ'' in the definition 37) is added to Σ' *with its epistemic status*. Suppose, for simplicity, that this material is only p itself. If p is asserted, we have only to add p, since p belongs to the speaker's world. If p is conditional, we may add either that p is possible, or that p is impossible (as in counterfactuals). Intuitively, we may not obtain q if we add the possibility of p, let alone the impossibility of p. To take possibilities and impossibilities into account, we must add three definitions. First, we must adapt the condition (34).

(44) Let $\Sigma = \{p_1 \dots p_n\}$ a set of propositions which constitute the premisses of p in a proof of $\Sigma \vdash p$, and let p be associated with the sentence X, we define $COND(\Sigma)$ to be:
(i) Σ itself if $IF(X) = assertion$,
(ii) $\{$ *Might* $p_1 \dots$ *Might* p_n $\}$ if $IF(X) = supposition$,
(iii) $\{$ $\neg$*Might* $p_1 \dots \neg$*Might* p_n $\}$ if $IF(X) = counterfactual$.

The $COND$ operator returns the initial set of propositions if we have an assertion, and a set of possibilities (resp. impossibilities) if we have a supposition (resp. a counterfactual). The next step is to define the interaction between the possibilities or impossibilities and the premises in the speaker's world. Using Veltman's notion (Veltman 1996) of *update*, we define the addition of *Might p*. In update models, an information state is a set of worlds. A proposition holds or is true in an information state iff it holds in each of the worlds which make up this information state. Let $\sigma + p$ denote the result of adding a proposition p to an information state σ. This operation amounts to consider only the subset of worlds in σ which contain p. If p was already true in σ, adding p to σ produces nothing. If p is incompatible with other information in σ, $\neg p$ holds in every world of σ. So $\sigma + p$ is the empty set, or absurd information state. If p is true in some worlds of σ and false in some others, $\sigma + p$ denotes the subset of worlds in σ where p is true. The case of possibility formulas is essentially the same.

(45) Let σ be an information state, the information state σ + *Might p* is:
(i) the absurd state if $\sigma + p$ is the absurd state,
(ii) σ itself if $\sigma + p$ is not the absurd state.

The addition of $\neg$*Might p* is similar. Adding $\neg$*Might p* to σ produces the absurd state if σ contains p or is compatible with p and σ itself if p is not compatible with σ. Finally, we devise a new definition of *donc*.

(46) $DONC_{p,q}(\Sigma_w, \Sigma'_{w_{speaker}})$ iff there exists a non-empty set $\Sigma'' \subseteq \Sigma$ such that:
(i) $\Sigma'' \cup \Sigma'$ is consistent,
(ii) $\Sigma \vdash p$,
(iii) $\Sigma' \cup \Sigma'' \vdash q$,
(iv) if σ is the minimal information set containing Σ', q holds in $\sigma + COND(\Sigma'')$.

This definition simply enriches the previous definition (37) for *donc* with the condition (iv). If p corresponds to an *if*-sentence, $COND(\Sigma'')$ will be a set of possibilities or impossibilities. The addition of this set to σ will produce σ or the absurd state of information. In the first case, q does not necessarily hold, because (iii) makes the truth of q depend of the truth of Σ' *and* Σ'' (which must be non–empty). In the second case, no proposition holds. An analogous reasoning applies to the counterfactual value of the *if*–sentence.

However, certain facts invite a more flexible analysis.[26] Consider:

(47) a. ?? Every rectangle is four–sided. If A is a rectangle, DONC A is four–sided

b. A is even. If B = 2A, ? DONC B is even

c. A is superior to B, if B is superior to C, ? DONC A is superior to C

These data suggest that *donc* is better when an asserted *factual* premise occurs in the reasoning. There are variations across speakers about the nature of the reasoning (logical, empirical, etc.) which licenses *donc* in such cases. There are also variations as to whether one asserted factual premise is sufficient. By and large, for *donc* to be tolerated in examples such as (47a–c) it is necessary that (i) the reasoning be sufficiently explicit, (ii) at least one asserted factual premise occur in the reasoning. This combination of preferences stems from the general constraint on *donc*, that the conclusion q be provable. We do not pretend to explain these facts, but we can dissipate some of the feeling of mystery which they evoke.

The reason why, other things being equal, (47a) is worse than (47b) and (47c) is that the only asserted premise is a rule, not a fact. While facts can be strictly idiosyncratic, rules cannot in general. Technically, this means that facts can be true in only one information state, and false outside, while rules must hold 'more generally'. We will not discuss the different options for measuring the degree of generality, and will simply assume that rules must hold in a set of information states which may not be a singleton. In consequence, if we update the same information state σ with a rule r and with a fact f, we have, in general, $\sigma + f \subset \sigma + r$. We note that deductions which point to facts, or *factual deductions*, entail the introduction of some fact in the premises: one may not deduce a fact only from a set of rules; the information must grow at some place, that is, the set of worlds compatible with the information states must decrease in size, down to the point where we keep only the worlds where the fact to be proven is true.

[26] In French, there is a strong difference between the sentential *donc*, which is in the initial position and the VP–internal one. In the discussion to come, we consider only the sentential *donc*, which is more problematic.

Consider (47a). The first premise is a rule. According to definition (46,iv), the following configuration must obtain: from the minimal information state containing the rule 'every rectangle is four–sided' ($= r$) updated by the conditional 'if A is a rectangle' ($=$ *Might* p), we can deduce 'A is four–sided' ($= r$). Let $\mathbf{0}$ be the minimal information state, the minimal information state containing r is $\mathbf{0} + r$. So, the update corresponding to (47a) has the form:

(47a) : $(\mathbf{0} + r) + Might\ p \vdash q$.

The update corresponding to the better examples is:

(47b), (47c) : $(\mathbf{0} + f) + Might\ p \vdash q$

In general, due to the difference we postulated between facts and rules, we have: $\mathbf{0} + f \subset \mathbf{0} + r$. So, the information on premises introduced in (47b) and (47c) is more precise than that introduced in (47a). We ascribe the observed difference to this fact. More generally, we submit that cases like (47b) or (47c) are more likely to be perceived as deductions than (47a) because their information growth resembles that of a factual deduction.

This hypothesis predicts that, in the case of *rule deductions*, that is, deductions which prove rules, it will be sufficient to have one asserted rule to improve the examples. The following data show that it is indeed the case.

(47) d. ?? If every equilateral triangle is equi–angled, and if every equi–angled triangle has three 60 degrees angles, DONC every equilateral triangle has three 60 degrees angles

e. Every equilateral triangle is equi–angled, if every equi–angled triangle has three 60 degrees angles, ? DONC every equilateral triangle has three 60 degrees angles

Summarizing, these data suggest that introducing *if*–sentences in a *donc*–reasoning is acceptable insofar as the growth of information in the premises mimics that of the same reasoning in its 'normal', i.e. non–conditional, form.

5.2. REEVALUATION PCs

Reevaluation PCs have the following characteristic: in a form $X\ C_{reev}\ Y$, the informational effect of X is reevaluated when Y is introduced. We will describe here only the two PCs *de toute façon* and *quoi qu'il en soit*, and simplify their GQ relations for shortness. The associated relations will be denoted by DTF and QQS respectively.

Let us first describe intuitively the behavior of these two PCs. In a form $p\ C_{reev}\ q$,

1. if C_{reev} = *de toute façon*, then,[27]
   (a) q entails p, as in *John hates Paul,* DE TOUTE FAÇON *he hates everybody*, or
   (b) there is some r such that q entails r regardless of p (whether p is true or false), and p, in some context(s), entails r, as in *John completely forgot the meeting,* DE TOUTE FAÇON *it was postponed.* Suppose that r is something like 'John could not make the declaration he had prepared for the meeting'. Then, the cancellation of the meeting entails r irrespective of John's presence or absence. Or, more marginally,[28]
   (c) there is some r such that p entails r in some context(s) and q blocks the derivation of r from p in the speaker's world, as in *John flunked his exam,* DE TOUTE FAÇON *he can take this exam once again.* If r is 'John will not get his diploma', q blocks the derivation of r since John has got another chance. This is done by constructing a situation of the speaker's world, $w_{speaker}$, which allows one to derive the negation of a premise which is used in the derivation of r in some other[29] world.
2. If C_{reev} = *quoi qu'il en soit*, there is some r such that either the above situation 1c obtains, or p entails r in some contexts and q entails $\neg r$ in the speaker's world, as in *Mary is smart,* QUOI QU'IL EN SOIT *she is too young for this position*, where we understand that a possible conclusion from 'Mary is smart', e.g. 'The company should hire her', is anyway contradicted by q: Mary is too young to be hired.

(48) $DTF_{p,q}(\Sigma_w, \Sigma'_{w_{speaker}}, \{r\})$ iff Σ' is non empty and:
(i) $\Sigma', q \vdash p$, or
(ii) $\Sigma, p \vdash r$ and $\Sigma', q \vdash r$, or
(iii) $\Sigma, p \vdash r$ and $\Sigma', q \vdash \neg p'$ for some p' in Σ.

(49) $QQS_{p,q}(\Sigma_w, \Sigma'_w, \{r\})$ iff:
(i) $\Sigma, p \vdash r$ and $\Sigma', q \vdash \neg p'$ for some p' in Σ, or,
(ii) $\Sigma, p \vdash r$ and $\Sigma', q \vdash \neg r$.

De toute façon and *quoi qu'il en soit* are inappropriate in some configurations of indirect opposition compatible with *but* or *mais.* E.g., in *Mary is smart* (⇒ 'she can be hired') *but she is rather young* (⇒ 'she should not be

[27] We ignore here extra factors connected with the difference between logical and generic entailment, and with restrictions on causal ordering.
[28] Speakers do not agree on this inferential structure.
[29] As remarked in note 22, worlds can be inconsistent. Should this be avoided, we ought to require that w be different from $w_{speaker}$, if we want $\vdash$ to be monotonic.

hired'), the two PCs are clumsy. This is explained by our hypothesis, since, in such cases, there is no pragmatically available interpretation in which the restriction *she is rather young* blocks the conclusion *she can be hired.* It is only in contexts where this interpretation emerges naturally that such sentences are felicitous.

6. Conclusion

While we have provided here only a very limited sample of problems, observations and techniques, we have made clear two points:

- PCs impose constraints on the construction and management of their arguments. They behave as complex relations on complex semantic entities. In some cases, their behavior can be seen as a generalized quantification on proof space.
- The observed constraints are essentially *hybrid.* In contrast with other domains in lexical semantics, one cannot rely only on feature and type systems to account for the phenomena. It is necessary to formulate the constraints at a level where inferential (global) and more local parameters, particular to each class of PCs, interact. However, this interaction is not just discourse–driven, which would take us back to some vague conception of contextual dependence. It is controlled by lexical informations associated with PCs, which determine how inference affects them. Moreover, taking into account detailed instructions allows us to distinguish between lexical items which belong to the same functional class, that is, to address the problem of *plesionymy* (Hirst 1995).

In this paper, we have not considered the relation of PCs to more general discourse phenomena. In most approaches to discourse, a distinction between content relations and discourse or rhetoric moves is used. Generally speaking, one can say that discourse coherence is achieved by connecting discourse units (e.g. sentences, speech acts, interactional moves) so that no unit remains unattached. Several attachment procedures have been proposed (see Polanyi 1985, Asher 1993, Traum & Heeman 1996, Roulet *et al.* 1985 for some discussions and examples). It should be clear that the relations we consider here are much more low–level and pertain to content rather than to global discourse organization. Although we use sometimes labels for rhetoric moves (e.g. explanation) to provide intuitive guidance, we are not concerned with discussing precise definitions of such labels. This 'minimalist' option is possible only because rhetoric relations are usually defined in terms of basic content relations, not the reverse.

However, this leaves two questions open.

(a) Can rhetoric relations always be eliminated in favor of content relations in the case of PCs ? Answering this question depends on further empirical studies. Note that this problem is in principle distinct from that of the identification of a generating subset of discourse relations, allowing one to derive their whole spectrum (Sanders et al. 1993).

(b) Are there specific attachment properties of PCs? In his 1993 book, Asher has generalized the open constituent approach of (Polanyi & Scha 1984). This allows in particular more flexibility for linking a sentence to a non immediately preceding one. Is this sentence–hopping phenomenon controlled by the PCs in some way?

We intend to address these questions in further research.

References

Allen J.F. (1981) An interval–based representation of temporal knowledge, 7th International Joint Conference on Artificial Intelligence, pp. 221–226.

Anderson A.R. , Belnap N.D. (1975) *Entailment. The Logic of Relevance and Necessity*, Princeton: Princeton University Press.

Anscombre J.C., Ducrot O. (1983) *L'argumentation dans la langue*, Bruxelles: Mardaga.

Asher N. (1993) *Reference to Abstract Objects in Discourse*, Dordrecht: Kluwer.

Austin J.L. (1962) *How to Do Things with Words*, Oxford: Clarendon Press.

Barwise J., Cooper R. (1981), Generalized quantifiers and natural language, Linguistics and Philosophy **4**, pp. 159–219.

Bierwisch M. (1980) Semantic structure and illocutionary force, in Searle J.R. , Kiefer F., Bierwisch M. (eds) *Speech Act Theory and Pragmatics*, Dordrecht: Reidel, pp. 1–37.

Blakemore D. (1987) *Semantic Constraints on Relevance*, Oxford: Blackwell.

Bonomi A., Casalegno P. (1993) *Only*: Association with focus in event semantics, Natural Language Semantics **2**, pp. 1–45.

de Cornulier B. (1985) *Effets de sens*, Paris: Editions de Minuit.

Ducrot O. (1980) *Les mots du discours*, Paris: Editions de Minuit.

Ducrot O. (1995) Les modificateurs déréalisants, Journal of Pragmatics **24**, pp. 145–165.

Elhadad M. (1993) Generating argumentative judgment determiners, in Proceedings of the 11th National Conference on Artificial Intelligence, Menlo Park: AAAI Press and the MIT Press, pp. 344–349.

Fernandez X. (1996) *Les particules énonciatives*, Paris: Presses Universitaires de France.

Ferrari A. (1995) *Connessioni. Uno Studio Integrato Della Subordinazione Avverbiale*, Geneva: Slatkine.

Gärdenfors P. (1988) *Knowledge in Flux*, Cambridge: MIT Press.

Gil A. (1995) *Textadverbiale in den Romanische Sprachen*, Bonner Romanistische Arbeiten **53**, Berne: Peter Lang.

Grieve J. (1996), *A Dictionary of French Connectors*, London: Routledge.

Grote B., Lenke N. (1995) Ma(r)king concessions in English and German, in Proceedings of the Fifth European Workshop on Natural Language Generation, May 1995, Leiden, Netherlands.

Glasbey S. (1993) Distinguishing between events and times: some evidence from the semantics of *then*, Natural Language Semantics 1, pp. 285–312.

Halliday M. A. K., Hasan R. (1976) *Cohesion in English*, London: Longman.

Hintikka J. (1962) *Knowledge and Belief*, Ithaca: Cornell University Press.

Hirst G. (1995) Near–synonymy and the structure of lexical knowledge, Working Notes of the AAAI Spring Symposium on Representation and Acquisition of lexical Knowledge: Polysemy, Ambiguity, and Generativity, 27–29 March 1995, Stanford University, pp. 51–56.

Hobbs J.R., Stickel M.E., Appelt D.E., Martin P. (1993) Interpretation as abduction,

Artificial Intelligence **63**, pp. 69–142.

Hybertie C. (1995) *La conséquence en français*, Paris: Ophrys.

Iatridou S. (1994) On the contribution of the conditional *then*, Natural Language Semantics **2**, pp. 171–199.

Jayez J. (1988) *L'inférence en langue naturelle. Le problème des connecteurs*, Paris: Hermès.

Jayez J., Rossari C. (1996) The semantic of pragmatic connectives in tree adjoining grammars. The French *donc* example, to appear in Abeillé A., Rambow O. (eds), Proceedings of the Tag+4 Conference, Stanford: CSLI.

Keenan E.L., Westerståhl D. (1997) Generalized quantifiers in linguistics and logic, in van Benthem J., ter Meulen A. (eds) *Handbook of Logic and Language*, Amsterdam: North-Holland, pp. 837–893.

Knott A. (1996) *A Data-Driven Methodology for Motivating a Set of Coherence Relations*, Ph.D. thesis, University of Edinburgh.

König E. (1981) The meaning of scalar particles in German, in Eikmeyer H.J., Rieser H. (eds) *Worlds, Words and Contexts*, Berlin: de Gruyter, pp. 107–132.

König E. (1986) Conditionals, concessive conditionals and concessives: areas of contrast, overlap and neutralization, in Traugott E.C., ter Meulen A., Reilly J.S. , Ferguson C.A. (eds) *On Conditionals*, Cambridge: Cambridge University Press, pp. 229–246.

Lakoff R. (1971) If's, and's and but's about conjunction, in Langendoen D.T. (ed) *Studies in Linguistic Semantics*, New York: Holt, Rinehart and Winston, pp. 114–149.

Lang E. (1984) *The Semantics of Coordination*, Amsterdam: John Benjamins (Translation from the German original *Semantik der koordinativen Verknüpfung*).

Lindström P. (1966) First-order predicate logic with generalized quantifiers, Theoria **35**, pp. 186–195.

Manzotti E. (1995) Aspetti linguistici dell'esemplificazione, Versus **70–71**, pp. 49–114.

Merin A. (1994) *Decision-Theoretic Pragmatics*, unpublished manuscript of an ESSLI advanced course, Copenhagen, 8–19 August 1994.

Moeschler J. (1989) Modélisation du dialogue, Paris: Hermès.

Morel M.A. (1996) *La concession en français*, Paris: Ophrys.

Nemo F. (1992) *Contraintes de pertinence et compétence énonciative. L'image du possible dans l'interlocution*, Thèse de Doctorat, Ecole des Hautes Etudes en Sciences Sociales, Paris.

Pasch R. (1989) Ueberlegungen zum Begriff des Satzmodus, Linguistische Studien **193 A**, pp. 1–88.

Polanyi L., Scha R. (1984) A syntactic approach to discourse semantics, in Proceedings of the 10th International Conference on Computational Linguistics, pp. 413–419.

Polanyi L. (1985) A theory of discourse structure and discourse coherence, in Proceedings of the 21st Annual Meeting of the Chicago Linguistic Society, pp. 306–322.

Reichman R. (1985) *Getting Computers to Talk Like You and Me*, Cambridge: MIT Press.

Rossari C. (1994) *Les opérations de reformulation*, Berne: Peter Lang.

Rossari C., Jayez J. (1997) Connecteurs de conséquence et portée sémantique, Cahiers de Linguistique française **19**, to appear.

Rothstein S. (1995) Adverbial quantification over events, Natural Language Semantics **3**, pp. 1–31.

Roulet E. et al. (1985) *L'articulation du discours en français contemporain*, Berne: Peter Lang.

Sanders T. J. M., Spooren W. P. M., Noordman L. G. M. (1993) Coherence relations in a cognitive theory of discourse representation, Cognitive Linguistics **4**, pp. 93–133.

Schiffrin D. (1987) *Discourse Markers*, Cambridge: Cambridge University Press, 2nd edition.

Schurz G. (1991) Relevant deduction, from solving paradoxes towards a general theory, Erkenntnis **35**, pp. 391–437.

Searle J.R , Vanderveken D. (1985) *Foundations of Illocutionary Logic*, Cambridge: Cambridge University Press.

Sperber D., Wilson D. (1986) *Relevance. Communication and Cognition*, Oxford: Basil Blackwell.

Sweetser E. (1990) *From Etymology to Pragmatics: Metaphorical and Cultural Aspects of Semantic Structures*, Cambridge: Cambridge University Press.

Thistlewaite P.B., McRobbie M.A., Meyer R.K. (1988) *Automated Theorem-Proving in Non-Classical Logics*, London: Pitman.

Traum D., Heeman P (1996) Utterance units and grounding in spoken dialogue, to appear in the Proceedings of ICSLP.

Tzouvaras A. (1996) Aspects of analytic deduction, Journal of Philosophical Logic **25**, pp. 581–596.

Veltman F. (1986) Data semantics and the pragmatics of indicative conditionals, in Traugott E.C., ter Meulen A., Reilly J.S. , Ferguson A.C. (eds) *On Conditionals*, Cambridge: Cambridge University Press, pp. 147–177.

Veltman F. (1996) Defaults in Update Semantics, Journal of Philosophical Logic **25**, pp. 221–261.

INTERLINGUAL REPRESENTATION OF COMPLEX PREDICATES IN A MULTILINGUAL APPROACH: THE PROBLEM OF LEXICAL SELECTION

MARÍA ÁNGELES ZARCO
Departamento de Filologia Francesa e Inglesa
Universidad de Cádiz (Spain)
Bartolomé LLompart s/n
11003 Cádiz, Spain

Abstract. In this paper we study the Syntagmatic, Paradigmatic and Transfer relations Complex Predicates (CPs) enter, within a Machine Translation (MT) approach. On the one hand, we describe the cross-linguistic validation of lexical semantic representations applied to the natural language realizations of the Italian predicative forms METTERE, PRENDERE and DARE, in their combination with Nouns, and to their equivalent realizations in English and Spanish. On the other, we suggest the use of such representations to stand for lexicalized verbal forms as well, so that paraphrasis phenomena are handled. Finally, we focus on the problem of lexical selection. We apply Mel'cuk Lexical Functions to solve co-occurrence restrictions and we suggest the need to establish a relationship among semantic and lexical formulae. We claim both representations can be related, since Lexical Functions can be defined on a conceptual basis, and we suggest the need for a new LF $Oper_{1-2}$. Our analysis has a multilingual application.

1. Introduction

Complex predicates (CPs) are a type of collocation formed by the combination of the structure V+N used in a non-idiomatic meaning (considerations on transparency are given on the next page) and restricted in commutability not only by grammatical and semantic valence but also by usage (Aisenstadt, 1978). We refer to structures such as "To give a kick", "To have a bath", "To take a risk", etc.

P. Saint-Dizier (ed.), Predicative Forms in Natural Language and in Lexical Knowledge Bases, 321–347.

We propose these structures enter into a three-dimensional research field within a Machine Translation approach: the Paradigmatic, Transfer, and Syntagmatic relations. From a Paradigmatic point of view, we deal with the problem of paraphrasis ("To take a risk" - "To risk"), Transfer relations answer for structural mismatches through languages (English: "To commit suicide" - Spanish: "Suicidarse") and, finally, the Syntagmatic level that refers to the problem of co-occurrence restrictions ("To commit a crime" - *"To make a crime") (Henceforth, the asterisk indicates that the sentences or phrases are not grammatical).
Our aim is to be able to give a kind of interlingual representation to solve, firstly, paraphrasis and structural transfer mismatches and, secondly, lexical selection.
To do so, we used the Corpus from the Istituto di Linguistica Computazionale (Pisa) and decided to draw Italian CP forms as a first step. Our choice on the Italian verbs METTERE, PRENDERE and DARE is due to the need to look for verbs that were part of a CP and that their number of realizations were of a considerable amount. We drew all the realizations where the verbs METTERE, PRENDERE and DARE in combination with a Noun and acting as a CP appeared, and paired them with the Spanish and English realizations with the same meaning. Second, we joined to our CPs for each language the corresponding lexicalized forms with the same meaning, so that we consider the problem of meaning representation across languages and within the same language.
As underlying representations, we have applied Jackendoff's (1990) Conceptual Structures (CSs) to relate CPs and lexicalized forms within a multilingual approach. This way, structural transfer relations and paraphrasis are handled, but we claim as well, that such CSs -Interlingua representations- are not enough, especially for language generation, since they do not account for lexical selection (Smadja & McKeown, 1991).
The interlingual formalism used to represent co-occurrence restrictions is Mel'cuk Lexical Functions, that, apart from codifying lexical selection, formalizes deep syntactic actants. We consider it to be a better and deeper lexical representation form compared to bilingual context-dependent rules (Danlos & Samvelian, 1992).
Thus, in our work we will be dealing with the meaning-surface realization relationship in a monolingual and multilingual basis, as interlingual representations stand for both CPs and lexicalized forms. Following Cattell (1984), we consider that both surface expressions encode the same semantic elements.
Our particular concern is to show how such interlingual representations codify the meaning of CPs and verbal forms, how we need some kind of codification to answer for lexical selection -one of the CP linguistic features

-, and how establishing a semantic-lexical formulae relationship seems to be appropriate. Our aim is to show that CSs and LFs are defined on the same bases, that they supply for mutual deficiencies and that it should be necessary to count on a new Lexical Function we called $Oper_{1-2}$

2. Linguistic Analysis of Complex Predicates

Complex predicates, defined, as stated above, as a type of collocation formed by the structure V+N used in a non-idiomatic meaning and restricted in commutability not only by grammatical and semantic valence, but also by usage, can be analysed from three different levels: the Syntagmatic, Paradigmatic and a new kind of paradigmatic relation, the Transfer relation, that relates a word or phrase in one language to a semantically equivalent word or phrase in another language (Evens, 1988). Let us examine them.

2.1. SYNTAGMATIC LEVEL

According to the first one - **the Syntagmatic**-, CPs are a type of collocation (Halliday, 1966; Aisenstadt, 1978; Allerton, 1984) formed by the structure V+N, characterized by the following features:

1. Most of verbs within a CP are empty of meaning, they seem to do only a little more than provide the verb-function and carry the signification of tense and number (R. Cattell, 1984:2). These verbs are used for syntactic support (support verbs) and can be defined in terms of their property to preserve the relationship between the first actant and the supported noun (deverbal noun) (Machonis, 1991). The constituents have a weakened grammatical meaning. Howarth (1993) classifies them as **delexical collocations**:
   To **make** an offer
   To **give** a kick
   To **take** a look
2. Some verbs within a CP have a metaphorical or very narrow specific meaning:
   To **exert** influence
   To **offer** an apology
   To **shrug** one's shoulders
   To **clench** one's teeth
3. They are structures with a variable degree of lexical selection. Commutability restrictions are conditioned by usage and they affect to just one constituent of the CP structure. The fewer the collocator

synonymous, the stronger the lexical restriction. Cowie (1986, 1991) and Howarth (1993) distinguish between **free collocations** and **restricted collocations**. The former refers to combinations that allow substitution of any of the elements. In contrast, restricted collocations are semi-fixed combinations and are characterized by the limited collocational range of their elements:

To **make** an offer/***do** an offer
To **take/have/give** a look
To **make/take/*do** a decision

4. The elements of such structure have a lexical dependence relation from the Noun (base) to the Verb (collocator) (Smadja & McKeown, 1991). It is the Noun, the main semantic layer, that selects a specific Verb which excludes other potential candidate verbs.
5. The reasons for lexical dependence are arbitrary and do not apply to categories but to lexical elements (Smadja, 1989; Mackin, 1978). Commutability restrictions on lexical co-occurrence are only partly explicable in semantic terms. "*In this type of case, the language simply seems to dictate, for no good semantic reason, that such and such a combination does or does not occur. The restriction is thus an arbitrary one*" (Allerton, 1984:28).
6. They do not form an opaque unit. Their meaning is compositional. The meaning of individual complex expressions (e.g. sentences) are functions of the meaning of their component basic expressions (e.g. words) (Hutchins & Somers, 1992). That is to say, their meaning is fully transparent (Cruse, 1986); It is made up as the sum of the meanings of their constituents, in contrast with the idiom that is an indivisible unit.
7. Following M. Wood (1986), lexical selection and compositionality criteria do not classify the three different combination groups - free phrases, restricted collocations and idioms- in three separate and well delimited areas. The idea of a continuum is suggested, where there are cases placed in the intersection sharing features from two different sets. According to this, we show how our examples vary in their grammatical changes depending on their closer-farther position to the free word combination group:

Some CPs admit **insertion of material** (count nouns):

To give a hug: He gave her a **big** hug
To make a mistake: He made a **serious** mistake
To commit a crime: He committed a **dreadful** crime

Some do not:

(mass nouns)

To commit suicide: *He committed **dreadful** suicide

To take part: *He took **important** part in those games

(count nouns)

To hazard a guess: ?He hazarded a wild guess

(The question mark shows this could be one, if any, of the very few adjectives that can be inserted between verb and noun for this complex predicate)

Some CPs occur in **Cleft sentences**:

To make a mistake: It was the first **mistake** that she **made**

but:

To commit suicide: *It was suicide that she committed

Complex predicates vary for **passivization**:

To commit a crime: A dreadful crime was committed in the 4th Avenue

but:

To commit suicide: *Suicide is committed mainly at night

Syntagmatic substitution is admitted by some CPs:

He made a mistake today and I made **it** yesterday

He committed a crime with a gun and I committed **it** with a knife

Firstly, he gave his brother a hug and, secondly, he gave **it** to his sister

but it is not admitted by some others:

*He committed suicide yesterday and I will commit **it** tomorrow

*He took part in the 1989th Games and I took **it** in the 1993rd.

8. Some of the Complex Predicates have a corresponding lexicalized verbal form in the same language. This feature is analysed in the following section.

2.2. PARADIGMATIC LEVEL

In relation to the **Paradigmatic level**, level where the different syntactic realizations common to the same semantic content are established, we

need to account for facts such as the relationship guaranteed by the System between CPs and their lexicalized forms. By lexicalization we refer to the meaning-in-form relation established when a set of meaning components is in association with a morpheme, making up the whole of the morpheme's meaning (Talmy, 1985). The Paradigmatic level will set the relationship between one morpheme's semantic makeup equivalent to that of a set of other morphemes in a syntactic construction. We will refer to the one morpheme's semantic makeup as the lexicalized form (term from McCawley, 1968) and will talk about incorporation (Gruber, 1965) when referring to the thematic role that does not have a syntactic realization.
As it is stated by Cattell (1984), there is an important relationship between the following pairs:

(1)
  a. Bill coughed
  b. Bill gave a cough

(2)
  a. Sue looked at the book
  b. Sue had/took a look at the book

The same relationship can be affirmed for the following pairs in Spanish:

Cometer una equivocacion	Equivocarse
Dar un abrazo	Abrazar
Tomar parte	Participar
Dar un paseo	Pasear

There could be differences in emphasis, but it is obvious that the following pairs are very close to each other in meaning: (the English gloss is placed between single quotes)

(1)
  a. El ha cometido una equivocación en el trabajo
  ('He has committed a mistake at work')
  b. El se ha equivocado en el trabajo
  ('He was wrong at work')

(2)
  a. El ha cometido una gravísima equivocación
  ('He has committed a very serious mistake')
  b. El se ha equivocado gravemente
  ('He was wrong very seriously')

(3)
  a. Hoy he cometido varias equivocaciones
  ('I have committed several mistakes today')
  b. Hoy me he equivocado varias veces
  ('I was wrong several times today')

They paraphrase each other and, if not absolute synonyms, they are very

similar in meaning, and, thus, related. Problems of the following type:
(1)

a. He cometido una gran equivocación
('I have committed a big mistake')

b. *Me he equivocado grandemente
('I was greatly wrong')

occur because such adjective has no lexically-corresponding adverb, even though the modifiers in the two types of construction are systematically related to each other (Cattell, 1984:9). We still have another example that casts doubt upon the one-to-one relationship between the two types of predication:
(1)

a. El ha cometido una gravísima equivocación

b. El se ha equivocado gravemente

but:

c. *El ha cometido una equivocación gravemente
('He has committed a mistake very seriously')

It seems that, even though close in meaning, CPs and lexicalized forms cannot be interchangeable for every environment. We assume, following Cattell, that, for modification, CPs are not a single unit and, thus, the nominals in the expressions act like a real noun, taking only modifiers relevant to nouns.
Such relational approach is based on the idea that if a grammar is expected to describe all the structures of a language, it should correlate those which are paraphrases of each other. (Boas, 1977)

2.3. TRANSFER LEVEL

And, finally, the **Transfer level** that, as mentioned, relates a word or phrase in one language to a semantically equivalent word or phrase in another language. It often happens that we have to use complex expressions in one language in order to express "concepts" which can be expressed in the other language by single lexical items (Tsujii, 1989).
We enumerate the following possible classification for the Spanish-English structural matchings where complex predicates are involved:
Example 1:

CP & Lexicalized Form - CP

Cometer una equivocacion	To make a mistake
Equivocarse	To make a mistake

CP & Lexicalized Form - Lexicalized Form

Dar puñaladas	To stab

Apuñalar	To stab

Example 2:

CP & Lexicalized Form - CP & Lexicalized Form

Dar un abrazo	To give a hug
Abrazar	To hug
Tomar parte	To take part
Participar	To participate

Example 3:

Complex Predicate - Complex Predicate

Cometer un crimen	To commit a crime

Example 4:

Lexicalized Form - Complex Predicate

Suicidarse	To commit suicide
Ducharse	To have a shower
Bañarse	To have a bath

As we can see from the examples above, we are dealing with cases that show a lack of structural correspondence among languages, examples that, according to B. Dorr's (1991) classification, are of **lexical divergences**. Examples 1 and 4 show a lack of structural matching since there is no English lexicalized form for the CP "to make a mistake" that corresponds to the Spanish "equivocarse", and the English form "to stab" can be expressed in Spanish either by a lexicalized form or by a complex predicate. On the other hand, example 4 shows a lack of lexicalized forms in English that corresponds to the Spanish verbal forms indicated. Example 2 shows a symmetrical structural matching, whereas we include example 3 as a token of the CP structural matching type with a lack of lexicalized forms in both languages.

3. Interlingual Representation of CPs and Lexicalized Forms

After having mentioned the three levels of analysis CPs enter, we propose to apply to them and their lexicalized forms Jackendoff's Conceptual Structures, considered as interlingual representations for MT, in a way to relate their common underlying meaning with their different surface realizations. The Conceptual Structure of a lexical item is defined as:

> (...) an entity with zero or more open argument places. The meanings of the syntactic complements of the lexical item fill in the values of the item's argument places in the meaning of the sentence (Jackendoff, 1990:24).

As far as CPs and lexicalized forms are concerned, the main contribution of such interlingual representation comes from the fact that the same structure stands for both of them, since, as it was said before, the meaning of CPs and lexicalized forms is similar. The key relies in how our author represents in the structure the implicit arguments that do not have a syntactic realization. This fact is achieved by indexation:

> (...) if a constituent of a verb's meaning is indexed, its semantic features appear as a selectional restriction; if a constituent is unindexed, its features appear as the content of an implicit argument (Jackendoff, 1990:55).

Thus, we chose three Italian light verbs with a theoretically high occurrence level in collocational structures.
We worked through 12590 occurrences of the verb *Mettere*, 10635 of the verb *Prendere* and 11188 of the verb *Dare*. Thus, our study focused on some part of the **Motion Event**, represented in the conceptual structure by the conceptual constituents $[_{event}GO_{loc}]$ in an agentive construction $[_{event}CAUSE([_{thing}])]$. The *loc* and *ident* subindices are **semantic field features** that designate the field in which the Event or State is defined. The basic motion Event consists of one object (the **Theme** in Gruber and Jackendoff's terminology; the **Figure** in Talmy's (1985)) moving with respect to something or someone (**Goal** in the verbs analized).
As I have said before, we paired the Spanish and English realizations similar to the Italian ones from the Corpus of the Istituto di Linguistica Computazionale (Pisa) and applied the same CS and AFF ("affect") function to a common meaning in the three languages.
According to Culicover and Wilkins (1986), Talmy (1985) and Jackendoff (1990), conceptual roles fall into two tiers: a **Thematic tier** dealing with motion and location, and an **Action tier** dealing with Actor-Patient relations. Thus, following Jackendoff (1990), we supplement our conceptual structures with the function $[AFF([], [])]$, an additional mainstream fuction alongside the thematic functions. Its first argument is the Actor, and the second one, the Patient.
In our attempt to explore meaning-surface relations, we matched the surface entities from the Corpus with the conceptual form that encodes their meaning. The classification achieved for METTERE and PRENDERE/ DARE constructions results from the different types of lexicalization patterns. All the examples registered with a similar behaviour and in any of the three languages are displayed for each subsection in our aim to set forth a class of meaning-in-form language patterns across languages. We do not feel the need to give a translation of the Spanish and Italian forms, since any complex predicate or verbal form corresponds to the equivalent English one.

3.1. METTERE

Lexicalized forms are classified according to the following patterns:

1. Verbs that incorporate the Theme: (These verbs are said to form a minor system of agentive verbs in English, where Figure and Path conflate with Motion (Talmy, (1985))
   Tappare . Ēgli mette il tappo nel ()
   Taponar > El pone el tapón en ()
   To Stopper > He puts the stopper on ()
   $[_{event}CAUSE([_{thing}]_A[_{event}$
   $GO_{loc}([_{thing}TAPPO/TAPON/STOPPER]_{\{A\}}$
   $[_{path}TO_{loc}([_{place}ON_{loc}([_{thing}]_A)])])])]$

   $[AFF([_{He}],[_{Stopper}])]$

   Subscript A distinguishes those constituents that are arguments from those that are not (implicit arguments, selectional restrictions). Jackendoff's proposal is that, through A-marking, lexical entries stipulate their conceptual arguments, but to leave to the linking rules to determine the syntactic positions to which the arguments link. The A-mark in brackets indicates its optionality and, thus, the same structure for complex predicates and lexicalized forms is used. For the predicate "to put the stopper on", the conceptual constituent: $([_{thing}TAPPO/TAPON/STOPPER]_A)$ is A-marked, and for the lexicalized form "to stopper", such constituent is an implicit argument, hence, unindexed.

   Another verb of this type is:

   - / Anillar / To Ring.
   Mettere l'anello in () / Poner anillos a () / To put a ring on ()

   (The meaning under analysis here is the equivalent to "mark (a bird) with a ring or clip for subsequent identification")
   The Italian realization does not have a lexicalized form.

2. Verbs that incorporate the Theme with an abstract spatial sense (Motion/Figure/Path):

Accentuare	Egli mette l'accento sul ()
Acentuar	El pone el acento en ()
To Stress	He puts the stress on ()

   $[_{event}CAUSE([_{thing}]_A,[_{event}$

$GO_{loc}([_{thing} ACCENTO/ACENTO/STRESS]_{\{A\}}$
$[_{path} TO_{loc}([_{place} ON_{loc}([_{thing}]_{A})])])])]$

$[AFF([_{He}],)]$

3. Verbs that incorporate the Goal: (According to Talmy (1985), these form a minor system of agentive verbs where Ground and Path are conflated together with Motion)

Archiviare	Egli mette qualcosa in archivio
Archivar	El mete algo en un archivo
To File	He puts something in a file

$[_{event} CAUSE([_{thing}]_{A}, [_{event}$
$GO_{loc}([_{thing}]_{A}, [_{path} TO_{loc}([_{place} ON_{loc}$
$([_{thing} ARCHIVIO/ARCHIVO/FILE]_{\{A\}})])])])])]$

$[AFF([_{He}], [_{Something}])]$

Verbs of this type are:

Imprigionare / Encarcelar / To Jail
Mettere () in prigione / Poner () en la cárcel / To put () into jail

- / Escenificar / To Stage
Mettere () in scena / Poner () en escena / To put () on stage

Intascare / - / To Pocket
Mettere () in tasca / Meter () en el bolsillo / To put () in a pocket

- / - / To Box
Mettere () in una cassa / Meter () en una caja / To put () in a box

Collocare / Colocar / To Place
Mettere () in un posto / Poner () en un sitio / To put () in a place

Inscatolare / Enlatar / To Can
Mettere () in scatola / Poner () en lata / To put () in a can

4. Verbs that incorporate the Goal within an identificational field (Motion/Ground/Path):

Chiarire	Egli mette qualcosa in chiaro
Aclarar	El pone algo claro
To Clarify	He makes something clear

$$[_{event}CAUSE([_{thing}]A[_{event} \\ GO_{ident}([_{thing}]A[_{path}TO_{ident}([_{place}IN_{ident} \\ ([CHIARO/CLARO/CLEAR]_{\{A\}})])])])]$$

$[AFF([_{He}], [_{Something}])]$
Verbs of this type are:

Evidenziare / Evidenciar / To Clarify
Mettere () in evidenza / Poner () en evidencia / To make () clear

Inginocchiarsi / arrodillarse / To Kneel down
Mettere () in ginocchio / Poner () de rodillas / To go down on () knees

Liberare / Liberar / To Liberate
Mettere () in libertá / Poner () en libertad / To set () free

Relazionare / Relacionar / To Relate
Mettere () in relazione / Poner () en relación / To put () in relationship

Ridicolizzare / Ridiculizar / To Ridicule
Mettere () in ridicolo / Poner () en ridículo / To make a fool of ()

Risaltare / Resaltar / To Emphasize
Mettere () in risalto / Poner () de relieve / To place emphasis on ()

3.2. PRENDERE / DARE

Lexicalized forms are classified according to the following patterns:

1. Verbs that incorporate the Goal (Motion/Ground/Path):

Alloggiare(i)	Egli prende alloggio
Alojarse	-
To lodge(i)	He takes lodging
Alloggiare(t)	Egli da' alloggio
Alojar	El da alojamiento
To Lodge (t)	He gives lodging

We regard the lexical entries of the lexicalized forms above as having two alternative conceptual structures, due to their intransitive behaviour and the possible double event/state reading. There are several linguistic tests that support such difference. Verbs with an event reading can go in "present continuous tense", "combine with an adverb", and answer for the question: "What did X do?". Whereas the state reading can be expressed in "simple present tense" and go with "a durative adverbial phrase". In the conceptual structure we encode the event reading with the function **GO**:

"He quickly lodged in a hotel"

and the state one with the function **INCH** $[_{event}INCH([_{state}])]$, that represents the two senses related in the inchoative relation (a change taking place whose final state is the state reading):

"He lodges in a hotel for three weeks"

The same alternative conceptual structure will be given when the intransitive lexicalized forms show a reflexive sense.

$$[_{event}CAUSE([_{thing}]_A[_{event} GO_{loc}([_{thing}]_A[_{path}TO_{loc}([_{place}IN_{loc} ([_{thing}ALLOGGIO/ALOJAMIENTO/LODGING]_{\{A\}})])])])]$$

$$[AFF([_{He}],[_{Someone}])]$$

$$[_{event}CAUSE([_{thing}]_A [INCH([_{state}BE_{loc}([_{thing}]_A[_{place}IN_{loc} ([_{thing}ALLOGGIO/ALOJAMIENTO/LODGING]_{\{A\}})])])])]$$

$$[AFF([_{He}]_\alpha,[\alpha])]$$

where the Actor is affected. Greek letters stipulate binding between conceptual positions, that is to say, they stipulate that two or more (it could be more than two in the conceptual structure) roles are satisfied by the same individual. Such coindexing plays an internal role to conceptual structure as opposed to Roman alphabet subscripts, that stipulate correspondence between syntactic and conceptual positions.

Verbs of this type are:

Arrischiarsi / Arriesgarse / To Risk(i)
Prendere un rischio / Tomar un riesgo / To take a risk
Rischiare / Arriesgar / To Risk (t)

Distanziarsi / Distanciarse / To Separate (i)
Prendere distanza / Sacar distancia / To get ahead of
Distanziare / Distanciar / To Separate (t)

Sedersi / Sentarse / To Sit (i)
Prendere posto / Tomar asiento / To take a seat
Fare sedere / Sentar / To Sit (t)

Formarsi	Qualcosa prende forma
Formarse	Algo toma forma
To Form (i)	Something takes form
Formare	Egli da' forma a qualcosa
Formar	El da forma a algo
To Form (t)	He gives form to something

$$[_{event} CAUSE([_{thing}]_A [_{event} GO_{ident}([_{thing}]_A [_{path} TO_{ident}([_{place} AT_{ident}([_{property} FORMA/FORMA/FORM]_{\{A\}})])])])])]$$

$$[AFF^{+}([_{He}], [_{Something}])]$$

The notation AFF $^{+}$ expresses that the second argument is positively affected.

$$[_{event} CAUSE([_{thing}]_A [INCH([_{state} BE_{ident}([_{thing}]_A [_{place} AT_{ident}([_{property} FORMA/FORMA/FORM]_{\{A\}})])])])])]$$

$$[AFF^{+}(, [_{Something}])]$$

The action tier indicates that something is affected by the Event without implicating any specific Actor.

Verbs of this type are:

Colorarsi / Colorearse / To Colour up (i)
Prendere colore / Tomar color / To take on a colour
Colorare / Colorear / To Colour (t)

2. Verbs that incorporate the Theme (Motion/Figure):

Baciare	Egli da' un bacio a qualcuno
Besar	El da un beso a alguien
To Kiss	He gives a kiss to someone

$$[_{event} CAUSE([_{thing}]^{\alpha}_A [_{event} GO_{poss}([_{thing} BACIO/BESO/KISS]_{\{A\}} [_{path} FROM_{poss}([_{thing} \alpha] TO_{poss}([_{thing}]_A))])])])]$$

$$[AFF^{+}([_{He}], [_{Someone}])]$$

Assistere	Egli da' assistenza a qualcuno

Asistir	El da asistencia a alguien
To Assist	He gives assistance to someone

$$[_{event} CAUSE([_{thing}]^{\alpha}_{A} [_{event} GO_{poss}([_{thing} ASSISTENZA/ASISTENCIA\ ASSISTANCE]_{\{A\}} [_{path} FROM_{poss}([_{thing} \alpha] TO_{poss}([_{thing})]_{A}))])])]$$

$$[AFF^{+}([_{He}], [_{Someone}])]$$

Verbs of this type are:

- / Citar / -
Dare appuntamento / Dar una cita / To make an appointment

Calciare	Egli da' un calcio a qualcuno
-	El pega una patada a alguien
To Kick	He gives a kick to someone

$$[_{event} CAUSE([_{thing}]^{\alpha}_{A} [_{event} GO_{poss}([_{thing} CALCIO/PATADA/KICK]_{\{A\}} [_{path} FROM_{poss}([_{thing} \alpha] TO_{poss}([_{thing})]_{A}))])])]$$

$$[AFF^{-}([_{He}], [_{Someone}])]$$

The AFF $^{-}$ notation expresses that the second argument is negatively affected.

Verbs of this type are:

Incolpare / Culpar / To Blame
Dare la colpa a / Echar la culpa a / To put the blame on

4. Lexical Functions: use and definitions

We have shown how Conceptual Structures (CSs) are a very convenient way of representing different realizations with a close meaning. As said before, we consider them as interlingual representations within an MT context for cross linguistic representations. B. Dorr (1991)'s description of UNITRAN explains how CSs are used in the analysis phase of the system. But, as stated above, the lexical dependence relation from the collocator to the base is an arbitrary one. The selection of the collocator by the base is not due to semantic resons only. Collocations are important for language generation, as one of its main problems is that of choosing the right word.

But generation-lexicon entries seem to ignore syntagmatic lexical relations (Smadja, 1991).
For example, in the corresponding CPs for the verbal forms **To blame** and **To separate** referred above, the base for each language selects a rather different collocator:

English: **To put** the blame on ()
Spanish: **Echar** la culpa a ()('To throw the blame to ()')
Italian: **Dare** la colpa a ()('To give the blame to ()')
English: **To get** ahead of
Spanish: **Sacar** distancia ('To draw distance')
Italian: **Prendere** distanza ('To take distance')

As the English glosses show, the Spanish verb "Echar" ('to throw') and the Italian "Dare" ('to give') are not translations of the English verb "To put", nor the Spanish verb "Sacar" ('to draw') or the Italian form "Prendere" ('to take') translate the English collocator form "To get". On the other hand, we see that the Spanish and Italian verbs for these examples are related in meaning, but they are not accurate translations of each other either, since the Spanish translation of the Italian verb "Dare" is "Dar" and "Tomar" for the Italian "Prendere".
The problem, thus, is that a compositional analysis of meaning does not produce acceptable translations, even though complex predicates are fully transparent, as opposed to idioms that are opaque. The selection of the forms "To put", "Dare" and "Echar", and "To get ahead", "Sacar" and "Prendere" does not obey to semantic reasons only, but to arbitrary ones. There is no other verb to replace them in those constructions.
According to Mackin (1978), collocations are not a problem for linguistic comprehension, but for production. The translation of the Italian form "Io non ti voglio dare la colpa" into English *"I do not want to give you the blame", may be understood within context, but the proper translation "I do not want to put the blame on you" will make the difference between high and low quality translations.
Following Church & Hanks (1990), it is a common practice in linguistics to classify words on the basis of their co-occurrence with other words. The aim, though, is to avoid bilingual context-dependent rules, like for our example:

To put (blame) - Echar
To put (blame) - Dare

and to achieve a sufficient degree of abstraction when making formal lexical statements (Halliday, 1966). Our claim in this paper is that such selection can be represented by Mel'cuk **Lexical Functions** (LFs) that, apart from specifying lexical items occurring together, could be defined on the same basis as Conceptual Structures: **Thematic Roles** (Zarco, 1994). We use the functions developed by Mel'cuk and Zolkovskij (1970), (1988) and Mel'cuk and Polguere (1987) as part of the Explanatory Combinatorial Dictionary (ECD).

Definition:

An LF is a function in the mathematical sense representing a certain extremely general idea, such as 'very', 'begin' or 'implement', or else a certain semantico-syntactic role. An LF **F** associates with a word **W** called its argument, or **Key Word**, the set of words or phrases which express -contingent on W- the meaning or role which corresponds to **F**. (Mel'cuk & Zolkovskij, 1970: 24)

Lexical Function (argument)=value

The set of LFs we will apply to our CPs are of the **semantic parameter** type, since we are dealing with words that appear together in the text. The following function Oper, as well as Func and Labor, are used to represent the syntactic link that semantically empty verbs establish between the name of a participant of the situation denoted by W, to the name of the situation itself, i.e., to W, that is to say, they could be applied to complex predicates.

Oper $_i$ (W $_o$)

The role of the first deep actant (and the grammatical subject) of the Oper verb is played by the **i-th** participant of the situation, while the second deep actant (or the first surface complement) is represented by **Wo**. Other actants, if any, denote further participants of the situation.

Oper $_1$ (io)

The basic inherent action or relation of **D1** as the grammatical subject towards **io** as the grammatical object, i.e. the term chosen to denote what D1 does to io.

io represents the key word (W) or argument of the Lexical Function.

$Oper_1(appologies)$ =To Offer

$Oper_1(orders)$ =To Give

Oper $_2$ (io)

The basic relation of **D2** as the grammatical subject to **io** as the main grammatical object.

$Oper_2(orders)$ =To Have

$Oper_2(visit)$ =To Receive

Mel'cuk (Mel'cuk & Zolkovskij (1970)), in his definition of Lexical Functions, talks about deep actants, such as D1 and D2, and surface syntactic functions (grammatical subject, grammatical object and marginal grammatical object). For Mel'cuk, subindices 1 and 2 denote deep syntactic structure actants. As we guess, they represent the difference in the actant-action relationship the following predicates show: "I give orders"/ "I have orders", but Lexical Functions are mainly described in semantic and syntactic terms, and we are not given an accurate definition of such actant-action relationship. The only information considered, we know by intuition from the given examples. With our study, we think we can give a more precise definition of actants and, thus, of some Lexical Functions.
After having paired our CSs and AFF functions with LFs for each complex predicate and lexicalized form, we realize there is some sort of relationship we can establish between both types of representation. For each type of LF, the conceptual representation seems to be constant:

To lodge (t) - To give lodging

$$[_{event}CAUSE([_{thing}]_A[_{event} GO_{loc}([_{thing}]_A[_{path}TO_{loc}([_{place}IN_{loc} ([_{thing}ALLOGGIO/ALOJAMIENTO/LODGING]_{\{A\}})])])])]$$

$$[AFF([_{He}],[_{Someone}])]$$

$Oper_1(lodging)$ =To give

To lodge (i) - To take lodging

$$[_{event}CAUSE([_{thing}]_A [INCH([_{state}BE_{loc}([_{thing}]_A[_{place}IN_{loc} ([_{thing}ALLOGGIO/ALOJAMIENTO/LODGING]_{\{A\}})])])])]$$

$$[AFF([_{He}]_\alpha,[\alpha])]$$

$Oper_2(lodging)$ =To take

To risk (t) - To put at risk

$$[_{event}CAUSE([_{thing}]_A[_{event} GO_{ident}([_{thing}]_A[_{path}TO_{ident}([_{place}IN_{ident} ([_{thing}RISCHIO/RIESGO/RISK]_{\{A\}})])])])]$$

$$[AFF([_{He}],[_{Someone}])]$$

$Oper_1(risk)$ = To put at

To risk (i) - To take a risk

$$[_{event} CAUSE([_{thing}]_A [INCH([_{state} BE_{ident}([_{thing}]_A [_{place} IN_{ident}([_{thing} RISCHIO/RIESGO/RISK]_{\{A\}})])])])]$$

$$[AFF([_{He}]_\alpha, [\alpha])]$$

$Oper_2(risk)$ = To take

To form (t) - To give form

$$[_{event} CAUSE([_{thing}]_A [_{event} GO_{ident}([_{thing}]_A [_{path} TO_{ident}([_{place} AT_{ident}([_{property} FORMA/FORMA/FORM]_{\{A\}})])])])]$$

$$[AFF^+([_{He}], [_{Something}])]$$

$Oper_1(form)$ = To give

To form (i) - To take form

$$[_{event} CAUSE([_{thing}]_A [INCH([_{state} BE_{ident}([_{thing}]_A [_{place} AT_{ident}([_{property} FORMA/FORMA/FORM]_{\{A\}})])])])]$$

$$[AFF^+(, [_{Something}])]$$

$Oper_2(form)$ = To take

To colour (t) - To give colour

$$[_{event} CAUSE([_{thing}]_A [_{event} GO_{ident}([_{thing}]_A [_{path} TO_{ident}([_{place} AT_{ident}([_{property} COLORE/COLOR/COLOUR]_{\{A\}})])])])]$$

$$[AFF^+([_{He}], [_{Something}])]$$

$Oper_1(colour)$ = To give

To colour up (i) - To take on a colour

$$[_{event} CAUSE([_{thing}]_A [INCH([_{state} BE_{ident}([_{thing}]_A [_{place} AT_{ident}$$

$([_{property} COLORE/COLOR/COLOUR]_{\{A\}}))]))]))]$

$[AFF^{+}(, [_{Something}])]$

$Oper_2(colour)$ = To take on

It seems that the LF Oper $_1$ corresponds to a conceptual structure of the type:

$[_{event} CAUSE([_{thing}]_A GO_{loc}([_{thing}]_A[\ldots]))]$,
whereas Oper $_2$ can be conceptually represented by:

$[_{event} CAUSE([_{thing}]_A[INCH([_{state} BE \ldots])])]$.

Even more, it is the AFF function that shows to be different for realizations apparently similar of the Prendere/Dare constructions:
To lodge (i) - To take lodging
$[AFF([_{He}]_\alpha, [\alpha])]$

To risk (i) - To take a risk
$[AFF([_{He}]_\alpha, [\alpha])]$

To form (i) - To take form
$[AFF^{+}(, [_{Something}])]$

To colour up (i) - To take on a colour
$[AFF^{+}(, [_{Something}])]$

Being obvious that Oper $_1$ represents the CP "Dare alloggio", "Dare forma" and "Dare colore", where 1 is Mel'cuk's D1 grammatical subject, it seems obvious as well that actants in the examples above have a different role from the Oper $_1$ lexical function, as their corresponding AFF functions show. As indicated, they could be represented by Oper $_2$, as the examples by Mel'cuk from the $Oper_2$ definition above suggest:
$Oper_2(orders)$ = To have
$Oper_2(visit)$ = To receive

Even though, in the "To take lodging" and "To take a risk" CPs, the AFF function shows an Actor that is affected, whereas in the "To take form" and "To take on a colour" CPs, there is no Actor at all, but Patient. The following examples support such statement. If we try to combine our verbal forms under analysis with a reflexive pronoun and a thematically

dependent (Wyner, 1994) adverbial form "sin dudarlo" (English: 'unhesitatingly'), we get different results:
(Intransitive "To lodge" does not mean "To lodge someone", but "To lodge oneself", so, we consider the following sentences as synonyms. The analysis is realized on the Spanish examples. We give the English translation between inverted comas " ")

(1)
a. El se alojó en Córdoba ("He lodged in Córdoba")
b. He took lodging in Córdoba
(For the last example, the English form is considered since there is no complex predicate form in Spanish for the case under study).

(2)
a. El se arriesgó en el trapecio ("He risked on the trapeze")
b. El tomó un riesgo en el trapecio ("He took a risk on the trapeze")

Both structures are synonyms. But let us see the following case:

(3)
a. El se coloreó a sí mismo con un lápiz ("He coloured himself up with a pencil")
b. El se coloreó con un lápiz ("He coloured up with a pencil")
c. *El tomó color con un lápiz ("He took on a colour with a pencil")

The complex predicate form "Tomar color" is not a synonym of the lexicalized form "Colorearse". The Spanish reflexive form has a double reading, one with an Actor and another one without it. (This is the reason why the lexicalized form "colorearse" can be combined with a reflexive form "a sí mismo", as example a) shows). The complex predicate is the syntactic realization of the latter. Let us consider the same example combined with the thematically dependent adverbial form:
d. El se coloreó sin dudarlo ("He unhesitatingly coloured up")
e. *El tomó color sin dudarlo ("He unhesitatingly took on a colour")

The complex predicate "Tomar color" cannot be combined with an adverbial that modifies the Actor, if there is no such actant in the structure. We get the same results for the CP "Tomar forma" ("To take form"):

(4)
a. El se formó a sí mismo sin dudarlo ("He unhesitatingly formed himself")
b. El se formó sin dudarlo ("He unhesitatingly formed")

c *El tomó forma sin dudarlo ("He unhesitatingly took form")

The fact that such adverbial cannot be combined with both complex predicates, whereas it is possible for the others two, proves the need to distinguish these realizations. We think Oper $_2$ cannot represent both and, thus, it is here that we consider the need for a new Lexical Function that we call Oper $_{1-2}$, different from Oper $_1$ and Oper $_2$, that we define later:

$Oper_1(lodging)$ = To give
$Oper_2(form)$ = To take
$Oper_{1-2}(risk)$ = To take

Thus, as for the verbs and CSs mentioned above, their lexical entries should be increased by the following lexical function information:

$Oper_1(tappo)$ =**Mettere//Tappare**
$Oper_1(tapon)$ =**Poner//Taponar**
$Oper_1(stopper)$ =**To Put//To Stopper**

The double slash (//) represents a merged LF, whose value includes the meaning of the function and the key word as well. The merged function is very close to Gruber's incorporation and it is applied to lexicalization patterns.

$Oper_1(accento)$ =**Mettere//Accentuare**
$Oper_1(acento)$ =**Poner//Acentuar**
$Oper_1(stress)$ =**To Put//To Stress**
$Labor_{1,2}(archivio)$ =**Mettere in//Archiviare**
$Labor_{1,2}(archivo)$ =**Poner en//Archivar**
$Labor_{1,2}(file)$ =**To Put in//To File**
$Labor_{1,2}(chiaro)$ =**Mettere in//Chiarire**
$Labor_{1,2}(claro)$ =**Poner en//Aclarar**
$Labor_{1,2}(clear)$ =**To Make//To Clarify**
$Oper_1(alloggio)$ =**Dare//Alloggiare(t)**
$Oper_1(alojamiento)$ =**Dar//Alojar**
$Oper_1(lodging)$ =**To Give//To Lodge(t)**
$Oper_{1-2}(alloggio)$ =**Prendere//Alloggiare(i)**
$Oper_{1-2}(alojamiento)$ =**//Alojarse**
$Oper_{1-2}(lodging)$ =**To Take//To Lodge(i)**

The LF $Oper_{1-2}$ is first described in Zarco (1994) as the function that describes the relation of **1-2** as the grammatical subject towards the key word as the grammatical object. The subindex **1-2** represents the constituent that corresponds to a thematic role **Theme** that is **Actor**, being optionally **Affected**.

$Oper_2(forma)$ =**Prendere//Formarsi**

$Oper_2(forma)$ =**Tomar//Formarse**
$Oper_2(form)$ =**To Take//To Form(i)**
$Oper_1(forma)$ =**Dare//Formare**
$Oper_1(forma)$ =**Dar//Formar**
$Oper_1(form)$ =**To Give//To Form(t)**
$Oper_1(bacio)$ =**Dare//Baciare**
$Oper_1(beso)$ =**Dar//Besar**
$Oper_1(kiss)$ =**To Give//To Kiss**
$Oper_1(assistenza)$ =**Dare//Assistere**
$Oper_1(asistencia)$ =**Dar//Asistir**
$Oper_1(assistance)$ =**To Give//To Assist**
$Oper_1(calcio)$ =**Dare//Calciare**
$Oper_1(patada)$ =**Pegar**
$Oper_1(kick)$ =**To Give//To Kick**

The use of LFs as a codifying way for lexical restrictions shows to be a very useful way for Machine Translation (Maxwell, 1993). LFs codify lexical restrictions that are not represented by CSs, this is their main advantage towards Jackendoff's representation. LFs are considered as interlingual representations, since they are not mere enumerations of lexical cooccurrences, but they give semantic, syntactic and thematic information as well. So, what is what makes us think of LFs as part of the interlingual representation in combination with CSs and not as being the best possible representation on their own? LFs are defined mostly in semantic terms, whereas CSs in conceptual ones. LFs offer a codification of the sort of relation established between a pair of words, but the representation of open argument places to be filled by the possible syntactic arguments of the lexical item is not achieved, since LFs are concerned with the relationship established by the pair only.
Even though, if the examples above are not convincing enough, let us consider the following fact. The relationship between lexicalized forms and CPs is not always a one-to-one relationship. We find examples where there are more than one collocator for the same base:
To Knock: To **Deal/Strike/Give** someone a blow.
Abofetear: **Dar/Propinar** bofetadas a alguien.
(To slap in the face: To give a slap in the face to someone)
To Object: To **Make/Raise** an objection

We ignore how such very little differences in meaning (we should talk about the notion of **use** as well) can be codified and represented by CSs. On the other hand, with the indices of LFs, such differences can be specified:
$Oper_{1(general)}(blow) = give$

$Oper_{1(specific)}(blow) = deal, strike$
$Oper_{1(general)}(bofetada) = dar$
$Oper_{1(specific)}(bofetada) = propinar$
$Oper_{1(general)}(objection) = make$
$Oper_{1(specific)}(objection) = raise$

5. Analysis of Results

1. Complex Predicates and their lexicalized forms are paraphrases of each other and, thus, should be related by a common meaning representation.
2. Conceptual Structures are a good example of interlingual representations for CPs and their corresponding lexicalized terms, not only within the same language, but across languages, since they conceptualize meaning in a very abstract way. Thus, the structure can be a common one. As in B. Dorr's proposal, they seem to be useful in the analysis phase of the system.
3. Conceptual Structures do not account for lexical selection, so it seems to be rather difficult for the system to choose the appropriate lexeme, when there is lexical divergence among the languages involved in the generation process. CSs deal with thematic roles but not with lexical elements and it has been referred before that lexical selection in CPs answers to arbitrary reasons as well.
4. LFs seem a good way for specifying lexical relations since they are not a mere enumeration of words occurring together in the text, but they codify semantic and syntactic information. They are interlingua representations as CSs, their level of representation is just less deep, since they deal with lexemes and their cooccurrence restrictions. They are a better representation form than bilingual context-sensitive rules.
5. The main problem shown by LFs is their definition (Heylen, 1993; Bloksma et al., 1993). We find references to subindices 1 and 2 (D1, D2), but they are not defined. After having applied CSs and LFs to our corpora, we have realized of the constant application of certain LFs with always the same CSs. That is to say, the function:
$[AFF([_{Actor}], [_{Patient}])]$, corresponds to the conceptual structure:
$[_{event}CAUSE([_{Actor}], [_{event}GO([_{Patient}]...)])]$. The function:
$[AFF([_{Actor}]_{\alpha}, [\alpha])]$, corresponds to the conceptual structure:
$[_{event}CAUSE([_{Actor}], [_{event}GO([_{Actor}]...)])]$. And the function:
$[AFF(, [_{Patient}])]$, corresponds to the conceptual structure:
$[_{event}CAUSE([_{thing}], [INCH([_{state}BE([_{Patient}]...)])])]$.

Obviously, if, according to Mel'cuk, LF subindices are deep structure actants, we need three different LFs for the AFF functions described above.

It is this part of our small contribution, a definition of some LFs in terms of thematic and action relations and the addition of a new Oper function, Oper $_{1-2}$, different to the Mel'cuk's Oper $_1$ and Oper $_2$, and supported by the different AFF function described above:

$Oper_1$

If for a CP formed by a Verb plus a Noun, the subject of such CP is undergone by the action role **Actor**, then such actant is D1 and should be expressed by $Oper_1$.

$Oper_2$

If for a CP formed by a Verb plus a Noun, the grammatical subject of the CP is not undergone by the action role **Actor**, then such actant is D2 and should be expressed by $Oper_2$.

$Oper_{1-2}$

If for a CP formed by a Verb plus a Noun, the subject of such CP is undergone by a constituent whose thematic and action role is (THEME-ACTOR/≺AFFECTED≻), such actant should be represented by: $Oper_{1-2}$.

6. Lexical entries could be formulated in terms of their syntactic category, their conceptual structure and lexical relations. For methodological purposes, conceptual and lexical formulae should be defined on a similar basis, so that both generalizations have a common nature when the linking principle applies.

Acknowledgements

We gratefully acknowledge Prof. N. Calzolari and the Istituto di Linguistica Computazionale, Pisa (Italy), that offered us the possibility to work in a Corpus with more than 15 million occurrences. We are very thankful for her advice and guidance. We gratefully acknowledge, as well, Patrick St. Dizier for his gentle help and suggestions on the general edition of this work, especially on LaTex.

References

Aisenstadt, E., Collocability Restrictions in dictionaries, in *Hartmann R.R.K. Dictionaries and their uses. Papers from the 1978 B.A.A.L. Seminar on Lexicography*, vol. 4, 71-74. 1978.

Allerton, D.J., Three (or four) levels of word cooccurrence restriction, in *Lingua*, 63: 17-40, 1984.

Bloksma, S.L., Heylen, D., Maxwell, K.G., Analysis of Lexical Functions, in *Collocations ET-10/75*, vol.1: 49-69, 1993.

Boas, H.U., Lexical entries for verbs in a contrastive lexicon english-german, in *Abraham*, 1977: 191-216, 1977.

Cattell, R., Composite Predicates in English, in *Syntax and Semantics*, 17, London: Academic Press, 1984.

Church, K.W., Hanks, P., Word Association Norms, Mutual Information and Lexicography, in *Computational Linguistics*, vol.16, n.3: 22-29, 1990.

Cowie, A.P., Collocational Dictionaries - A Comparative View, in *Fourth Joint Anglo-Soviet Seminar*, Murphy (ed.), London: British Council: 61-69, 1986.

Cowie, A.P., Multiword Units in Newspaper Language, in *Perspectives on the English Lexicon: A Tribute to Jacques Van Roey*, Granger (ed.), Cahiers de l'Institut de Linguistique de Louvain, CILL 17. 1-3: 101-116, 1991.

Culicover, P., Wilkins, W., Control, PRO, and Projection Principle, in *Language*, 62: 120-153, 1986.

Danlos, L., Samvelian, P., Translation of the Predicative Element of a Sentence: category switching, aspect and diathesis, in *TMI-92: Proceedings of the Fourth International Conference on Theoretical and Methodological Issues in Machine Translation*, Montreal: 21-34, 1992.

Dorr, B., Conceptual Basis of the Lexicon in Machine Translation, in *Lexical Acquisition. Exploiting on-line resources to build a lexicon*, Zernik, Uri (ed.), London:LEA, 1991.

Evens, M.W., *Relational Models of the Lexicon*, Cambridge, Cambridge University Press, 1988.

Gruber, J.S., *Studies in Lexical Relations*, Doctorate Thesis, MIT., 1965.

Halliday, M.A.K., Lexis as a linguistic level, in *In Memory of J.R. Firth*, E.E. Bazell, J.C. Catford, M.A.K. Halliday, R.H. Robins (eds.), Longman, Green & Co., 1966.

Heylen, D., Lexical Functions and Machine Translation, in *Collocations ET-10/75*, vol.1: 153-159, 1993.

Howarth, P., A Phraseological Approach to Academic Writing, in *Language, Learning and Success: studing through English*, G. Blue (ed.), London: Macmillan, 1993.

Hutchins, W.J., Somers, H.L., *An Introduction to Machine Translation*, London: Academic Press, 1992.

Jackendoff, R., The Status of Thematic Relations in Linguistic Theory, in *Linguistic Inquiry*, 18: 369-411, 1967.

Jackendoff, R., *Semantic Structures*, Cambridge, The MIT Press, 1990.

Jayaseelan, K.A., Complex Predicates and θ-Theory, in *Syntax and Semantics*, W. Wilkins (ed.), 21, New York: Academic Press: 91-111, 1988.

Machonis, P., The support verb 'make', in *Computational Lexicography: Proceedings of the COMPLEX Conference*, F. Kiefer (ed.), Research Institute for Linguistics, Hungarian Academy of Sciences: 141-153, 1991.

Maxwell, K.G., Translating Collocations with Lexical Functions, in *Collocations ET-10/75*, vol.1: 101-115, 1993.

McCawley, J.D., Lexical Insertion in a Transformational Grammar without Deep Structure, in *Papers from the 4th Regional Meeting of the Chicago Linguistic Society*, B. Darden, C-J.N. Bailey & A. Davison (eds.), Depart. of Linguistics, University of Chicago, Chicago, 1968.

Mel'cuk, I. & Polguere, A., A Formal Lexicon in Meaning-Text Theory, in *Computational Linguistics*, vol.13: 261-275, 1987.

Mel'cuk, I. & Zolkovskij, A., Towards a functioning 'Meaning-Text' Model of Language, in *Linguistics*, 57: 10-47, 1970.

Mel'cuk, I. & Zolkovskij, A., The Explanatory Combinatorial Dictionary, in *Relational Models of the Lexicon*, M. Evens (ed.), C.U.P.: 41-74, 1988.

Smadja, F.A., Macrocoding the lexicon with co-occurrence knowledge, in *Lexical Acquisi-*

tion. Exploiting on-line resources to build a lexicon, Zernik, Uri (ed.), London: LEA: 165-189, 1991.

Smadja, F. & Mckeown, K., Using Collocations for Language Generation, in *Comput. Intell.*, 7: 229-239, 1991.

Talmy, L., Lexicalization Patterns: Semantic Structure in Lexical Forms, in *Language Typology and Syntactic Description 3: Grammatical Categories in the Lexicon*, T.Shopen, Cambridge, C.U.P.: 57-149, 1985.

Tsujii, J., Machine Translation in Natural Language Understanding, in *Literary and Linguistic Computing*, vol.4, n.3: 214-217, 1989.

Wood, M.M., *A definition of Idiom*, Bloomington: Indiana University Linguistics Club, 1986.

Wyner, A.Z., *Boolean Event Lattices and Thematic Roles in the Syntax and Semantics of Adverbial Modification*, Unpublished PhD. dissertation, Cornell University, 1994.

Zarco, M.A., *Las Estructuras Conceptuales y las Funciones Léxicas en el ámbito de la Traducción Automática: Elementos relacionables del Lexicón*, Doctorate Thesis, University of Cadiz, Spain, 1994.

Zuluaga, A., *Introducción al estudio de las expresiones fijas*, Cirencester (UK): Studia Romanica et Linguistica, Verlag Peter D. Lang, 1980.

THE SEMANTICS OF EVENT-BASED NOMINALS

FEDERICA BUSA
Volen Center for Complex Systems
Department of Computer Science
Brandeis University
Waltham, MA 02254

Abstract

In this paper, I examine the set of agentive nominals and argue that, in spite of their heterogenous behavior, we can find generalizations by characterizing their semantic properties in terms of the *stage-level* and *individual -level* distinction. This parallel with the domain of predicates provides a useful semantic classification of the class of agentive nouns and explains a range of syntactic phenomena as well as issues of semantic interpretation.

1. Introduction

Parsons (1990) has provided substantial motivation for the necessity of adopting more complex logical forms involving events to describe the semantics of English sentences. A more controversial position is adopted in Generative Lexicon theory, (cf. Pustejovsky, 1991a, 1991b, 1995), which argues in favor of a richer structure for all lexical items. In particular, similarly to verbs and adjectives, also nominal forms involve events in their representation. This move is not only consistent with an event-based semantics for other categories of the language, but it is supported by a variety of phenomena including type coercion, adjectival modification, and, among others, also agentive nominals. In this paper, I present an event-based treatment of agentive and instrument nouns, showing that a framework for lexical semantics that provides a fine-grained view of event structure and a systematic method for interpreting how events are encoded in the lexicon makes interesting generalizations about the semantic properties of these nominal types. In particular, I will show that the semantic representa-

P. Saint-Dizier (ed.), Predicative Forms in Natural Language and in Lexical Knowledge Bases, 349–374.

tion of agentive nominals requires explicit reference to the well-known distinction between *stage-level* and *individual-level* predicates (Carlson, 1977). The former refer to episodic or temporary properties of an individual, as in *John is tired/available*, the latter make reference to properties that hold irrespectively of specific situations, as in *John is tall/intelligent.* Agentive nominals are generally taken to correspond to a set of deverbal nouns which are morphologically derived by *-er* affixation (e.g., *baker*, *winner*, *runner*), a process which is sensitive to the argument structure of underlying verbal forms (cf. Keyser and Roeper, 1984, Moortgat, 1987, Sproat, 1985, Fabb, 1984, Grimshaw, 1990, Rappaport and Levin, 1992). This theoretical position gives a prominent role to morphological processes which are expected to account for both a syntactic generalization, i.e. there is a category change from a verb to a noun, and a semantic generalization, i.e. the derived noun refers to an individual who performs the action denoted by the verbal stem. In the discussion below, I will show that by studying the properties of agentive nouns from the broader perspective of event semantics, we can discover interesting semantic properties and capture generalizations that would otherwise remain unnoticed if we were to focus uniquely on the availability of an external argument in the representation of a verbal form from which an agentive noun is derived.

The analysis presented below is based on the assumption that morphological operations are an instance of creative use of language, which should consistently derive new forms for which we can find independent motivation in the language. This methodology can be viewed analogously to an account of the semantics of causation involving both lexical causatives (e.g. *kill*) as well as phrasal causative (e.g. *make someone die*). In the rest of this paper, I will show that the class of agentive nominals comprises a larger set than simply *-er* nominals and motivate two broad semantic classes, namely *stage-level* and *individual-level* nominals. The general picture that emerges is that *stage-level* and *individual-level* nominals can be broadly distinguished as the result of how the defining event is quantified. The analysis is couched within the framework of Generative Lexicon theory (Pustejovsky, 1995), which is well suited for expressing quantificational information about events in the lexicon, for capturing fine-grained semantic distinctions within each class of nouns, and for explaining compositionally how certain nominals may shift between a *stage-level* and an *individual-level* interpretation in different contexts.

2. The Basic Distinction

Nouns which denote participants in an event fall under two general classes according to whether or not they require the context to make reference to the event which defines the individual. This is illustrated in the sentences in (1) and (2), (cf. Pustejovsky, 1997, for the use of the symbol "!" to indicate semantically odd sentences):

(1) When we bumped into the violinist$_i$ on Sunset Boulevard,
a. he$_i$ was playing a serenade on a street corner.
b. he$_i$ was selling hot-dogs on a street corner.

(2) When I saw a pedestrian$_i$ on Sunset Boulevard,
a. he$_i$ was about to get run over by a truck.
b. !he$_i$ was selling hot-dogs on a street corner.

In (1), we can felicitously refer to *violinist* without the individual being engaged in the activity of "playing the violin". Conversely, reference to *pedestrian*, in (2), is felicitous only if the individual is walking at the time of reference, hence the oddness of (2b). Following the solution suggested in Pustejovsky (1995), the behavior of these two types of nominals can be captured by viewing it as the analogue of the *individual-level* and *stage-level* distinction in the nominal domain. Nouns such as *violinist* are treated as *individual-level agentive nominals* (ILNs), and nouns such as *pedestrian* as *stage-level nominals* (SLNs).

Similarly to (1) and (2) above, the examples in (3)-(4) show that *pedestrian* and *rioter* require the matrix predicate and the event of walking or rioting to occur within the same temporal interval, (cf., (4)).[1] The same does not hold with *carpenter* and *senator*, (cf., (3)):

(3) a. A carpenter was arrested by the police
(arrested while building furniture, while doing something else)
b. A senator was sick.
(sick while in a meeting, while doing something else)

(4) a. A pedestrian was run over yesterday.
(run over while walking, !!while doing something else)
b. A rioter was arrested yesterday.
(arrested while demonstrated, !!while doing something else)

The distinction between ILNs and SLNs is also supported by additional grammatical properties which these nominals appear to share with *stage-level* and *individual-level* predicates. Temporal and locative modifiers, which are licensed with stage-level but not with individual-level predicates, are available with SLNs, (cf. (5c-d), (6c)) while they are restricted with ILNs, (cf. (5c-d), (6d)):

[1]As pointed out in Pustejovsky (1995), there are cases where reference to the individual outside the defining event is possible, e.g. *Hey, you're the passenger from my flight yesterday!*, as long as there is an appropriate situation which is existentially quantified.

(5) TEMPORAL MODIFIERS
a. John is available in the afternoon. (*stage-level*)
b. !!John is intelligent in the afternoon. (*individual-level*)
c. Rush-hour pedestrians are often in a bad mood. (SLN)
("pedestrians walking at rush hour")
d. !Rush-hour smokers are often in a bad mood. (ILN)
(! "smokers smoking at rush-hour")

(6) LOCATIVE MODIFIERS
a. John drives in France. (*stage-level*)
b. !!John is a linguist in France. (*individual-level*)
c. Watch out, there is a pedestrian on the crosswalk! (SLN)
("pedestrian is walking on the crosswalk")
d. Watch out, there is a violinist on the crosswalk! (ILN)
(? "violinist is playing on the crosswalk")

With SLNs, both temporal and locative modifiers refer to the event which defines the individual. Conversely, ILNs do not license temporal modifiers, and locative modifiers are only interpreted as making reference to the location of the individual, but not of the event (cf. Randall, 1982).

Similarly, frequency adverbials and frequency adjectives are available with *stage-level* predicates and SLNs, respectively, but not with *individual-level* predicates and ILNs:

(7) a. John is frequently intelligent/tall. (*individual-level*)
b. John is frequently available/drunk. (*stage-level*)
c. John only gives a discount to frequent customers. (SLN)
d. !I Frequent violinists have callous fingers. (ILN)

Note, however, that ILNs such *drinker* and *smoker* pattern with SLNs, as shown in (8), below:

(8) John is a frequent/moderate smoker/drinker.

The examples above can be accounted for by drawing appropriate semantic distinctions within the class of ILNs. Finally, an additional distinction between ILNs and SLNs is discussed in Gupta (1980) and Carlson (1982). Nominals such as *passenger* and *batter* can either make reference to the individuals involved in the events or to the events themselves; ILNs can only make reference to the individuals, as illustrated below:

(9) a. Twenty customers came in today.
b. I have seen eight doctors in the last month.

In (9a), the number of customers may be twenty or less since the same individual might have visited the business more than once. In (9b), the number of doctors is exactly eight. The distinction between SLNs and ILNs is also motivated by additional grammatical facts including their distribution with complements, coercion predicates, when-clauses, and temporal adverbials, as discussed in greater detail in Busa (1996).

In the rest of this paper, I discuss how the type and the quantificational properties of the events which define agentive nouns account for the facts presented above. Within recent work in lexical semantics, Rappaport and Levin (1992) have observed that the set of agentive nominals splits into two classes: *non-event agentive nouns* and *event agentive nouns*. The former class comprises nominals which do not presuppose the occurrence of the defining event, namely instruments, e.g. *can opener*, as well nominals that denote occupations, e.g. *teacher*. The class of *event agentive nouns*, instead, presuppose the occurrence of the event that characterizes the individual, as with expressions such as *builder of the house*. The analysis in this paper, differs in that all agentive nominals are characterized in terms of events, irrespectively of whether or not the event which defines an individual is interpreted as occurring or having occurred (cf. Chierchia, 1995, for individual-level predicates). The different interpretations emerge as the result of the quantification as well as of the event type of the action that characterizes the agentive noun. Before turning to the specific components of the account, the next section outlines the framework of Generative Lexicon theory which provides a model of lexical representation that makes the desired distinctions. In particular, lexical items are provided with internal semantic structure which makes explicit quantificational information and supports compositional operations of sense extension.

3. Elements of Lexical Representation

A lexical representation in GL is viewed as a rich system of semantic relations, along similar lines outlined in Moravcsik (1990), as well as Jackendoff's *Conceptual structure* (cf. Jackendoff, 1990). The *qualia*-based representations, in GL, provide an adequate framework for making explicit the relations between individuals and events as the result of how qualia roles are interpreted. In a GL lexical entry, events are encoded in the TELIC and the AGENTIVE roles. More specifically, the TELIC role encodes the characterizing function of an individual, while the AGENTIVE expresses the mode of coming into being of the entity being defined. In other words, different qualia roles induce a different interpretation on the event that they encode. By viewing qualia roles as introducing different forms of quantification, we can exploit their properties to account for the interpretation of the two classes of agentive nouns.

As discussed in (Moravcsik,1990, Pustejovsky, 1991), the TELIC role has a generic interpretation, in that it specifies what is, when properly used, the particular function(s) of the object being defined. In other words, the TELIC role imposes a particular interpretation to the relation that it encodes. As mentioned above, ILNs such as *violinist* and *smoker* do not require the defining event of playing the violin or smoking to occur at the time of reference, since these events denote a "stereotypical" property of the individual. Given the interpretation of qualia roles, the TELIC role provides a natural place for encoding this type of event-based information. The representation of nominals such as *violinist* or *smoker* is given below:

(10)

violinist	
FORMAL =	**"person"**
TELIC =	**"play violin"**

smoker	
FORMAL =	**"person"**
TELIC =	**"smoke"**

The representation of SLNs involves different components, since an individual is a *pedestrian* or a *passenger* if the defining event is occurring. In other words, performance of the event brings about an individual as the agent of it. This information can be naturally encoded in the AGENTIVE role:

(11)

pedestrian	
FORMAL =	**"person"**
AGENTIVE =	**"walk on street"**

passenger	
FORMAL =	**"person"**
AGENTIVE =	**"travel on vehicle"**

Encoding events in different qualia roles has consequences for how that event is interpreted. In particular, qualia roles introduce different quantificational properties on the events that they encode. As proposed in Pustejovsky (1995), the AGENTIVE role introduces existential quantification, while the TELIC role is interpreted as a modal operator. As an example, consider a nominal such as *knife*, whose TELIC role expresses the persistent property that this object is for cutting. This linguistic knowledge does not entail that every knife cuts or that a particular knife will be cutting in any given circumstance. Thus, the TELIC role induces a modal interpretation of the event that it encodes, such that, for an individual for which the TELIC is defined, we interpret that event as a definitional property which is true irrespectively of actual situations. We can make more explicit these quantificational properties by refining the representations in (10) and (11) above. The representation of ILNs and SLNs is summarized below, where the type of the lexical item is expressed in the argument structure, α is the defining event, D-E indicates a default event which is already quantified in the lexicon, either existentially, (i.e. AGENTIVE), or "generically", (i.e. TELIC):

$$(12)\quad \begin{bmatrix} \text{ILN} & \\ \text{ARGSTR} = & \begin{bmatrix} \text{ARG1} = \mathbf{x}\text{:human} \\ \ldots \end{bmatrix} \\ \text{EVENSTR} = & \begin{bmatrix} \text{D-E1} = \mathbf{e}{:}\tau \\ \ldots \end{bmatrix} \\ \text{QUALIA} = & \begin{bmatrix} \text{FORMAL} = \mathbf{x} \\ \text{TELIC} = \alpha(\mathbf{e},\mathbf{x}) \end{bmatrix} \end{bmatrix} \begin{bmatrix} \text{SLN} & \\ \text{ARGSTR} = & \begin{bmatrix} \text{ARG1} = \mathbf{x}\text{:human} \\ \ldots \end{bmatrix} \\ \text{EVENSTR} = & \begin{bmatrix} \text{D-E1} = \mathbf{e}{:}\tau \\ \ldots \end{bmatrix} \\ \text{QUALIA} = & \begin{bmatrix} \text{FORMAL} = \mathbf{x} \\ \text{AGENTIVE} = \alpha(\mathbf{e},\mathbf{x}) \end{bmatrix} \end{bmatrix}$$

The analysis presented in this paper is aimed not only at motivating the usefulness of an event-based analysis for distinguishing two broad classes of nominals but also at capturing the semantic distinctions found within each class. Thus, for the set of ILNs it is necessary to define precisely the conditions that restrict the interpretation of the event encoded in the TELIC, in order to account for the availability of frequency adjectives with *smoker* and *drinker* but not with *violinist* or *sculptor*. Similarly, the representation for SLNs, in (12), will have to be refined to reflect how these nouns constrain the surrounding context are the result of their semantic type.

4. The Modal Force of ILNs

Above, I have argued that ILNs do not require the event that defines the individual to occur at the time of reference, hence the property of being a *violinist* or a *smoker* holds irrespective of any specific situation of playing the violin or smoking. The class of ILNs, however, is not fully homogenous since *smoker* and *violinist* differ in important ways. Consider, for instance, the constrasts below:

(13) a. John is a violinist but doesn't play violin anymore.
b. John is an alcoholic but doesn't drink anymore.

(14) a. !!John is a smoker but doesn't smoke anymore.
b. !!John is a drinker but doesn't drink anymore.

Whereas a *smoker* refers to an individual who has the habit or the inclination to smoke with some of regularity, a *violinist* can be minimally characterized as an individual who has the ability of playing the violin. Consider the sentences in (15) with their approximate glosses:

(15) a. John is a violinist.
"John has the ability/knows how to play the violin"
b. John is an alcoholic.
"John has the chronic diseased caused by heavy drinking"

c. John is a smoker.
"John has the habit of smoking"
d. John is a husband.
"John is married to someone"
e. John is a doctor.
"John is competent to perform doctor-like activities"

Adapting some of the data on the interpretation of *when* clauses (Declerk, 1988) to the present concerns, we can approach a generalization:

(16) a. A person is a violinist when he knows how to play the violin.
b. A person is an alcoholic when he has a chronic diseased from the habit of heavy drinking.
c. A person is a smoker when he has the habit of smoking.
d. A person is a a doctor when he is competent to perform doctor-like activities

Note that with the ILNs above the interpreation of *when* is that of a conditional operator. These provide a step towards an abstraction for describing the semantic import of agentive nous, as shown in the paraphrases below:

(17) a. A person is an α when he HAS THE ABILITY to ψ.
b. A person is a β when he HAS THE HABIT of φ.
c. A person is a a γ when he HAS THE COMPETENCE to ϕ
d. A person is a a δ when he IS IN THE STATE OF χ

The common element is that all the ILNs above involve a conditional expression whereby the antecedent, namely a stative predicate that necessarily holds of the individual, provides the set of conditions under which the characterizing event may take place.

These properties can be explicitely encoded as predicates in the representation of the noun indicating different modes of coming into being of the defining event and hence of the individual performing it. In other words, the difference between the variety of ILNs can be captured by what type of stative predicate is expressed in the AGENTIVE role: in the case of *violinist* is the modal *ability*, and in the case of *smoker* is the partly extensional predicate *habit*. The resulting qualia structure representations for the two nominals are given below,:

(18)
$$\begin{bmatrix} \textbf{violinist} & \\ \text{FORMAL} = & \textbf{“person”} \\ \text{TELIC} = & \textbf{[1]=“play violin”} \\ \text{AGENTIVE} = & \textbf{“ability to [1]”} \end{bmatrix} \qquad \begin{bmatrix} \textbf{smoker} & \\ \text{FORMAL} = & \textbf{“person”} \\ \text{TELIC} = & \textbf{[1]=“smoke”} \\ \text{AGENTIVE} = & \textbf{“habit to [1]”} \end{bmatrix}$$

Nominals such as *doctor* and *husband* encode different stative predicates: with *doctor*, or *lawyer* there is an additional sense that the *ability* to perform the "doctor-activities" and "lawyer-activities" are recognized upon obtaining a certain degree, i.e. a *capacity*,[2] whereas *husband* encodes a non-dispositional predicate, which can be broadly characterized as an *agreement*. Since *ability*, *habit*, *capacity* and *agreement* carry modal force, the interpretation of the defining event of different nominals crucially depends on the modal force associated with the dispositional or non-dispositional predicate in their representation.

Without detailing the specific semantic import of all these stative predicates, I will informally hint at the properties of *habit* and *ability*, (cf. Busa, 1996, for details), which differ in crucial ways. An event which is interpreted relative to an ability is fully intensional in that it may or may not occur. Conversely, when we interpret an event relative to *habit*, an extensional component is introduced. More specifically, the events that constitute the past occurrences up to the present are extensional (e.g. AGENTIVE), while there is intensional component comprised of possible events that are likely to occur in the future (e.g. TELIC). This difference predicts the contrast in (13)-(14) above, as well as distinct behavior with different adjectives, as shown below:

(19) John is a former violinist/musician/professor. ($sense_1$)
$\Rightarrow$ John is a violinist/musician/professor. ($sense_2$)

(20) John is a former smoker/drinker.
$\not\Rightarrow$ John is a smoker/drinker.

In addition, adjectives that make reference to an ability, such as *natural* or *talented* are ruled out with *smoker* and *drinker*:

(21) a. !!a natural smoker/drinker
b. !!a talented smoker/drinker

Conversely, adjectives that make reference to the occurrence of a habit are licensed with *smoker* but not with *violinist*:

(22) a. habitual/frequent/moderate drinker/smoker
b. !!habitual/frequent/moderate violinist/alcoholic

In (22b), a frequency adjective is not licensed since it could only modify the state denoted by *ability*. In (22a), instead, given the extensionalized events associated with the semantics of *habit*, frequency adjectives are licensed.

Finally, the explanation for why agentive nouns in predicative position are not licensed with PP modifiers follows from the analysis presented above:

(23) a. !John is a violinist at home.
b. !John is a smoker in bars.

[2]Cf. Von Wright (1963) who call *capacities* second order abilities.

Chierchia (1995) argues that the semantic ill-formedness of the expressions in (23) is explained by the generic interpretation of the predicative nominal, which carries the constraint that the individual be located anywhere. Thus, the examples in (23) are ruled out because of two competing if not contradictory locations. According to the analysis presented here, the explanation is consistent with Chierchia's, given that a stative predicate holds irrespectively of any spatial or temporal restriction introduced by the locative PP.

ILNs such as *violinist* appear to have the same distribution as SLNs under certain readings, where they are licensed with locative and temporal modifiers, as in (24)-(25) below:

(24) a. John is a violinist in France.
b. John was a violinist last year.

(25) a. John is an employed violinist in France.
b. John was an employed violinist last year.

The sentences in (24) are acceptable under the occupational sense of *violinist*. I argue that ILNs do not change class, i.e. to SLNs, but rather, as the result of further specification of the context, they acquire an extensional reading similar to *smoker*, discussed above. Compare, for instance, the occupational reading of *violinist* in (26b) with the nominal *smoker*:

(26) a. !John$_i$ is an employed violinist but he$_i$ does not play anymore.
b. !John$_i$ is a smoker but he$_i$ does not smoke anymore.

Under the occupational reading of *violinist*, the interpretation of the defining event is restricted to those situations in which the individual is employed. As argued in Busa (1996), the semantics of *employed* introduces a different modal base, which makes the expression *employed violinist* similar to extensional ILNs such as *smoker*, (cf. (26b)). The interpretation is obtained by viewing the modifier *employed* as "wrapping" the nominal head within another level of quantification, namely the conditions associated with the individual being employed. This involves adding another level of modality corresponding to *instrumental modality* (cf. Kratzer, 1981, Krifka et al., 1995), for which the modal base is given in terms of purpose. Without detailing the specifics of how the adjective *employed* restricts the modal base, the compositional interpretation of *employed violinist* is derived by annotating the information associated to *violinist* with the information introduced by *employed* which is a stage-level predicate expressing that an individual carries out a particular event for the purpose of earning money. In this section I have shown that the distinctions between the set of ILNs can be captured in terms of different stative predicates in their AGENTIVE role. In turn, the stative properties of dispositional and non-dispositional predicates is responsible for the persistency associated with ILNs, which makes them analogous to individual-level predicates.

5. The Existential Force of SLNs

Although the differences between ILNs and SLNs can be broadly stated in terms of quantificational properties of the defining event, the representation in (12), above, is not adequate. An explicit semantics for SLNs has to account for two crucial facts. In particular, SLNs such as *pedestrian* or *passenger* constrain the interpretation of the matrix predicate to overlap with the defining event. Given such a constraint, it should be expected that the set of SLNs includes nominals that establish different temporal relations with the matrix predicate. Consider the examples below:

(27) a. John's widow is getting married next week.
 b. !John married his future widow (Asher, p.c.)

(28) a. The fugitive is now in jail.
 b. !The fugitive is planning to escape tonight.

(29) a. John's murderer was caught.
 b. !John$_i$'s murderer will kill him$_i$ tomorrow.
 c. !!John$_i$ killed his future murderer.

(30) a. I met John's fiancée.
 b. !John$_i$'s fiancée is happily married to him$_i$.
 (cf. John$_i$'s wife is happily married to him$_i$.)

The nominals in (27)-(30) belong to the class of SLNs since they require the context to be temporally related to the event characterizing the individual. Thus, in (27)-(29) the temporal relation between the defining event and the matrix can be roughly stated as **before**, and in (30), as **after**. More specifically a *widow* is someone who has lost her husband, a *fugitive* is an individual who has escaped and is free, a *murderer* is someone who has murdered someone else, a *fiancée* is a person who will possibly marry another person.

A solution that accounts uniformly for the different interpretations of the nominals above exploits the properties of the defining event, which is already part of their representation. In other words, the transitional properties of the event description defining the individual are responsible for the discourse effects that are produced with certain SLNs. More specifically, assuming the traditional distinction between eventualities (Bach, 1981, Dowty, 1979, Mourelatos, 1978), *pedestrian*, *fugitive*, and *murderer* differ in that *walk* is an *activity*, *escape* an *achievement*, and *murder* an *accomplishment*.

In the following sections, I will describe how we can achieve a uniform treatment of SLNs which differ on the type of event that they encode, by observing that what appears to be a restriction that SLNs impose on the matrix predicate can be stated as a dependence of the nominal on the external situation (i.e. matrix predicate). I will discuss how this property, which I call *predicative dependence*, is part of the semantic type distinguishing SLNs from other nominals and how it accounts for the different temporal relations that SLNs establish with the surrounding context.

5.1. THE COMPLEX TYPE OF PREDICATIVELY DEPENDENT EXPRESSIONS

The assumption that lexical items are structured semantic entities, whose representation determines grammatical behavior, requires an explicit statement of how the properties of words are a reflection of different underlying semantic types which vary in their complexity. To illustrate this point, consider the examples below:

(31) A woman hit a pedestrian crossing the street.
a. [$_S$ A woman$_i$ [$_{VP}$ hit a pedestrian$_j$] [$_{VP}$ e$_i$ crossing the street]].
b. [$_S$ A woman$_i$ [$_{VP}$ hit [$_{NP}$ a pedestrian$_j$ e$_j$ crossing the street]].

(32) A woman hit a violinist crossing the street.
a. [$_S$ A woman$_i$ [$_{VP}$ hit a violinist$_j$] [$_{VP}$ e$_i$ crossing the street]].
b. [$_S$ A woman$_i$ [$_{VP}$ hit [$_{NP}$ a violinist$_j$ e$_j$ crossing the street]].

(33) A woman hit a pedestrian riding a bike.
a. [$_S$ A woman$_i$ [$_{VP}$ hit a pedestrian$_j$] [$_{VP}$ riding a bike]].
b. !! [$_S$ A woman$_i$ [$_{VP}$ hit [$_{NP}$ a pedestrian$_j$ e$_j$ riding a bike]].

The sentences in (31) and (32) have two interpretations which vary according to the binding of the trace. This ambiguity, however, is absent in (33), where the trace cannot be bound to *pedestrian* if the event which is predicated of the individual cannot be interpreted as coreferential with the defining event. Assuming that the subject is VP internal (Koopman and Sportiche, 1991), and that the *while*-clause is adjoined to VP, the predicative dependence of SLNs can be viewed in terms of binding theory (Chomsky, 1981). In all of the sentences above, the pronoun is c-commanded by one of the antecedents, however, in (33b) although *pedestrian* c-commands the trace, the trace cannot be bound to it. The conclusion is that the temporal overlap which is required between the SLN and an external event appears to hold for the minimal c-command domain of the nominal. In turn, such an event has to be consistent with the event that defines the individual. The relation between a SLN and the surrounding context can be stated in terms of the constraints on anaphors:

Predicative Dependence
A relation which requires an event in the representation of a lexical item

to be anaphorically bound to the predicate in its smallest c-command domain.

In order to show how this relation is built into the type of SLNs, I will digress into the theory of semantics types adopted in GL theory. In Pustejovsky (1995, 1996) it is observed that in addition to simple or *unified types* such as *rock*, *chair*, *violinist*, language expresses types that can be treated as underlyingly relational. Consider the following examples:

(34) a. First John was late for lunch$_i$ (`event`),
b. and then he didn't even want any of it$_i$ (`phys_object`).

(35) a. I will lend you that thick book$_i$ (`phys_object`),
b. you will find it$_i$ quite instructive (`information`).

Nominals such as *lunch* and *book* present a challenge to conventional treatments of polysemy which rely on an enumeration of senses either by multiple listings of the same lexical entry or by meaning postulates. Within a view of lexical semantics where the rules of composition are built into underspecified representation, the properties of nominals displaying systematic polysemy are treated as having an underlying *complex type*. As outlined in Pustejovsky (1995, 1997), a complex type or *dot object* is a Cartesian type product of n distinct types which characterizes the interpretation of lexical items that carry multiple senses simultaneously. Viewed as a complex type, the nominal *book* denotes `physical_object.information`, and *lunch* denoted `physical_object.event`. The types alone, however, do not fully determine the semantics of the dot object, since they would appear to be contradictory in absence of a coherent structuring. This structuring is provided by a relation $\mathcal{R}$, which for *book* $\mathcal{R}$ expresses *containment*, namely the physical_object contains the information. With *lunch*, $\mathcal{R}$ is different in that the type refers to `food` as well as its the specific function (TELIC), namely *eat*. This is shown below:

(36) a. $\lambda x.y[book(x : \texttt{phys\_object}, y : \texttt{information}){:}hold_F(x, y)...]$
b. $\lambda x.e[lunch(x : \texttt{phys\_object}, e : \texttt{event}):$
$food_F(x) \wedge \lambda w \lambda e[eat_T(e, w : human, x)..]$

Note that with *book* both types are related in the formal, such that the nominal appears to be relational, whereas with *lunch* the two types are distributed throughout different qualia roles.

The relation $\mathcal{R}$ not only provides a coherent structuring of apparently contradictory types, but imposes restrictions on their expressibility, thus accounting for the range of syntactic expressions available with a dot object. A nominal such as *door*, which is of type `aperture.phys_object` is expected to pattern with nominals whose type is `aperture`, (i.e. *entrance*), as well as nominals whose type is `phys_object`, (i.e. *board*):

(37) a. the door to the kitchen (`aperture.phys_object`)
b. the entrance to the kitchen (`aperture`)
c. !!the board to the kitchen (`phys_object`)

(38) a. the wooden door (`aperture.phys_object`)
b. !!the wooden entrance (`aperture`)
c. the wooden board (`phys_object`)

Further motivations and more specific mechanisms concerning dot objects are discussed in Pustejovsky (1995, 1997). To see how this discussion is relevant to the analysis of SLNs, consider the difference between *passenger* and *man* in (39):

(39) a. Once on the road (and especially if the passengers were all *regulars* and *masculine*), the schedule meant nothing. (Brown Corpus)
b. !!....(and especially if the men were all *regulars* and *over 20*).....

The nominal *passenger*, in (39a), allows distinct adjectives to modify different components of its meaning. The adjective *regular* modifies the event of traveling, while the adjective *masculine* adds information relative to the gender of the individuals. On the contrary, *man*, in (39b), does not carry event-based information that would license modification by a frequency adjective. This behavior is analogous to the polysemous nominals *lunch*, *book*, and *door*, discussed above.
The complex type of *passenger* is `individual.event`, where the event is not the defining event, i.e. walk for *pedestrian*, but rather the event on which the noun is predicatively dependent. In intuitive terms, the relation $\mathcal{R}$ between the two types can be viewed as establishing an anchoring for the individual and the defining event in the external situation. This is shown below, where $\mathcal{R}_{PD}$ is the relation of *predicative dependence* relating the individual and the external event, α is the defining event, and *in* expresses containment between two events:

$$(40)\quad \begin{bmatrix} \text{SLN} & \\ \text{ARGSTR} = & \begin{bmatrix} \text{ARG1} = \mathbf{x{:}human} \end{bmatrix} \\ \text{EVENSTR} = & \begin{bmatrix} \text{D-E1} = \mathbf{e_1{:}process} \\ \text{E2} = \mathbf{e_2{:}}\tau \\ \text{REST} = \mathbf{in(e_2,e_1)} \end{bmatrix} \\ \text{QUALIA} = & \begin{bmatrix} \text{FORMAL} = \text{SLN}(\mathbf{x,e_2}) \\ \text{AGENTIVE} = \alpha(\mathbf{e_1,x}) \end{bmatrix} \end{bmatrix}$$

The complex relation $\mathcal{R}_{PD}$ is a constraint on the well-formedness of an expression applied to an individual involved in an event. In other words, both types of the dot object make reference to an existentially quantified event - e.g., the event that defines the individual - to which they are related structurally.

Adopting the generalization that event types largely determine syntactic behavior, (cf., Levin 1993 and references therein), the analysis of SLNs as complex types allows a uniform account for the different temporal relations established with the context. Thus, rather than specifying, for each nominal, whether the defining event occurs before or after the event denoted by matrix verb, we can maintain the generalization that the matrix predicate is intepreted to occur within the temporal interval in which the defining event occurs by relying on the internal structure of events. This solution is not only theoretically more elegant, but it is necessary to properly account for the data, since for a *widow* the state that results from having lost the husband holds at the time of reference. Similarly, a *fugitive* denotes an individual who has escaped in the past and for which the state of being free holds at the time of reference. Our generalization can be maintained by adopting a view of the event structure representation which treats events as complex objects. I address these issues in the next section.

5.1.1. *The Role of Event Headedness*

In GL theory, where transitions are represented as complex event structures (cf., Pustejovsky, 1988, Moens and Steedman, 1988, Pustejovsky 1995) different types of transitions are distinguished minimally by the notion of head. The head of an event structure is expressed by the operator *, which designates relative prominence of one subevent, and is motivated by a number of grammatical facts (e.g. modifier scope, adjuncts, grammatical alternations, and so on). In particular, the set of nominals associated with achievements, namely *escapee* and *fugitive*, versus *winner* and *discoverer* are interpreted differently. Nominals for which the defining event is a right-headed transition, i.e. *escape*, require that the matrix predicate is interpreted to overlap with the final state. Conversely, the interpretation of nominals for which the defining event is a left-headed transition, i.e. *murder*, can be compared with the interpretation of the perfect aspect. If we were to assume an analysis in which *fugitive* established a precedence relation between an event of escaping and the state of being in prison, then (41b) would be predicted to be fully acceptable:

(41) a. The fugitive/escapee has not yet been found.
b. !The fugitive is in jail.
(c.f. The fugitive is now in jail.)

Similarly, a simple precedence relation would not explain the contrast in (42) below:

(42) a. The winner of the Boston marathon$_i$ had crossed the finish line when it started to rain.

b. !The winner of the Boston marathon$_i$ had crossed the finish line when it$_i$ was called off.

Consider, first, a variety of nominals associated with *right-headed transitions*, namely *fugitive*, *widow*, *winner* or *discoverer*:

(43) a. The fugitive is driving a red Alfa.
b. John met a widow.
c. The winner of the Boston Marathon was rude to the press.

In the above examples, the matrix predicate is interpreted as occurring within the interval in which the resulting states "be free", "have lost the husband" and "having won" hold. Given the configurational properties of events and arguments as expressed in the qualia, the underlying representation of the event descriptions associated with the nominals above is presented in (40) below, where the individual denoted by the agentive noun corresponds to the external argument in the relation of the headed event:

(44) RIGHT-HEADED TRANSITIONS

a. $\begin{bmatrix} \alpha \\ \ldots \\ \text{FORMAL}^* = \alpha\_\mathbf{result}(\mathbf{e_2,x}) \\ \text{AGENTIVE} = \alpha\_\mathbf{act}(\mathbf{e_1,y,x}) \end{bmatrix}$ ⇒ UNACCUSATIVE- SLN *widow, dead*

b. $\begin{bmatrix} \alpha \\ \ldots \\ \text{FORMAL}^* = \alpha\_\mathbf{result}(\mathbf{e_2,x}) \\ \text{AGENTIVE} = \alpha\_\mathbf{act}(\mathbf{e_1,x,y{:}location}) \end{bmatrix}$ ⇒ UNACCUSATIVE- SLN *fugitive, escapee*

c. $\begin{bmatrix} \alpha \\ \ldots \\ \text{FORMAL}^* = \alpha\_\mathbf{result}(\mathbf{e_2,x}) \\ \text{AGENTIVE} = \alpha\_\mathbf{act}(\mathbf{e_1,x,y}) \end{bmatrix}$ ⇒ CAUSATIVE- SLN *winner, discover*

Nominals such as *winner*, *discoverer*, *widow* and *dead* have a similar representation, although the configuration between subvents and arguments varies. As the result of sharing the same headedness specification, for all these SLNs the operator *in* relates the headed subevent with the matrix predicate, thus producing the correct interpretation for the examples in (43). The differences between these nominals, interestingly, appear to correlate with the causative/unaccusative distinction. The nominals *escapee*, *fugitive* and *widow* incorporate the semantics of unaccusative verbs, while *winner* and *discoverer* incorporate the semantics of causative right-headed transitions (e.g. *achievements*). The interpretation of nominals such as *murderer* and *author*, requires additional parameters, given that the final state makes reference to the individual who died or the entity which has been produced/written. Consider the representation in (45), which presents the configurational properties of subevents and arguments for a causative left-headed transition (e.g. *accomplishment*) such as *murder*, or *write a book*:

(45) CAUSATIVE LEFT-HEADED TRANSITIONS

a. $\begin{bmatrix} \alpha \\ \ldots \\ \text{FORMAL} = \alpha\_\mathbf{result}(\mathbf{e}_2,\mathbf{y}) \\ \text{AGENTIVE}^* = \alpha\_\mathbf{act}(\mathbf{e}_1,\mathbf{x},\mathbf{y}) \end{bmatrix}$ $\Rightarrow$ *murderer, author*

Given a configuration such as (45), the agentive noun makes reference the external argument in the headed subevent. However, if the derivation were uniquely determined by the initial subevent, the interpretation of *murderer* would be "the x who is murdering", which does not reflect the perfect reading of the event, namely "the x who has murdered". For accomplishments of the type shown in (45), I assume that the representation involves an additional state that asserts the persistence of the "having done", as discussed in Parsons (1990) for the perfect aspect.

Finally, there is a small set of nominals which are correlated to a transition that has not culminated. This set includes *applicant*, *laureando* (ABD), *candidato* (candidate), *cresimando* (confirmee), and *fiancee*, among others. The agent in these cases is defined relative to an ongoing activity which might or might not culminate in a certain state, thus the contrast in (46) below:

(46) a. The candidate delivered a very good speech.
b. !The candidate who was elected yesterday will not raise taxes.

With these nominals, the matrix predicate is included in the initial process, and no information about the occurrence of the final state is given. The distinctions presented above indicate that the interpretation of agentive nominals, be they derived or lexically encoded, is not uniquely determined by the properties of argument structure or event structure in isolation. Rather, the interpretation of the nominal as well as the availability of the morpheme *-er* with productive forms are jointly determined by argument structure, event structure and qualia. More specifically, *-er* seems to be correlated with the availability in the AGENTIVE role of the external argument in the headed relation. The lack of nominals such as *arriver* and *dier* is accounted by this fact, namely the headed relation is the final state, which predicts the availability of the form *arrivee*.

A general observation is in order at this point. There are a number of cases that appear to be apparent contradictions to the generalization above:

(47) a. The fugitive has been caught and he is currently being interrogated.
b. That pedestrian has travelled to New York several times.

c. The murderer was born in Ohio.

In all the above cases the matrix predicate is in the perfect tense and carries a persistence which allows us to reverse the conditions on the interpretation of SLNs. A more detailed explanation is beyond the scope of this paper, for details see Busa (1997).

In this section, I have motivated a variety of SLNs, according to the type of the event which defines an individual. Although there are potentially various ways of formalizing the set of SLNs, I have argued that any representation ought to account for the dependence that these nouns establish with the matrix predicate. In particular, this behavior can be viewed as the result of anaphoric-like properties of SLNs, whose underlying semantic type is complex. Intuitively, this view poses the question of the role that events play in the representation of nominals, in that SLNs display a behavior that appears to be *in-between* that of individuals and events or situations. This means that our framework should involve a rich set of relations between semantic types, which in GL are expressed in terms of complex types or dot object.

6. Agentive Nominals and Coercion

There are a number of examples where the behavior of agentive nominals appears to contradict the generalization that the stage-level versus individual-level distinction holds in the domain of nominals. Consider the following sentences, with the predicate *become*:

(48) a. John became a violinist at the age of five. (ILN)
b. !John became intelligent. (*individual-level predicate*)
c. John became drunk immediately. (*stage-level predicate*)
d. !John become a passenger/pedestrian last night. (SLN)

The contrast in (48) might suggest that SLNs do not behave like stage-level predicates at all, but rather, in composition with *become*, they have the same distribution as individual-level predicates. Conversely, both stage-level predicates and ILNs are licensed. Without detailing the semantic properties of *become*, it can be viewed as construing a transition expressing that a given resulting state obtains. ILNs are licensed with *become* with the interpretation "John developed the ability to play the violin at five", namely *become* makes reference to the beginning of the dispositional state, i.e. *ability* in the AGENTIVE role. With stage-level predicates such as *drunk*, *angry*, and so on, which contain a default event expressing the coming into being of the state of the individual, *become* makes explicit the causal relation between subevents. Under the current treatment, individual-level predicates are uniquely defined by the FORMAL role containing a stative predicate for which no mode of coming into being is specified. In this case, there is no anchoring for *become* into an initial subevent. The oddness of (48d) follows from the observation that a SLN is uniquely defined by the event which occurs, and in this case the event is an *activity*. In addition, given the temporal interpretation between events, *become* would be construed as occurring within the interval in which the individual is α-ing. The data in (48) indicate that individual-level predicates are ruled out because they lack an AGENTIVE role, and SLNs are ruled out because they lack a resulting state, which are available respectively with ILNs and stage-level predicates. This predicts also that SLNs encoding a transition should be licensed with *become*, as shown below:

(49) a. John became a winner.
b. John became a widower.

Similarly, SLNs are not available as subjects of *begin* in saturated raising constructions, (cf. (50)) below:

(50) a. The violinist/lecturer began.
b. !The passenger/participant/tenant began.

The verb *begin* is an aspectual predicate selecting the initial point of an event, and expressing that the subevent of beginning temporally precedes the event which begins. For nouns such as *violinist*, by coercion, an event which satisfies the selectional properties of the predicate can be recovered in the TELIC role. Conversely with *passenger*, the only event which would be available for coercion violates the restriction on the interpretation between events. In order to satisfy the constraints associated with SLNs, the act of beginning should be included within the event that characterizes the individual; however, *begin* carries its own temporal constraints which are incompatible with those introduced by a SLN, namely the composition fails since both lexical items impose incompatible constraints on the temporal relation between events.

A similar set of facts accounts for the behavior of agentive nouns with the predicate *enjoy*, (cf., (51)):

(51) a. John enjoyed the violinist.
b. !John enjoyed the pedestrian/fugitive.

Again, *violinist* by coercion provides an appropriate event that can be reconstructed (cf. Pustejovsky, 1995). Conversely, since the SLN is predicatively dependent on the matrix verb, there is no event that can be reconstructed. Compare, however, (52b) to the fully grammatical sentence with the SLN *visitor*:

(52) John enjoyed his visitor.

The sentence in (52) above is interpreted as John enjoyed the visitor who visited him. This sort of reciprocity allows an event to be reconstructed, thus satisfying the requirements associated with SLNs.

An interesting and unexpected consequence of this analysis, in which coercion extensionalizes the event defining an ILN and certain properties of SLNs are stated in terms of constraints on anaphors, concerns certain instances of metonymy, discussed in Jackendoff (1992) and Nunberg (1995):

(53) a. Ringo squeezed himself into a tight parking space.
b. Ringo was hit in the fender.

Without detailing all the issues associated with the interpretation of the examples in (53) above, the sentences make reference to Ringo's car. The properties which are relevant to the present discussion concern the fact that both (53a-b) require that Ringo is driving at the time, given the ungrammaticality of (54):

(54) a. !!Mary saw Ringo (=Ringo's car) squeezed into a tight parking space.
b. !!Ringo (=Ringo's car) was hit in the fender while he was on stage.

These examples indicate that certain principles accounting for the interpretation of agentive nominals in different contexts are actually quite general and find independent motivation in a set of phenomena which would appear to be largely unrelated. Clearly, this is a simple observation that just skims the surface of the issues involved in handling metonymy and requires future research and further substantiation. The observation is meant to simply point out that, given the appropriate machinery, the type of processes which affect the lexicon and syntax may also be properties of other elements.

7. Remarks on Instruments

Having defined the general properties of ILNs, in this section I will show that nominals denoting *instruments* are formally similar to ILNs as the result of their lexical representation. However, given a number of observations concerning the ontological status of instruments, I argue that important semantic distinctions should be made.

Morphologically derived nominals denoting *instruments* appear to have the same distribution as ILNs denoting animate agents, in that frequency adjectives are not licensed and complements are either bare plural phrases, or compound modifiers:

(55) a. distributore di caramelle/*della caramella (*instrument*)
distributor of candies/of the candy (i.e. candy machine)

b. professore di letteratura/*della poesia/letteratura (*agent*)
professor of literature/of the poem/literature

(56) a. apriscatole/apribottiglie
can opener/bottle opener
b. brick mason

Given the similarity in grammatical behavior, as well as the productivity of the morpheme *-er*, the general class of ILNs should also include *-er nominals* denoting instruments. Focusing on the role of productive morphology, the generalization usually assumed is that the morpheme *-er* applies to an external argument which is interpreted as the agent of an event, irrespectively of whether the type of that argument is **human** or **physical_object** (cf. Randall, 1982, Keyser and Roeper, 1984, Fabb, 1984, Sproat, 1985, Moortgat, 1987, Grimshaw, 1990, Rappaport and Levin, 1992).

Recent analyses (Fabb, 1984, Sproat, 1985, Rappaport and Levin, 1992) have suggested a variety of solutions for the treatment of instruments relative to the set of agentive *-er*-nominals. Sproat (1985), for instance, argues that instruments should be treated as "lexical idioms", since he observes a lack of systematicity in the argument role that *-er* picks out. Adopting a semantic view crucially based on theta-roles, Sproat illustrates a number of cases where *-er* picks out an argument that is not properly an *actor*. Consider, for instance, the noun *respirator*, which does not correspond to the ACTOR role of the verb *respirate*, but denotes a machine that aids to respirate. Similarly, the instrument *propeller*, in (57a), is not related to *ship*, in the same way *writer* is related to *book* in (57b):

(57) a. the propeller of the ship;
b. the writer of the book;

The difference that Sproat notices is that *book* corresponds to the THEME of *write*, while *ship* is not be the THEME of *propel*, since it is the possessor of the propeller. I argue that this is insufficient evidence to say that *instruments* are lexical idioms, since the problems noticed by Sproat seem to derived from the inadequacy of theta-roles (cf. Dowty, 1991) to fully capture fine-grained semantic distinctions between argument types.

In Rappaport and Levin (1992), the behavior of *instruments* appears heterogeneous. They distinguish between *event* and *non-event* nominals according to whether or not the nominalized form licenses a complement, and therefore identify two corresponding types of instruments. Whereas nominals such as *opener*, *wiper*, and so on, are analyzed as non-event instruments, the nominals below are taken to be event instruments since they license a complement:

(58) a. A protein ... a potent <u>inducer</u> of new blood vessel growth.

b. Rather they show that sound is a later elicitor of reaching behavior...
c. Woks have always been conservers of cooking oil.....

According to Rappaport and Levin, evidence for the conclusion that the *-er* nominals in (58) qualify as instruments is provided by the availability of the following sentences:

(59) a. We used this protein to induce new blood vessel growth.
b. We induced new blood vessel growth with this protein.

If the availability of constructions such as (59a) and (59b) are to be taken as evidence that (59a) makes reference to an instrument, than that instrument should be *protein* and not *inducer*. An even more important drawback is that a test such as (59) does not make the right predictions. Consider the sentences below:

(60) a. We used the train to go to work yesterday.
b. We went to work yesterday with the train.

(61) a. John used a fake moustache to hide his scar.
b. John hid his scar with a fake moustache.

According to the test provided by Rappaport and Levin, there should nominals such as *goer* and *hider*, associated with (60) and (61), which denote instruments.

One crucial question which should be asked concerns the definition of the criteria for stating that a noun denotes an instrument. Clearly, although we think of instruments as particular types of tools, it is possible to construe an instrumental reading for speech-acts as well as events. These, however, are meaning shifts with which I am not concerned here. If the properties of a noun are crucially part of its semantic type, which distinguishes it from other objects in the world, then it is hard to see how the *-er* nominals in (58) above qualify as instruments in the same way as *knife*, *chair*, *can opener*, *coffee grinder*, *respirator*, *propeller*, and so on. Nominals such as *inducer* and *respirator* do not share the grammatical behavior of instruments:

(62) a. The knife/coffee grinder in on the second shelf.
b. !!The inducer/elicitor/conserver is on the second shelf.

(63) a. A knife is an instrument for cutting.
b. !!A conserver is an instrument for conserving.

(64) a. John opened the can with a good knife.
b. !!John conserved the oil with a good conserver.

The nominals *conserver* or *inducer* should best be treated on the par of other nouns which Rappaport and Levin classify as nonagentive abstract nouns, such as those in (65) below:

(65) a. Anger is a great <u>defuser</u> of pent-up emotions.
b. Powerful emotions are the certain <u>levellers</u> of ordinary feelings.

The nominals in (65), in fact, are similar to those in (58) since the following constructions are available:

(66) a. Our emotions were defused by anger.
b. Ordinary feelings are levelled with powerful emotions.

Without detailing the properties of these nouns, which display the behavior of stage-level nominals, contrary to Rappaport and Levin's claim, they do not fall into the class of instruments.

Pursuing the view that productive morphological processes should derive new forms for which there is independent motivation in the lexicon, *-er* nouns denoting instruments should be formally similar to nominals such as *knife* or *chair* which are instruments specified for a particular function. As also noted by Kastovsky (1977), instruments are named according to a regular function that they serve and not to one single action, which makes them formally similar to ILNs denoting individual agents. This observation points to a crucial distinction between natural kinds and artifacts. For artifacts, but not for natural kinds, the functionality of the object plays a crucial role, as shown below:

(67) a. !a good rock;
b. a good chair;

Whereas a good chair is one that is comfortable to sit on, it is not clear what a good rock is, unless the expression is uttered by a climber for whom a rock acquires a very specific functionality, i.e. to climb on.

In GL, the functionality is specified in the TELIC role, which is defined for all nouns whose type is `artifact_tool`:

$$(68)\quad \begin{bmatrix} \textbf{artifact\_tool} \\ \ldots \\ \text{QUALIA} = \begin{bmatrix} \text{FORMAL} = \textbf{phys\_obj(x)} \\ \text{TELIC} = \textbf{R(e}_2\textbf{,x)} \\ \text{AGENTIVE} = \textbf{make(e}^T\textbf{,z:human,x)} \end{bmatrix} \end{bmatrix}$$

The general type of an `artifact_tool` given in (68) specifies that the object has been created for a certain purpose that is left underspecified above. Without detailing the mechanisms that involved in the definition of *unified types* in GL (cf. Pustejovsky, 1995, for details), suffice it to say that an instrument such as *knife* is a subtype of `artifact_tool`, with its TELIC role specified:

(69)
$$\begin{bmatrix} \textbf{knife} & \\ \dots & \\ \text{QUALIA} = & \begin{bmatrix} \text{FORMAL} = \textbf{artifact\_tool(x)} \\ \text{TELIC} = \textbf{cut(e}_2\textbf{,x,y:stuff)} \\ \text{AGENTIVE} = \textbf{make(e}_1\textbf{,z:human,x)} \end{bmatrix} \end{bmatrix}$$

The relation in the AGENTIVE role embedded in the TELIC can be viewed as the antecendent of a conditional expression stating that for every cutting event there is some individual using that knife. Alternatively, it can be viewed as restricting a contextual variable (cf. Chierchia, 1995) to all those situations in which someone uses the knife. Consider now, how the representation of the noun *propeller* follows precisely from it being an *artifact_tool* with a specific purpose:

(70)
$$\begin{bmatrix} \textbf{propeller} & \\ \dots & \\ \text{QUALIA} = & \begin{bmatrix} \text{FORMAL} = \textbf{artifact\_tool(x)} \\ \text{CONSTITUTIVE} = \textbf{part\_of(x,y:artifact\_tool)} \\ \text{TELIC} = \textbf{propel(e}_2\textbf{,x,y:artifact\_tool)} \\ \text{AGENTIVE} = \textbf{make(e}_1\textbf{,z:human,x)} \end{bmatrix} \end{bmatrix}$$

This discussion is meant to show that the formal similarity between ILNs and instruments is the result of the characterizing event being encoded in the TELIC role in both cases. One important distinction between instruments and human agents is that the "agentivity" which is represented in the TELIC of artifacts is different from that of ILNs, in that for artifacts it is clearly understood as being mediated by an individual using the instrument. I will not discuss these issues here, since drawing the appropriate distinctions involves formal considerations that are beyond the scope of this paper (cf., Busa 1996, for details).

In this section, I have shown that the set of nominals denoting instruments is fully homogeneous and their representation is consistent with that of the general type of **artifact_tools**. Although *-er* nouns denoting instrument and ILNs share a similar representation, namely the defining event is encoded in the TELIC role, I argue that instruments are not properly agentive nouns. This distinction is desirable, in that the ontological status of the two nominal types is crucially different. If we treated all instruments as agentive nominals, we be forced to conclude that *chair* is an agentive nominal also, since its semantic type is no different from the instrumental interpretation of a nominal such as *opener*. The distinctions discussed above suggest that if instruments are formally similar to ILNs, then there should be other nominals which are formally similar to SLNs. In particular, for nominals such as *laundry*, *grocery*, and so on (R. Jackendoff, p.c.) the conditions associated with the interpretation of SLNs appear to hold, since they are definable relative to situated events of washing or being on the shelf of a supermaket. This certainly does not satisfactorily capture the properties of their, but certainly indicates that the stage-level versus individual-level properties are not uniquely associated with human agents.

8. Conclusion

Focusing on the distinctions between nouns such as *violinist* or *smoker*, on the one hand, and *pedestrian*, *passenger* or *winner*, on the other, I have argued that the semantic properties of agentive nominals bear directly on a number of semantic issues, and most prominently the stage-level versus individual-level distinction, which allows us discover predictable patterns of behavior within the class of agentive nominals. Extending the proposal made in Pustejovsky (1995), I have shown that by focusing on the role of events in the representation of nominals, the properties of a variety of agentive nouns can be explained by how the event defining the individual is quantified. In turn, quantification is derived from the lexical entry, by identifying the modes of explanation for a particular individual, i.e. *qualia structure*. Thus, to say that an ILN is defined relative to the TELIC role, is to say that the event which characterizes the individual is understood relative to its functionality, and hence to a modal base which is determined by the nature of that particular activity. Similarly, to say that a SLN is defined relative to the AGENTIVE role, is to say that what "explains" that individual is the performance of the defining event. Focussing on the role of events to account for the interpretation of agentive nouns constitues a promising are a reseach into the semantics of productive affixes. In particular, the set of *-ee*-nominals appears to be often associated with right-headed transitions, and the set of *-er* nominals with left-headed transitions, processes, and states. Finally, the formal similarity of ILNs and nominals denoting instruments poses the question of the existence of other nominals in the language which are formally similar to SLNs. The fact that this is indeed the case strengthen the usefulness of the distinction, which branches out to a much larger class of nouns than the set of morphologically derived *-er* nominals.

References

Bach, E. (1981). "On Time, Tense, and Aspect: An Essay in English Metaphysics," in P. Cole (ed.), *Radical Pragmatics*, Academic Press, New York.

Barker, C. (1995). "Episodic *-ee* in English," in *Aquilex Workshop on Lexical Rules*, Cambridge, UK, August 9-11.

Busa, F. (1996). *Compositionality and the Semantics of Nominals*, Ph.D. Dissertation, Brandeis University.

Busa, F. 1997. *Aspectual Nominals*, ms., Brandeis University.

Carlson, G. (1977). *Reference to Kinds in English*, Ph.D. Thesis, University of Massachussetts at Amherst.

Carlson, G. (1982). "Generic Terms and Generic Sentences," in *Journal of Philosophical*

Logic, vol. 11.
Chierchia, G. (1995). "Individual-Level Predicates as Inherent Generics", in Carlson G. and F. Pelletier, (eds.) *The Generic Book*, The University of Chicago Press.
Chomsky, N. (1981). Lectures on Government and Binding, Dordrecht: Foris Publications.
Declerk, R. (1988). "Restrictive When-Clauses," in *Lingustics and Philosophy*, 11.
Dowty, D. R. (1991). "Thematic Proto-Roles and Argument Selection," in *Language*, 67.
Fabb, N. (1984). *Syntactic Affixation*, Ph.D. Dissertation, MIT.
Grimshaw, J. (1990). *Argument Structure*, MIT Press.
Gupta, A. (1980). *The Logic of Common Nouns*, Yale University Press.
Jackendoff, R. (1983). *Semantics and Cognition*, MIT Press.
Jackendoff, R. (1990). *Semantic Structures*, MIT Press.
Jackendoff, R. (1992). "Mme. Tussaud Meets the Binding Theory," *Natural Language and Linguistic Theory*, 21.
Kastovsky, D. (1977). "Word-Formation, or: at the Crossroads of Morphology, Syntax, Semantics, and the Lexicon," in *Folia Linguistica*, X, 1/2.
Katz, J., and J. Fodor. (1963). "The Structure of a Semantic Theory," in *Language*, 39.
Keyser, J. and Roeper, T. (1984). "On the Middle and Ergative Constructions in English", in *Linguistic Inquiry*, 15.
Koopman, H. and D. Sportische (1991). "The Position of Subjects," *Lingua*, 85.
Kratzer, A. (1981). "The Notional Category of Modality," in H.-J. Eikmeyer and H. Reiser, (eds.), *Words, Worlds, and Contexts: New Approaches to Word Sematnics*, De Gruyter.
Kratzer, A. (1995). "Stage-Level and Individual-Level Predicates," in Carlson G. and F. Pelletier, (eds.), *The Generic Book*, The University of Chicago Press.
Krifka, M., F. Pelletier, G. Carlson, A. ter Meulen, G. Chierchia, and G. Link (1995). "Genericity: An Introduction," in Carlson G. and F. Pelletier, (eds.), *The Generic Book*, The University of Chicago Press.
Levin, B. 1993. *Towards a Lexical Organization of English Verbs*, University of Chicago Press, Chicago.
Moens, M. and Steedman (1988). "Temporal Ontology and Temporal Reference," *Computational Linguistics*, Vol. 14.
Mourelatos, A. P. D. (1978) "Events, Processes and States," *Linguistics and Philosophy*. Also published in P. Tedeschi, and A. Zaenen, (eds.), *Syntax and Semantics: Tense and Aspect*, vol. 14, Academic Press, 1981.
Moortgat, M. (1987). "Compositionality and the Synatx of Words", in Groenendijk, J., D. de Jong, and M. Stockhof (eds.), *Foundations of Pragmatics and Lexical Semantics*, Foris Publications.
Moravcsik, J.M. (1990). *Thought and Language*, Routledge.
Nunberg, G. (1995) "Transfers of Meaning," *Journal of Semantics*, 12.
Parsons, T. (1990). *Events in the Semantics of English*. MIT Press.
Pustejovsky, J. (1988). "The Geometry of Events," in *Studies in Generative Approaches to Aspect*, Carol Tenny (ed.), Lexicon Project Working Papers 24, MIT, Cambridge, MA.
Pustejovsky, J. (1991a). "The Generative Lexicon", in *Computational Linguistics*, 17.
Pustejovsky, J. (1991b). "The Syntax of Event Structure," in *Cognition*, 41.
Pustejovsky, J. (1995). *The Generative Lexicon*, MIT Press.
Pustejovsky, J. (1997). "The Semantics of Lexical Underspecification," in *Folia Linguistica*.
Randall, J. (1982). *Morphological Structure and Language Acquisition*. Ph.D. Dissertation, University of Massachusetts at Amherst.
Rappaport M. and B. Levin. (1992). "*-er* Nominals: Implications for the Theory of Argument Structure", in Stowell, T. and E. Wehrli (eds.), *Syntax and Semantics*, vol. 26, Academic Press.
Sproat, R. (1985). *On Deriving the Lexicon*, Ph.D. Dissertation, MIT.
Von Wright, G. *Norm and Action: A Logical Inquiry*, Routledge and Kegan Paul, London.

Index

Text, Speech and Language Technology

1. H. Bunt and M. Tomita (eds.): *Recent Advances in Parsing Technology.* 1996
ISBN 0-7923-4152-X
2. S. Young and G. Bloothooft (eds.): *Corpus-Based Methods in Language and Speech Processing.* 1997 ISBN 0-7923-4463-4
3. T. Dutoit: *An Introduction to Text-to-Speech Synthesis.* 1997 ISBN 0-7923-4498-7
4. L. Lebart, A. Salem and L. Berry: *Exploring Textual Data.* 1998
ISBN 0-7923-4840-0
5. J. Carson-Berndsen, *Time Map Phonology.* 1998 ISBN 0-7923-4883-4
6. P. Saint-Dizier (ed.), *Predicative Forms in Natural Language and in Lexical Knowledge Bases.* 1999 ISBN 0-7923-5499-0

KLUWER ACADEMIC PUBLISHERS – DORDRECHT / BOSTON / LONDON

www.ingramcontent.com/pod-product-compliance
Ingram Content Group UK Ltd.
Pitfield, Milton Keynes, MK11 3LW, UK
UKHW020132290926
15471UKWH00003B/440
* 9 7 8 9 4 0 1 7 2 7 4 7 1 *